Mean & Std. Deviation "σ" — p.109, p.42, 138, p.93
Normal Curve formula — p.108/109
68% w/in 1STD of Normal Curve — p.108

Cointoss p. 80/(69, 89
Picking 4 or w/o REPLACEMENT p. 79/80/81 close #'s
ORDER Dependent (Permutation) p.97 UnImportant (Combination) p.88

p.138 Estimate of Mean if 'σ' is Known
p.143 " " " " " " " unKnown

Multiply for "AND"
Add for "OR"

p.108
68%, 96%, 99%,

p.101 Bell curve History

CONTENTS OF APPENDICES

p.143 Find interval Range when mean known

Rolling Dice:

ODDS "11" on 1 pr. Dice:
$(1,5)$ or $(5,1) = (2)\left(\frac{1}{6}\right)^2 = \frac{2}{36}$

ODDS "7" on 1 pr. Dice.
$(4,3)(3,4)(5,2)(2,5)(6,1)(1,6) = 6\left(\frac{1}{6}\right)^2 = \frac{6}{36}$

Actually a Binomial Dist.
$$C_{n\,k}\, p^k (1-p)^{n-k} = C_{4\,2}\left(\frac{1}{2}\right)^2\left(\frac{1}{2}\right)^2$$

SAUNDERS PACKAGED PROGRAM FOR STATISTICS

Also available for use with *STATISTICS:*

Gilbert and Kurland: *Student Study Guide for STATISTICS* — $3.95

Audio Tapes for *STATISTICS* — $100.00

Computer Card Deck for Student Exercises — free upon adoption

Student Workbook for Computer Exercises — free upon adoption (indicate quantity required)

Coin toss: Ttl # Events = 2^n

ODDS any Event (# combinations) $\binom{n}{r} = \dfrac{n!}{r!(n-r)!}$

(See p. 90)

$\left. \begin{array}{ll} HHHH & HHTT \\ HTHT & HHHT \\ HTTH & HTHH \\ HTTT & HHTH \end{array} \right\} \times 2 = 16$ Diff in toss

i.e., 2 Tails in 4 tosses $= \binom{4}{2} = \dfrac{4!}{2!(4-2)!} = 6$; ODDS of 2 Tails $= 6 \times \left(\dfrac{1}{2}\right)^4 = \dfrac{6}{16}$

$_4C_2$

Coin toss — Normal Distrib. $\mu = np$; 100 tosses $\mu = 100 \times .5 = 50$

$\sigma = \sqrt{npq}$ [$q = 1-p$] $\sigma = \sqrt{100(.5)(1-.5)} = 5$

50 55 x
0 1   z

{See p. 93 (see prob. #21 p. 121)}
Binomial Approx. to Normal

$z = \dfrac{x - 50}{5}$; @ z=1 x=55

(Binomial Approx. to Normal :
$z = \dfrac{x - np}{\sqrt{np(1-p)}}$

statistics

NORMA GILBERT, Ph.D.
Department of Mathematics
Drew University

Odds of 2 people in A room having the same Birthday :

1st person   2nd person   3rd person

$\dfrac{365}{365}$ ✱ $\dfrac{364 \text{ days avail.}}{365}$ not same ✱ $\dfrac{363 \text{ days avail. not same}}{365}$ ✱ — — —

∴ If 21 people in Rooms, odds = $\dfrac{.52}{}$
" 22 " " " " , odds = $\dfrac{.49}{}$

Check to see if Wtd. Dice:

Straight ODDS = 230 3's in 1200 tosses

$_{1200}C_{230} \dfrac{1}{6}^{230} \dfrac{5}{6}^{770} = $ OVERFLOW

Binomial Distr.

W. B. SAUNDERS COMPANY
Philadelphia, London, Toronto

Toss 1200 times, '3' comes up 230x

$z = \dfrac{p - \hat{p}}{\sigma p} = \dfrac{\frac{230}{1200} - \frac{200}{1200}}{\sqrt{\dfrac{p(1-p)}{n}}} = \dfrac{.025}{\sqrt{\dfrac{\frac{1}{6}(5/6)}{1200}}}$

$= 2.3$
Bad Dice!

0   2.3
99%

Normal Distribution:
$\mu = 1200 \times \frac{1}{6} = 200$
$\sigma = \sqrt{npq} = \sqrt{1200 \frac{1}{6} \frac{5}{6}} = 12.9$

200 213

$z_{230} = \dfrac{230 - 200}{12.9} = 2.33$
Bad Dice

W. B. Saunders Company: West Washington Square
Philadelphia, PA 19105

1 St. Anne's Road
Eastbourne, East Sussex BN21 3UN, England

1 Goldthorne Avenue
Toronto, Ontario M8Z 5T9, Canada

Library of Congress Cataloging in Publication Data
Gilbert, Norma.
Statistics.
1. Mathematical statistics.   I. Title.
QA276.G45   519.5   75-22732
ISBN 0-7216-4127-X

Front cover is a reproduction of an original lithograph by Steve Holmes, and is published with the permission of the artist.

Statistics   ISBN 0-7216-4127-X

Last digit is the print number: 9   8   7   6   5   4   3   2

preface

This book is intended for students who need to understand how statistical decisions are made but who have little mathematical background (a year or two of high school algebra is sufficient). So many statistics texts aimed at this group have appeared that some explanation—other than the pleasure I've had in writing it—is needed of why another appears. Here are several reasons:

First, many formulas appear in any statistics book. Some texts emphasize how to use a formula. Learning how to use it is necessary, but I have tried to explain also where a formula comes from, why it should be used, and when it should not be used. Explanations are often intuitive and informal, but a student should get a grasp of why a decision is reached rather than just learn what is reached.

Second, in some texts the notation gets pretty involved. In this one it doesn't. (The exception is Chapter 16; by then a student is well accustomed to statistics. Some instructors will prefer to omit this chapter.)

Third, the book is suited to self-paced ("Keller Plan") study as well as to the usual lecture approach: An introductory section at the beginning of each chapter explains its scope and where emphasis should be put. Over 300 examples are worked out in detail. The exercises at the end of each chapter are preceded by a list of vocabulary and symbols which help a student review the chapter. The answers to odd-numbered problems are given in detail, in the belief that it does no good to know that the probability of an event, for example, is .40 if the student doesn't know how .40 is derived.

Last, some texts avoid probability, but an understanding of statistics without it is impossible. Others dive into probability immediately. Many students have trouble with this topic, however, so I have begun with descriptive statistics and delayed probability until Chapter 5; it is then used in every succeeding chapter.

Chapters 1 through 8, part of 9, and 10 should ordinarily be taken in sequence; material from the remainder of Chapter 9 and Chapters 11 through 17 can be selected according to the interests and speed of the students and the length of the course.

I am especially indebted to Professor David Moore of Purdue University, who read the manuscript at three different stages of its development and who made many splendid suggestions. Special thanks are due to others who read and criticized the manuscript: Stephen A. Book, California State College, Dominguez Hills, California; Albert Liberi, Westchester Community College, Valhalla, New York; Stanley M. Lukawecki, Clemson University, Clemson, South Carolina; Leon Gleser, Purdue University, West Lafayette, Indiana; and Judith Tanur, State University of New York at Stony Brook, Stony Brook, New York.

I am also indebted to Margo Hall (Drew University, Madison, New Jersey, 1974), who helped sort out several different drafts; to John Miano (Drew University, 1975), who suggested interesting exercises; to all the Drew students in Mathematics 3 who have used the original draft of the manuscript; and to John Snyder of W. B. Saunders Company for his cooperation and encouragement.

NORMA M. GILBERT

contents

contents

chapter one

what mathematical background is needed?

*Most of this chapter should be familiar stuff. Note that * is used for multiplication, rather than × or ·: 2 * 3 = 6. Summation notation (Sections 1.4 and 1.5, pages 4–7) may be new to you. You will use the remaining parts of this chapter so frequently, however, that you will be foolish indeed if you let anything slip by without understanding it. Even students who are good at mathematics are sometimes confused about whether multiplication or addition is carried out first (2 * 3 + 4 = 6 + 4 or 2 * 7?) and how parentheses are used to change the usual order.*

1.1 INTRODUCTION

Many students approach statistics with apprehension, having resolved after junior year in high school never to take another mathematics course. But be of good cheer for two reasons. The background in mathematics which you need is remarkably slight. And in your high school courses you may have learned how to solve quadratic equations, for example, without understanding **why** you were learning to do this. Now those of you who are concerned about using mathematics are studying statistics because a knowledge of this material is necessary background for your study of psychology, economics, zoology, or sociology. You will find an almost immediate application of the mathematics you learn to a subject of interest to you, and this delightful situation—in contrast to your past experience—will give most of you a new and splendid ability to cope with mathematics.

Try always to gain some perception of the general arguments that produce results rather than accept them blindly. An idiot can be trained (or a computer can be programed) to plug numbers into formulas; you must learn when and why a particular formula is used. Occasionally an explanation, in order to be mathematically proper, would require that you know calculus; in these cases the explanation is purely intuitive or is omitted. Fair warning will be given when such occasions arise.

There are very few mathematical tidbits that you need to read this book. One unusual notation needs comment: multiplication of numbers will be indicated by *:

$$4 * 2 = 8$$

This is done partly because this notation is often used on computer print-outs and you should become accustomed to it, and partly so that 4.2 in decimal notation will not be confused with the product 4 * 2. Multiplication of x by y will ordinarily be indicated by juxtaposition: xy, rather than $x * y$; similarly, fd rather than $f * d$. But neither a dot ($f \cdot d$) nor a cross ($f \times d$) will ever be used to indicate f times d.

Note especially that if any number is multiplied by zero, the product is 0:

$$7 * 0 = 0 \qquad 0 * 2/3 = 0$$
$$0 * 7 = 0 \qquad \sqrt{2} * 0 = 0$$
$$-3.8 * 0 = 0 \qquad 0 * x = 0$$

> A special notation will be used to warn you of traps that have been set for you: an unusual twist, a problem that cannot be solved, or some special reason to be wary.

1.2 PRE-TEST

If you can solve the following problems correctly, you are very well prepared mathematically for this book. If you have trouble with some of them, study this chapter with care. Problems 13–20 need familiarity with Σ (summation) notation. Do not lose heart if you are not yet familiar with it; read Section 1.4, page 4, with care. The answers are at the bottom of page 3.

1. $4.38 + .12 + .02 =$

2. $48 - 51 =$

3. $6.21 * .01 =$

4. $5 * 0 =$

5. $\dfrac{27 * 75}{25 * 9} =$

6. $3 * 5^2 + 5 =$

7. $3 * (5^2 + 5) =$

8. $4 - 10^2/20 =$

9. Evaluate $x^2 y + \dfrac{z}{y}$ if $x = 2$, $y = 3$, and $z = 6$.

10. Find $\overline{X}$ if $X = 10$, $s = 2$, $n = 25$, $z = 15$, and $z = \dfrac{X - \overline{X}}{\dfrac{s}{\sqrt{n}}}$.

11. $\sqrt{493} =$ (2 decimal places; use Table 1, Appendix C.)

12. $\sqrt{.00212} =$ (3 decimal places; use Table 1, Appendix C.)

In Problems 13–17, X values are 3, 5, 8, 2, 0, 3, 0; corresponding Y values are 2, 1, 3, −1, 0, 2, 4.

13. $\Sigma X =$

14. $\Sigma(X - Y) =$

15. $\Sigma X^2 =$

16. $(\Sigma X)^2 =$

17. $\Sigma(X - 2) =$

18.

Y	f	$\Sigma fY =$
3	2	
4	7	
5	2	

19. A set of five X values is given. Write a formula which says, "Subtract 3 from each X value, square each new number, add the new set, and then take the square root of the sum."

20.   A set of 10 Y values is given. Translate into English the directions given in the formula $\dfrac{\Sigma(2Y+5)^2}{10}$.

1.3   THE ORDER OF OPERATIONS

The word "operations" here refers to raising to powers, multiplication or division, and addition or subtraction. The rules agreed on by mathematicians are that these are carried out in the order given below when more than one operation is to be performed — that is,

> (a)   First raise to powers.
> (b)   Then multiply or divide.
> (c)   Then add or subtract.

Example 1

$2 * 3 + 4 = ?$

$2^2 * 3 + 1 = ?$

$3 - 12/2^2 = ?$

$2 * 3 + 4 = 6 + 4 = 10.$

$2^2 * 3 + 1 = 4 * 3 + 1 = 12 + 1 = 13.$

$3 - 12/2^2 = 3 - 12/4 = 3 - 3 = 0.$

But suppose you don't want to do things in that order? What if, for example, 2 and 3 are first to be added and then the result squared? Parentheses ( ), square brackets [ ], or curly brackets { } are used, and a new rule is added:

> If there are parentheses or brackets, simplify what is inside them first before continuing.

The notation $3(4 + 2)^2$ is used instead of $3 * (4 + 2)^2$.

Example 2

$(2 + 3)^2 * 4 = ?$

$(2^2 * 3 - 2)^2 * 4 = ?$

$2(3^2 + 4 \div 2) - 3(2^3 - 4)^2 = ?$

$(2 + 3)^2 * 4 = 5^2 * 4 = 25 * 4 = 100.$

Answers:
1. 4.52; 2. −3; 3. .0621; 4. 0; 5. $\dfrac{3 * 3}{1 * 1} = 9$; 6. 80; 7. 90; 8. −1; 9. 14; 10. 4; 11. 22.20; 12. .046; 13. 21; 14. $\Sigma X - \Sigma Y = 21 - (2 + 1 + 3 - 1 + 0 + 2 + 4) = 21 - 11 = 10$; 15. $9 + 25 + 64 + 4 + 0 + 9 + 0 = 111$; 16. $21^2 = 441$; 17. $\Sigma X - \Sigma 2 = 21 - 7 * 2 = 7$; 18. $6 + 28 + 10 = 44$; 19. $\sqrt{\Sigma(X - 3)^2}$; 20. Multiply each Y value by 2, add 5 to each product, then square each number in the new set; next add the squares, and finally divide the sum by the number of scores.

$$(2^2 * 3 - 2)^2 * 4 = (4 * 3 - 2)^2 * 4 = (12 - 2)^2 * 4 = 10^2 * 4 = 100 * 4 = 400.$$

$$2(3^2 + 4 \div 2) - 3(2^3 - 4)^2 = 2(9 + 4 \div 2) - 3(8 - 4)^2$$
$$= 2(9 + 2) - 3 * 4^2$$
$$= 2 * 11 - 3 * 16 = 22 - 48 = -26.$$

(You won't have anything even as complicated as this one for a long time.)

1.4 Σ NOTATION

You will soon need to learn seven Greek letters. The first of these is the Greek (capital) S, called **sigma,** and written Σ. We shall use Σ as an abbreviation for the word **sum** or for the phrase "the sum of _____."

If X values are 1, 2, 3, 5, then $\Sigma X = 1 + 2 + 3 + 5 = 11$. Add **all** the X values.

If the weights W of five students are 110, 100, 160, 200, 190, then $\Sigma W = 110 + 100 + 160 + 200 + 190 = 760$.

Remember the rules for order of operations when Σ (standing for addition) is combined with exponents, multiplication, or parentheses.

Example 1

If Y values are 4, 3, 1, 6, 2, find (a) ΣY, (b) ΣY^2, (c) $(\Sigma Y)^2$.

(a) $\Sigma Y = 4 + 3 + 1 + 6 + 2 = 16.$

(b) $\Sigma Y^2 = 4^2 + 3^2 + 1^2 + 6^2 + 2^2 = 16 + 9 + 1 + 36 + 4 = 66.$

(c) $(\Sigma Y)^2 = 16^2 = 256.$

Example 2

Two columns of numbers are given in the box below.

(a) $\Sigma fX = ?$

(b) $\Sigma fX^2 = ?$

(c) $(\Sigma fX)^2 = ?$

(d) $\Sigma (fX)^2 = ?$

f	X	fX	fX^2
2	−1	−2	2
3	0	0	0
1	1	1	1
2	2	4	8
		+3	11

(a) $\Sigma fX = ?$

Multiplication is carried out before addition; the notation ΣfX means "multiply each f value by the corresponding X value, and then add the results." A new column fX is written next to the two we already have:

$$\Sigma fX = 2 * -1 + 3 * 0 + 1 * 1 + 2 * 2$$
$$= -2 + 0 + 1 + 4 = 3$$

(b) $\Sigma fX^2 = ?$ $fX^2 = (fX)X = X(fX)$, so another column is added and its entries are determined by multiplying corresponding entries in the X and fX columns.

$$\Sigma fX^2 = \Sigma X(fX) = -1 * -2 + 0 * 0 + 1 * 1 + 2 * 4 = 2 + 0 + 1 + 8 = 11$$

Yes, it would be possible to square X and then multiply by f, resulting in columns headed f, X, X^2, fX^2. Two things are wrong with this—one invisible and one visible. The invisible one is that in statistics when we need ΣfX^2 we shall usually want ΣfX as well; the visible reason is that the f, X, fX, fX^2 route to finding ΣfX^2 involves multiplying adjacent columns, while the f, X, X^2, fX^2 route to finding ΣfX^2 means multiplying numbers in the first and third columns to determine the fourth. The human eye is not tremendously successful at playing leapfrog with columns.

(c) $(\Sigma fX)^2 = ?$ What is inside the parentheses is evaluated first.
$(\Sigma fX)^2 = (-2+0+1+4)^2 = 3^2 = 9$.

(d) $\Sigma(fX)^2 = ?$
$\Sigma(fX)^2 = (-2)^2 + 0^2 + 1^2 + 4^2 = 4+0+1+16 = 21$.

There are three rules for working with summations:

> **Rule 1.** $\Sigma(X + Y) = \Sigma X + \Sigma Y$. (This rule only makes sense if the number of X values is the same as the number of Y values.)
>
> **Rule 2.** If k is a constant, $\Sigma kX = k\Sigma X$.
>
> **Rule 3.** If A is a constant, $\Sigma A = nA$, where n is the number of values that are being added.

Example 3

X values: 4, 1, 6, 3; corresponding Y values: 2, 3, 7, −1. (a) $\Sigma(X+Y) = ?$ (b) $\Sigma 2X = ?$ (c) $\Sigma(X-2) = ?$ (d) $\Sigma 5 = ?$ (e) $\Sigma(X-Y) = ?$

(a) $\Sigma(X+Y) = \Sigma X + \Sigma Y = (4+1+6+3) + (2+3+7-1) = 14 + 11 = 25$. (Rule 1)

(b) $\Sigma 2X = 8+2+12+6 = 28$, but also
$= 2\Sigma X = 2(4+1+6+3) = 2 * 14 = 28$. (Rule 2, which says a constant can be factored out)

(c) $\Sigma(X-2) = (4-2) + (1-2) + (6-2) + (3-2) = 2 + (-1) + 4 + 1 = 6$, but also $= \Sigma X - \Sigma 2 = (4+1+6+3) - 4 * 2 = 14 - 8 = 6$.

(d) $\Sigma 5 = 4 * 5 = 20$. (Rule 3) It is assumed that the number of values being added (here, 4) is known from the context of the problem. A more sophisticated summation notation, using subscripts, can be adopted, but we shall not need it.

(e) $\Sigma(X-Y) = \Sigma[X+(-1)Y]$
$= \Sigma X + \Sigma(-1)Y$ (Rule 1)
$= \Sigma X + (-1)\Sigma Y$ (Rule 2)
$= \Sigma X - \Sigma Y$
$= (4+1+6+3) - (2+3+7-1)$
$= 14 - 11$
$= 3$.

1.5 FORMULAS

A number of formulas will be developed in this course; they simply give shorthand directions for carrying out operations. Study the following examples, and learn how to translate from English into mathematical language and vice versa.

Example 1

Translate into English: $\Sigma(X - 4)$.

Subtract 4 from each X value; then add the new set.

Example 2

Write a formula which gives directions to subtract 4 from each X value, square each number in the new set, and then add the squares.

$$\Sigma(X - 4)^2$$

Example 3

Write a formula which gives directions to subtract 4 from each X value, add the new set of values, and square the sum.

$$[\Sigma(X - 4)]^2$$

Now test yourself on these two. Remember the order of operations!

Example 4

Translate into English: $\Sigma(X + 3)$.

Example 5

Write a formula which gives directions to subtract 5 from each X value, square each number in the new set, add the squares, and divide the sum by 3.

Answers

4. Add 3 to each X value, then add the new set of scores.

5. $\dfrac{\Sigma(X - 5)^2}{3}$

If there are three X values: 3, 5, 7, then the operations you have carried out are the following:

	Example 1	Example 2	Example 4	Example 5	
X	$X - 4$	$(X - 4)^2$	$X + 3$	$X - 5$	$(X - 5)^2$
3	−1	1	6	−2	4
5	1	1	8	0	0
7	3	9	10	2	4
	$\overline{3}$	$\overline{11}$	$\overline{24}$		$\overline{8}$

Example 3: $[\Sigma(X - 4)]^2 = 3^2 = 9.$

Example 5: $\dfrac{\Sigma(X - 5)^2}{3} = \dfrac{8}{3}.$

Often you will be given X values and an f corresponding to each X value, as shown in the first two columns below (f will be used in later chapters for the **frequency** with which X values appear); the other four columns will be needed in Examples 6–9.

(1)	(2)	(3)	(4)	(5)	(6)
X	f	fX	$X-5$	$f(X-5)$	$f(X-5)^2$
3	1	3	−2	−2	4
5	4	20	0	0	0
7	2	14	2	4	8
		37		2	12

Example 6

Translate into English: $\Sigma fX = 37$.

Multiply each X value by the corresponding f and add the products; the result is 37 (Column 3).

Example 7

Write a formula which gives directions to subtract 5 from each X value, multiply the new set by the corresponding f, and add the products; the sum is 2.

$$\Sigma f(X-5) = 2$$

Note: Even if you are accustomed to using functions, do not confuse $f(X-5)$ as used here with functional notation; f will be used in this book only for **frequency** and never for a function.

Again, try two examples yourself.

Example 8

Translate into English: $\Sigma f(X-5)^2 = 12$.

Example 9

Write a formula which gives directions to subtract 5 from each X value, multiply the new set by the corresponding f, add the products, and then square the sum; the result is 4.

Answers

Subtract 5 from each X value, square each of the numbers in the new set, multiply each square by the corresponding f, and add the products (Column 6).
$[\Sigma f(X-5)]^2 = 2^2 = 4$ (Column 5).

1.6   SQUARE ROOTS AND SQUARES

In working out statistical problems, it will often be necessary to find a square root. Always remember two things: (1) you can never find the square root of a negative number, and (2) the square root is never negative.† $\sqrt{4} = 2$, **not** ±2. Use Table 1, Appendix C, to find a square root if you don't have a calculator.

---

†Many students are misled by the reasoning "if $x^2 = 4$, then $x = \pm\sqrt{4} = \pm2$, so $\sqrt{4} = \pm2$." The underlined part of this reasoning is wrong; $\sqrt{4} = +2$.

Example 1

$$\sqrt{34.5} = ? \qquad \text{(2 decimal places)}$$

Find 3.45 in the *n* column of Table 1, Appendix C. Since 34.5 is between 10 and 100, look across the row to the $\sqrt{10n}$ column, finding 5.874. Multiply by .1; $\sqrt{34.5} = 5.874$. Rounding to 2 decimal places, $\sqrt{34.5} = 5.87$.

Example 2

$$\sqrt{.0126} = ? \qquad \text{(3 decimal places)}$$

Find 1.26 in the *n* column of Table 1. Since .0126 is between .0100 and .0999, look across the 1.26 row to the $\sqrt{n}$ column, finding 1.122. Multiply by .1; $\sqrt{.0126} = .1122$. Rounding to 3 decimal places, $\sqrt{.0126} = .112$.

Example 3

$$134^2 = ? \qquad \text{(Use Table 1, Appendix C.)}$$

Find 1.34 in the *n* column, and discover that $1.34^2 = 1.7956$. Since 134 is between 100 and 999, multiply by 10,000; $134^2 = 17,956$.

1.7 GREEK LETTERS

You have already met Σ, the Greek capital S. You will need six other Greek letters; come meet them now all at once if you like. (If you prefer, wait until you bump into them later on.) Σ is a capital letter; all the others are lower case.

Greek letter	Name	Pronounced like	English equivalent
μ	mu	a kitty's mew	m
σ	sigma	stigma $-$ t	s
α	alpha	as in romeo	a
β	beta	"abate" in reverse	b
ρ	rho	fish egg; ignore the h	r
χ	chi	kie to rhyme with tie	$-$(German ch)

1.8 VOCABULARY AND SYMBOLS

$\star$

summation Σ

$\mu, \sigma, \alpha, \beta, \rho, \chi$

1.9 EXERCISES

1. $\dfrac{1}{2} \star \dfrac{3}{4} =$

2. $\dfrac{1}{2} + \dfrac{2}{3} =$

3. $26 - 34 =$

4. $3 - 4 - 6 + 2 + 4 - 3 + 6 =$

5. $4 \star 3 + 5 =$

6. $6 + 2 \star 3^2 =$

7. $(3^2 + 1)^2 \star 0 =$

8. $(2^2 + 3)^2 \star 2 =$

9. $(3^2 - 7)^3 * (4^2 - 7 * 2)^2 =$ 10. $(\sqrt{16} - 3 * 2) * (2^2 - 5) =$

11. $(9 + 4^2) * (3^2 - 5^2) =$

12.   Evaluate: $xy + z$, where $x = 2.3$, $y = 3.0$, and $z = 4.1$.

13.   Evaluate: $y\left(x + \dfrac{1}{yw}\right)$, where $x = 2$, $y = 3$, and $w = 0$.

14.   Evaluate: $(-3)^2$.

15.   Evaluate: -3^2.   16.   Subtract $+4$ from -18.

In Exercises 17–24, use the following X and Y values:

X	Y
3	2
4	1
1	0
2	1
5	-3

17. $\Sigma X =$ 18. $\Sigma X^2 =$

19. $(\Sigma X)^2 =$ 20. $\Sigma(X - 2) =$

21. $\Sigma(X - Y) =$ 22. $\sqrt{\dfrac{\Sigma(X - 3)^2}{4}} =$ (1 decimal place)

23. $\Sigma XY =$ 24. $(\Sigma X)(\Sigma Y) =$

In Exercises 25–29 use the following values of y and f:

y	f
2	4
4	2
5	4

25. $\Sigma fy =$ (What column should be added to the table above?)

26. $\Sigma fy^2 =$ (Hint: add column fy^2 to the table above Exercise 25.)

27. $(\Sigma fy)^2 =$ 28. $\bar{y} = \dfrac{\Sigma fy}{\Sigma f} =$

29.   If $\bar{y} = 3.6$, $\Sigma f(y - \bar{y}) =$ (Hint: add columns $y - \bar{y}$ and $f(y - \bar{y})$ to the table above Exercise 25.)

30. $\sqrt{844} =$ (1 decimal place)

31. $\sqrt{.00132} =$ (3 decimal places)

32.   Solve for x: $3 = 2x - 3$.

33.   Solve for x: $.06x + .05(4000 - x) = 240$.   (Hint: First multiply both sides of the equation by 100.)

34.   Solve for $\bar{X}$ if $z = 8$, $X = 3$, $s = 1$, and $n = 16$: $z = \dfrac{X - \bar{X}}{\dfrac{s}{\sqrt{n}}}$

35.   The scores (X) of 10 students on the Miller Personality Test are as follows: 22, 21, 16, 26, 23, 27, 23, 18, 23, 31.
      (a)   Write a formula which says: "Add the scores of all 10 students and divide the sum by the number of students; label the result $\bar{X}$ (read 'X bar')."

(b)   Translate into a formula the following directions: "Subtract 23 from each score, square each result, sum the squares, and then divide by one less than the number of scores. Take the square root of the result and label it *s*."

(c)   Carry out the directions given in (b) to one decimal place.

36.   The heights (*H*) in inches of six sophomore men are 75, 71, 74, 67, 68, 71.

(a)   Write a formula which says "Subtract 71 from each height, add the new set of numbers, divide the sum by the number of students measured, and then add 71 to the quotient; label the sum $\overline{H}$ (read '*H* bar')."

(b)   Translate into a formula the following directions: "Subtract $\overline{H}$ from each height, square each of the new values and then find their sum; divide the sum by one less than the number of men whose height has been measured, take the square root of the quotient and label the result *s*."

(c)   Carry out the directions given in (a) and in (b) to one decimal place.

37.   Eight defendants have been found guilty and are sentenced to terms (*T*) of 6, 10, 8, 1, 3, 1, 4, and 15 years in prison, respectively.

(a)   Translate into English: $\overline{T} = \dfrac{\Sigma(T-7)}{8} + 7$.

(b)   Translate into English: $s^2 = \dfrac{\Sigma(T-\overline{T})^2}{7}$.

(c)   Carry out the instructions given in (a) and in (b) to one decimal place.

38.   In June, 1975, observations (*O*) are made of 14 robins, 5 wrens, 17 starlings, and 12 sparrows on a certain plot of land. Based on previous experience in other years, the expected counts (*E*) on this plot are 10 robins, 8 wrens, 20 starlings, and 10 sparrows.

(a)   Translate into English: $X^2 = \Sigma \dfrac{(O-E)^2}{E}$.

(b)   Compute X^2 to one decimal place.

ANSWERS

1. $\dfrac{3}{8}$ 3.   −8

5.   17   7.   0

9.   8 * 4 = 32   11.   25 * −16 = −400

13.   Not defined: $\dfrac{1}{0}$ is meaningless; if you were tempted to hope that $\dfrac{1}{0} = 0$, take a peek at division. You know $\dfrac{8}{2} = 4$ because 8 = 2 * 4 (that is, division is defined in terms of multiplication: $\dfrac{x}{y} = z$ only if *x* = *yz*). But $\dfrac{1}{0} = 0$ implies that 1 = 0 * 0, which in turn implies 1 = 0 − an unseemly state of affairs if ever I saw one. Did you get caught in the trap this time? If, in this course, you find a need to divide by 0, check your work: you won't be tempted to do so again, so go back and find your mistake.

15.   −9   17.   15

19. $15^2 = 225$.

21. $\Sigma X - \Sigma Y = 15 - 1 = 14$ or $1 + 3 + 1 + 1 + 8 = 14$

23.   6 + 4 + 0 + 2 − 15 = −3   25.   8 + 8 + 20 = 36

27. $36^2 = 1296$.

29. $4 * -1.6 + 2 * .4 + 4 * 1.4 = 0$.   31.   .036

33. $x = 4000$.

35. (a) $\bar{X} = \dfrac{\Sigma X}{10}$ (b) $s = \sqrt{\dfrac{\Sigma(X - 23)^2}{9}}$ (c) $\bar{X} = 23.0, s = \sqrt{\dfrac{168}{9}} = 4.3$

37. (a) 7 is subtracted from each term and the new set is summed; the sum is divided by 8 (the number of prisoners), and 7 is added to the quotient. The result is labeled $\bar{T}$.

(b) The number $\bar{T}$ is subtracted from each term, the differences are squared, and then the squares are added. The sum is divided by 7, and the resulting number is labeled s^2.

(c) $\bar{T} = \dfrac{-8}{8} + 7 = 6$.

$$s^2 = \frac{0^2 + 4^2 + 2^2 + (-5)^2 + (-3)^2 + (-5)^2 + (-2)^2 + 9^2}{7} = \frac{164}{7} = 23.4.$$

chapter two

introduction to statistics

An introductory course in statistics consists of two parts: descriptive statistics and statistical inference. Here you will find a brief discussion of both, so you may have some idea of where you are going. To understand statistical inference, you will need to know the difference between a population and a sample. Most of this chapter deals with types of data, and how to present data in tables and in graphs.

2.1 WHAT IS STATISTICS?

The word "statistics" is used with two different meanings:

1. **Statistics are classified facts (especially, numerical facts) about a particular class of objects.** You have certainly heard of accident statistics (56,400 deaths by motor vehicle in the United States in 1969), baseball statistics (Ted Williams had a batting average of .406 in 1941), and statistics on students entering college (the average SAT score of freshmen entering Drew University this year is 602). In this definition, statistics is a plural noun: "Statistics **are . . . .**" This is the meaning of the word when used by the general public.

   The term **descriptive statistics** is used for the part of this course that deals with the presentation of numerical facts (data), in either tables or graphs, and with finding numbers which summarize data, for example, by giving the center and the spread about the center. ("My average on quizzes in Economics is 86, and my grades range from 74 to 96.")

2. **Statistics is the area of science that deals with the collection of data on a relatively small number of cases so as to form logical conclusions about the general case;** the conclusion you draw is called a **statistical inference.** In this meaning, statistics is a singular noun; the question at the beginning of this section ("What **is** statistics?") should shout loudly and clearly that this is the meaning of particular interest in this course. You are familiar with Presidential polls: 2,000 voters are asked their preference for President. On the basis of these results, claims (and remarkably accurate claims) are made about how 75,000,000 voters will cast their ballots on election day. A conclusion is made about how a large group will react from knowledge of reactions of a small part of the group.

   In general, can you draw conclusions about the whole if you can gather only part of the information needed to answer a question? The answer, of course, is "yes"—otherwise, few people would study statistics—but the "yes" is qualified by many limitations. In descriptive statistics—definition 1 above—errors may occur in recording data or in arithmetic or in rounding off computations, but otherwise the results are exact. In statistical inference, however, the situation is very different. As Wallis and Roberts write in *The Nature of Statistics,* "Statistics is a body of methods for making wise decisions in the face of uncertainty."† In order to make wise decisions, you must learn what kinds of information and how much of it are needed, what kind of conclusions can be drawn, and how accurate they are likely to be.

2.2 POPULATIONS AND SAMPLES

A main part of statistics is learning how to make wise decisions about a large group (a **population**) after studying detailed information about a small part of it (a **sample**). A population consists of **all** the individuals or objects in a well-defined group about which information is needed to answer a question. "Population" as used in statistics does not necessarily refer to all the individuals in a particular community. It may be the percent of carbon monoxide in the air at 12 noon each day of the year in each station of the U.S. Weather Bureau, or it may be those residents of Arkansas in 1976 who are registered voters.

We shall often be interested in measures or scores of different members of a population. We might be interested in the number of American soldiers who died in Viet Nam each year since 1964, or the amount Ms. Jonas pays for groceries each week in 1975. But, in either case, different members of the population may have the same measures or scores. Thus, if a population consists of the IQ of five students A, B, C, D, E, the scores these five students receive on a standardized IQ test may be 105, 120, 120, 115, 148, respectively; two scores are the same, even though the five students are different.

A sample is a relatively small group scientifically chosen so as to represent the population. Data for the sample are secured, but it is the whole population that the statistician is interested in. We'll look later (Chapter 8) into how a sample is scientifically chosen.

If you know the IQ of every seventh grader in the junior high school in Madison, New Jersey, do you have a population or a sample? It is impossible to answer until you know the question that is being asked. You have a population if the question is "What is the average IQ of seventh graders in Madison Junior High School at the present time?" You have all the relevant information about each member of the group about which the question is raised. You have a sample, however, if the question you are trying to answer is "What is the average IQ of present seventh graders in New Jersey?" or "What is the average IQ of seventh graders in Madison Junior High School over the past ten years?" What if the question is "What is the average IQ of present seventh graders in Madison Junior High School who are taking typing?" Here you have a population (you know the IQ of every member of the group about whom the question is asked)—plus some irrelevant information which you would ignore, namely, the IQ of seventh graders not taking typing.

Example 1

What is the population and what is the sample in the following statistical problem?

---

†Wallis, W. A., and Roberts, H. V., *The Nature of Statistics* (Glencoe, Illinois: The Free Press, 1965), p. 11.

A doctor tests a new drug on 100 patients with leukemia, chosen at random. After six years, 20 patients are alive. What proportion of all leukemia patients will be alive after six years of treatment with this drug?

The sample consists of the 100 patients on whom the new drug is tested. The population is all leukemia patients now living whose disease has been identified; the population about which the doctor wishes to draw conclusions consists of all leukemia patients treated with the drug now **or in the future.** Often, the population which you can sample and the population about which you hope to draw conclusions will differ, but the error made by extending conclusions about the sampled population to a larger group is often hard to estimate.

Example 2

"Sullivan Co. orders 10,000 light bulbs from a manufacturer who claims the mean life of his bulbs is at least 1,000 hours. Sullivan Co. tests 15 bulbs from the shipment. Using the results of these tests, should the shipment be accepted?" Describe the population and sample.

Here the population consists of 10,000 bulbs; the sample, of the 15 bulbs which are tested.

Example 3

"In a random sample of 1,000 Democrats, 400 favor capital punishment, whereas 250 of a random sample of 500 Republicans favor capital punish-ment. Find the difference in the proportions of all Democrats and Republicans who favor capital punishment." Describe the populations and samples.

Two different populations (one consisting of all Democrats, the other of all Republicans) are to be compared after studying two samples (1,000 Democrats and 500 Republicans). For the comparison to be meaningful, the terms "Democrat" and "Republican" should be carefully defined.

2.3   TYPES OF DATA

Data may be classified as categorical, ranked, or metric.

Categorical Data.   Individuals are simply placed in the proper category, and the number in each category is counted. Each item must fit into exactly one category.

Example 1

Sex of Essex College Students
1975

Sex	Number
Male	706
Female	678

In Example 1, categories are listed in a column. Sometimes information is classified in two ways at once. Then a row listing categories in the second set is added. Be espe-cially careful with your tally if you are making a two-way classification.

Example 2

Eye Color of Essex College Students by
Class 1975

		Class		
Color	FR	SOPH	JR	SR
Blue	124	86	82	98
Brown	150	170	135	136
Green	15	4	7	2
Hazel	103	85	95	92

In the first example there are two categories; in the table above there are 16 (brown-eyed sophomores, for example). If a pink-eyed albino freshman turns up at Essex College, another category would be added to the column under *Color,* and four more to the table; the line "Pink 1 0 0 0" would be added.

What's wrong with the following categories for types of publications? "Newspapers, books, magazines, *The New York Times.*"†

Ranked Data   have order among the categories. If six horses race, the statement "bets were paid off on A, B, and C but not on D, E, and F" presents two categories (winners and losers). The statement "A won, followed by B, C, D, E, and F in that order" presents ranked data. Note that there are not necessarily equal intervals or differences between ranks. A may win by a nose over B, but E may come in ten lengths ahead of F.

Metric Data   involve measurement. To continue the horse race analogy, the times the six horses take to run the race are metric data. Units are assigned (seconds or minutes for the horse race), with equal intervals between the units. This makes it possible to carry out arithmetical operations. With ranked data, we can only remark "A came in ahead of B." With metric data, we can say "A's time was 62.3 sec., B's 62.4 sec.; the difference is .1 sec."

Other examples of metric data are: incomes of residents of Milwaukee, times for rats to solve a given maze, and SAT scores of students who applied for admission to Cornhill College in the class of 1979.

In general, separate techniques are used for the different types of data. Much of the data dealt with in this book will be metric. Techniques developed for metric data can seldom be used with categorical or ranked data. At other times, techniques especially suited to ranked or categorical data will be presented. Most of these could be applied to metric data, but they usually are not since more powerful methods specific to metric data are preferred.

2.4   PRESENTATION OF DATA IN A TABLE

Eighty-two students obtain the following scores on the Miller Personality Test:

22	22	20	27	30	23	29	21	26
21	23	25	29	18	22	31	30	28
16	28	33	25	23	31	23	18	24
26	25	17	22	25	28	19	24	20

Scores continued on following page

---

†*The New York Times* is a newspaper and therefore fits into two categories.

23	26	21	31	25	24	33	29	20
27	21	25	28	24	23	25	30	27
23	26	22	24	17	33	26	24	19
18	33	25	28	31	29	27	28	24
26	24	22	26	24	18	21	29	22
31								

These are raw data (that is, data which have not been numerically organized). It is difficult to digest such a mass of figures, and indigestion becomes more acute as the number of figures is increased. A **tally** is made by listing the different scores in order from the smallest to the largest and adding a vertical tally mark as each score appears until a multiple of 5 is reached; then a horizontal tally mark is made through the previous four.

Score	Tally	Frequency
16	|	1
17	||	2
18	||||	4
19	||	2
20	|||	3
21	﬈﬈﬈	5
22	﬈﬈﬈ ||	7
23	﬈﬈﬈ ||	7
24	﬈﬈﬈ ||||	9
25	﬈﬈﬈ |||	8
26	﬈﬈﬈ ||	7
27	||||	4
28	﬈﬈﬈ |	6
29	﬈﬈﬈	5
30	|||	3
31	﬈﬈﬈	5
32		0
33	||||	4
		82

A **score** is any relevant numerical measurement (any piece of metric data), regardless of how obtained or the kind of unit used. After the scores are tallied, the number in each score class is written in a column labeled **frequency.** We shall commonly use X or Y for scores, f for frequency. A **frequency distribution** is a record of the number of scores that fall in each score class. The frequency distribution made from the 82 scores on the Miller Personality Test would start, then, as follows:

X	f
16	1
17	2
18	4
.	.
.	.
.	.

Each frequency distribution must include classes of observations (X's) and the frequency in each class (f's). Each score must fit into exactly one class.

Ignore the middle column of tally marks and look at the frequency distribution for scores on the Miller Personality Test; compare it with the raw data given at the beginning of this section. The frequency distribution is easier to assimilate; it is easy to determine the lowest score (16) and the highest score (35); it is easy to determine how many people scored 27 (4). There are still 18 different score classes. It is often advantageous to **group** data by combining score classes.

Here are two ways in which this may be done:

(a)			(b)	
X	f		X	f
16–17	3		16–18	7
18–19	6		19–21	10
20–21	8		22–24	23
22–23	14		25–27	19
24–25	17		28–30	14
26–27	11		31–33	9
28–29	11			82
30–31	8			
32–33	4			
	82			

As the number of score classes decreases, the information given becomes easier to assimilate. Grouped frequency distribution (b) above can be absorbed in about 5 seconds, while (a) takes longer, and the original distribution longer still. But simplicity has a price: detailed information is lost by grouping. (b) tells us there are 7 scores between 16 and 18, but we don't know how these 7 are distributed; the original data show 1 score of 16, 2 of 17, and 4 of 18. So grouping is not an unmixed blessing; a compromise must always be found between clarity and loss of information.

The **class width** is the difference between the lowest score that fits into one class and the lowest score that fits into the next higher class. (It is not the difference between the largest and smallest scores in one class.) In (a) above, each class is of width 2, and in (b) each has width equal to 3.

The **midpoint** of a class is the average (later you will learn to be fancier and call this the "mean") of the smallest and largest scores that could go into a class. The midpoint of the 16–17 class in (a) above is $\dfrac{16 + 17}{2} = 16.5$; the midpoint of the 16–18 class in (b) is $\dfrac{16 + 18}{2} = 17$.

The **lower class limit** is the smallest score that could go into that class; the **upper class limit** is the largest. In the first class in (b) above, the class limits are 16 and 18.

Class boundaries are halfway between the upper class limit of one class and the lower class limit of the next.

Example 1

X	Class Boundaries	Y	Class Boundaries
16–18	15.5, 18.5	.1–.3	.05, .35
19–21	18.5, 21.5	.4–.6	.35, .65
22–24	21.5, 24.5	.7–.9	.65, .95

Note that the last digit in a class boundary is 5, and the position of the decimal point in relation to this 5 tells you the unit of measurement of the scores. Thus, .5 implies scores are measured to the nearest unit, as in the X scores above; __5 implies measurement to the nearest tenth of a unit, as in the Y distribution. The class boundaries overlap, but this does not mean any score fits in more than one class. (Why?)

Example 2

Hourly Wages of Steel Workers in Pennsylvania

Wages (in dollars)	Class boundaries
0–2.99	0, 2.995
3–5.99	2.995, 5.995
6 and over	5.995, ∞

Note __5 in the class boundaries, since wages are measured to the nearest cent.

The upper limit of the last class is written as a "lazy 8" (∞), which you have probably learned to read as "infinity." It simply means every wage greater than or equal to $6 is included in this class. Such a class is called **open-ended.** It has no (upper) class limit and no midpoint.

Next, look at the lower limit of the first class. An hourly wage cannot be negative. The lower class boundary, then, is 0 rather than −.005.

2.5 HOW TO GROUP SCORES

Here are some guidelines for grouping.

1. Every score must fit into exactly one class.
   This guideline is never ignored.

2. Use 10 to 20 classes.
   This guideline is ignored often—but only for a good reason. Use as few as 5 classes if you are willing to sacrifice a lot of information in order to gain extreme simplicity; use many more than 20 classes if you are presenting masses of information to a sophisticated audience.

3. Classes should be of the same width.
   As we shall see in Chapters 3 and 4, arithmetic is simplified if this is the case. Also, this is the conventional method of grouping.

Sometimes it is impossible for all classes to be of the same size. A psychologist who times 40 rats in a maze, but removes any rat who has not found its way through the maze in 3 minutes, might present his data as follows:

Time (min.)	f
0 and under 1	10
1 and under 2	18
2 and under 3	4
3 and over	8

The width of the open-ended class cannot be determined.

4. Consider customary preferences in numbers (we count by 5, 10, 25 . . . most easily), and choose your classes to reflect this: 100–124, 125–149, . . . is better than 103–127, 128–152, . . .

Computations will be easier if the class width is an odd number. And most people are much more comfortable with the grouping in (a) than with that in (b):

(a)	(b)
100–124	100–122
125–149	123–145
150–174	146–168

The class width in (a) is 25; in (b) it is 23.

The last three guidelines are not firm rules; they are ignored if the purpose for which the data are used so demands.

A mass of raw scores is organized by using the following procedures:

1. Tally the scores.

2. Determine the range (= the difference between the highest and lowest scores.)

3. Select the class width and the number of classes so that the four guidelines are followed and so that (class width) * (number of classes) covers the range. Once you estimate the number of classes, you can divide the range by this number to approximate the class width.

4. Determine the class limits; that is, state which scores are to be included in each class.

5. Make a table showing the class limits and the frequency in each class. In any table, always include a title, give a date if meaningful, and give the source of data if you did not collect it yourself.

Example 1

The lowest of 100 scores is 10 and the highest is 45. How should they be grouped in about 10 classes?

The range is $45 - 10 = 35$. Should the class width be 4? $10 * 4 = 40$ covers the range, but 4 is an even number and guideline 4 is not followed. So we might have a class width of 3 and then would need 12 classes; $12 * 3 = 36$ covers the range. Classes will be 10–12, 13–15, . . . , 43–45.

Example 2

About 15 classes are to be used to group 300 scores which range from 207 to 592. How should classes be chosen?

The range is $592 - 207 = 385$. $\frac{385}{15} \approx 25$ (read $\approx$ as "approximately equals"); 25 is an odd number and a convenient choice for class width, but we need 16 classes ($16 * 25 = 400$ is greater than 385, but $15 * 25 = 375$ is less). Classes: 207–231, 232–256, . . . , 582–606? The "nice" class width of 25 is hardly apparent; a better choice is 200–224, 225–249, . . . , 575–599.

2.6 GRAPHICAL PRESENTATION OF DATA

To many people, data are more meaningful if presented graphically rather than in a table. Every graph, just like a table, should include a title, a date if this is meaningful, and the source of data if you did not collect it yourself.

Bar Charts. These are used for categorical data. Categories are usually shown on the horizontal axis, and frequency, proportion, or per cent is shown on the vertical axis, unless the names of the categories are long. The bars are separated from each other to emphasize the distinctness of the categories. The bars must all be of the same width (so the area of each bar is proportional to the number in that category). Labels are needed on both axes.

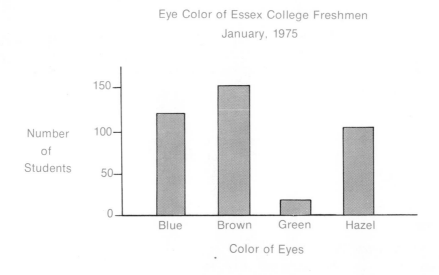

Eye Color of Essex College Freshmen
January, 1975

Histograms. A **histogram,** used for metric data, is similar to a bar chart except that the bars are not separate, and classes are ordered on the horizontal axis, with scores increasing from left to right. The bars are adjacent to emphasize that metric data is on a continuous scale. Either the midpoint or the class boundaries of each class are shown.

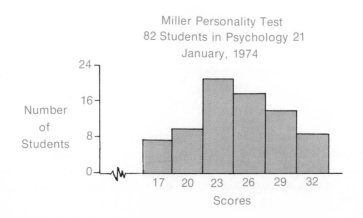

Miller Personality Test
82 Students in Psychology 21
January, 1974

A histogram cannot be used if one of the classes is open-ended.

Frequency Polygons. A **frequency polygon** is a line graph made by connecting the mid-points of the bars in a histogram by straight lines. If it makes sense, a class of zero frequency is assumed at each end to bring the eye down to the axis again.

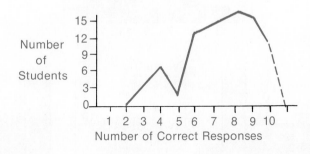

Correct Responses on Ten-Question T-F Test

Note: Should the dotted line on the accompanying figure be filled in as part of the graph, or should it have been omitted? It should have been omitted, since including it would imply that 11 is a possible score. The eye is not brought down to the axis on the right end of this frequency polygon.

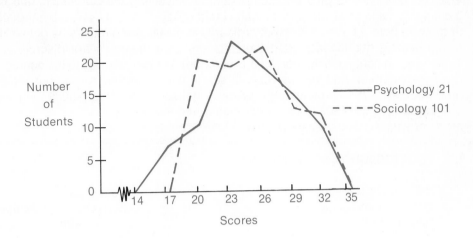

Miller Personality Text

Students in Psychology 21 and Sociology 101

A histogram is very confusing if two distributions are shown on the same graph; a frequency polygon is usually preferable.

Symmetrical and Skewed Distributions in Graphs. A histogram (or a frequency polygon) may be **symmetrical** (if you cut it vertically down the middle, one half can be flipped over exactly onto the other half), **skewed to the left** (has a longer tail on the left), or **skewed to the right.** Study these examples:

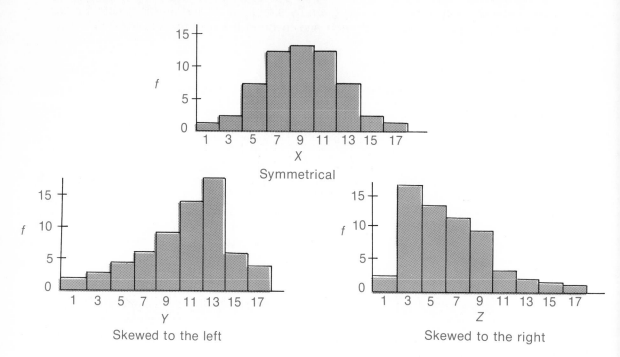

Symmetrical

Skewed to the left   Skewed to the right

Discrete and Continuous Variables.   In the distributions you have been considering, the variable (scores on the Miller Personality Test, for example) has taken on a finite number of values (16, 17, 18, . . . , 33). Such a variable is called **discrete.** A variable is called **continuous** if it can take on all the values of a continuous scale. For example, if X is any number between 1 and 4 (such as 1.42 or $\sqrt{3}$ or 3.9117), then X is a continuous variable. It is hard to give practical examples involving measurements: the height of a child is a continuous variable, in that a child passes through every intermediate height in growing from 3 feet to 4 feet; but ordinarily in measuring a child's height we measure to the nearest quarter inch, and therefore our measurements are discrete variables. Time flows on continuously (continuous variable), but is measured to the nearest second (discrete variable). And yet, as you will see in later chapters, continuous distributions are of tremendous theoretical importance in statistics. The graph illustrating a continuous distribution will be a curve rather than a frequency polygon.

2.7 VOCABULARY AND SYMBOLS

descriptive statistics	frequency
statistical inference	frequency distribution
population	class width
sample	midpoint of a class
categorical data	lower class limit
ranked data	upper class limit
metric data	class boundaries
raw data	open-ended class
tally	range
score	$\approx$

bar chart   skewed to the left
histogram   skewed to the right
frequency polygon   discrete variable
symmetrical distribution   continuous variable

2.8   EXERCISES

1.   The time each Brothers College student waits in line for dinner is measured on each of the nights February 1 through February 15. For which of the following questions do you have information about the population, and for which only about a sample (that is, for which do you have all data necessary to answer the question, and for which do you have only part of the data needed)?

What is the average time waiting in line
      (a)   for dinner on February 3?
      (b)   for dinner on Saturdays in February?
      (c)   for dinner during the second semester?
      (d)   for all meals on February 3?
      (e)   for dinner on February 1–7?

2.   For those questions in Exercise 1 for which a sample is given: Is the sample representative? Can you suggest a better method for choosing the sample?

3.   For each of the following, describe the population. If time and cost were of no concern, could you collect information about the entire population?

      (a)   You are trying to decide whether a coin is fair (that is, equally likely to come up heads and tails when tossed). You toss it a large number of times, counting the number of heads and of tails that come up.
      (b)   A doctor has a new treatment for coreopsis. He tests it on all people who have this disease in 1974.
      (c)   An economist is interested in the number of unemployed men over 17 in the United States who are seeking employment in January.

4.   Assume that you know the income from interest and dividends of each resident of Bangor, Maine. Give examples of situations in which you have information about (a) a population, (b) a sample.

5.   Describe the population(s) and sample(s):

      (a)   An economist working for the Bureau of Labor Statistics knows the percentage of workers unemployed in Newark, New Jersey, last month. This month he surveys 5000 Newark workers to discover whether the unemployment rate has changed.
      (b)   A political scientist is investigating the difference in the proportion of registered Republicans among well-to-do and poor voters in Nassau County, New York. He finds that among 1000 voters in Nassau County whose families have incomes over $16,000 the proportion of registered Republicans is 25.3 per cent, while this proportion is 22.0 per cent in 1200 families whose income is less than $8,000.
      (c)   Four identical packets of tomato seeds are treated with three different chemicals; then each of these packets plus a fifth of untreated seeds Is planted. The average yield per plant of 35 different plants from each of the four packets is determined to see whether chemical treatments affect yield.

(d) It is known that 1-year-old Eskimo dogs gain in weight, on the average, 2 pounds per month. A special diet supplement, HELTHPUP, is given to 35 1-year-old Eskimo dogs, and their gains in weight are measured to find out whether HELTHPUP affects their weight.

(e) The registrar at Peterson University knows that the grade point average of married students has been .3 higher than that of unmarried students in the past 4 years. This semester he compares the average GPA of 40 married and 50 unmarried students to see if there is still the same difference.

6. Classify as categorical, ranked, or metric data:

(a) At an art show, a judge lists his preferences for paintings.

(b) A chemist makes 10 different solutions by adding 100 cc. of sulfuric acid to 10, 20, . . . , 100 cc. of water, and then measures the boiling point of each solution.

(c) Dresses are inventoried by style number and size.

(d) Four hundred firemen are queried about their ethnic backgrounds.

(e) Ten bars of chocolate are rated by an expert taster.

(f) Thirty students are given IQ tests.

(g) Consumption of alcohol per capita in 1976 is determined for each of 46 countries.

7. (a) Make a tally for the following scores:

126	132	121	149	130	139	127	136	138	129
121	134	139	135	128	123	133	136	124	130
127	136	132	126	145	139	131	133	142	131
134	130	141	144	136	124	136	136	133	128
123	125	139	145	148	141	126	145	138	139
133	147	136	134	132	142	149	122	131	139
130	139	136	148	132	147	121	124	148	133
139	127	147	124	148	135	142	142	133	142
121	146	145	148	127	136	130	144	143	124
148	140	136	136						

(b) What is the smallest score? the largest? How many times does 139 appear?

8. Make a two-way classification (by sex and political affiliation) of the members of the Rhodes Political Union (R = Republican, D = Democrat, I = Independent, M = Man, W = Woman):

	1	2	3	4	5	6	7	8	9	10	11	12	13	14	15	16	17	18	19	20	21	22	23	24	25	26	27	28	29	30	31	32
R	x				x		x										x					x		x							x	
D								x		x				x				x	x	x					x		x		x			
I		x	x	x	x		x			x		x	x	x		x		x						x		x		x		x		x
M	x	x			x	x	x		x	x				x	x		x	x		x						x		x	x	x		
W			x	x			x				x	x	x			x			x			x	x	x	x		x			x	x	x

9. Criticize each of the following groupings of the scores which appear in Exercise 7. Which guidelines (p. 18) are not followed?

(a)		(b)		(c)	
X	f	X	f	X	f
120–123	7	121–126	16	120–122	6
124–127	13	126–130	15	123–125	8
128–130	8	131–135	17	126–128	9
131–135	18	136–140	22	129–131	9
136–139	21	141–145	14	132–134	13
140–143	9			135–137	13
144–147	10			138–140	11
148–151	8			141–143	8
				144–146	7
				147–149	11

10. (a) How should the scores in Exercise 7 be grouped into classes of width 5?
     (b) What are the class boundaries and the midpoint of the class which includes the lowest scores?

11. Follow the guidelines to find the class width and the number of classes for each of the following distributions. What should be the class limits for the class with the lowest scores?

	Number of scores	Highest score	Lowest score	Approximate number of classes
(a)	10,000	208	22	20
(b)	600	737	112	14
(c)	120	114	52	10

12. (a) The diameters of a sample of 120 1/2-inch bolts are measured, and found to range from .4954 in. to .5072 in. How should the data be grouped?
     (b) What are the class boundaries of the two classes with the smallest scores?
     (c) What are the midpoints of these two classes?

13. The registrar at Dixon College computes grade-point averages with A = 4.0, B = 3.0, C = 2.0, D = 1.0, F = 0.0. The grade-point averages of 490 students range from 0.42 to 4.00. How should these averages be grouped in about 15 classes? (Note: It is impossible for a student to have a grade-point average greater than 4.00.)

14. The highest grade on an hour exam in statistics is 97 and the lowest is 62; 100 students took the exam. How should the grades be grouped in classes whose width is 5? How many classes are needed?

15. Find the class boundaries, class width, and midpoint of the **first** and **second** classes in the following distributions. Let X stand for the midpoint of each class; find ΣfX where possible. (Add columns for X and fX to those shown.)

(a)
Trials for Rats to Learn Discrimination

Number of Trials	Number of Rats
1–3	7
4–6	5
7–9	3
10–12	3

(b)
Sale Prices of American Cars, 1969

Price	Per Cent
$1,500–2,499	28
$2,500–3,499	49
$3,500–4,499	17
$4,000 and over	06

(c)

Heights of 188 American Males

Height (in inches)	Number
60–64	10
65–69	150
70–74	120
75–79	8

16.   Use a histogram and a frequency polygon (if suitable) to present graphically each of the distributions shown in Exercise 15.

17.   Illustrate each of the following with whichever and of the following are suitable: a bar chart, a histogram, and a frequency polygon.

(a)

Suits Worn in Madison Ave. Agencies
on November 30

Type of Suit	Number
Gray flannel	1800
Gray tweed	1200
Gray serge	400
Other	200

(b)

Hourly Wages of N.J. Construction Workers

| | Per Cent | |
Wages	1960	1970
$1 but less than $3	20	5
$3 but less than $5	70	60
$5 but less than $7	10	35

18.   A student (who is about to flunk the course) wishes to illustrate the information given in the table at the left below with a frequency polygon. He comes up with the object shown at the right. Find two major and a number of (relatively) minor errors in

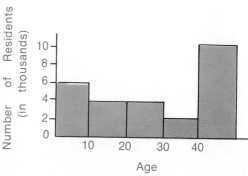

Ages of Residents of Milford, N.H.
Sept. 1, 1974

Age	Number of Residents (in thousands)
0, but less than 10	6
10, but less than 20	4
20, but less than 30	3
30, but less than 40	3
40 or over	10

his work. ("Major" means "so fundamental it cannot be easily patched up.") List his errors, and indicate which are the major ones.

ANSWERS

1. Sample: b, c, d. Population: a, e.

3. (a) The population consists of the results of **all** tosses of the coin, and cannot be determined (b) "Does the new treatment help all who have coreopsis now or in the future?": Since the population includes those who will suffer from this dread disease in the future, he cannot now collect all necessary information. "Does the new treatment help those who now have coreopsis?": Complete information is available—if you can recognize coreopsis. (c) Population: All unemployed men over 17 in the United States seeking employment in that January. Complete information could be determined by a census.

5. (a) Population: those in the labor force in Newark, N.J., this month; sample: 5,000 chosen from the Newark labor force this month. (b) Two populations: registered Republicans in Nassau County whose families make over $16,000 and under $8,000, respectively. He does badly, however, in characterizing these as "well-to-do" and "poor" without giving family size: a family with 14 children and an income of $16,500 may have fewer luxuries than a single person with an income of $7,000; samples: 1,000 voters from the first population and 1,200 from the second. (c) Four populations: all seeds identical to those in the packets and planted in the same soil and grown under the same conditions that have been treated with one of the chemicals or left untreated; samples: 35 plants from each population. (d) Population: all 1-year-old Eskimo dogs with HELTHPUP in their diets; sample: 35 such dogs. (e) Two populations: married and unmarried students, respectively, at Peterson University this semester; samples: 40 students from the first population and 50 from the second.

7. (a) X column: 121 to 149; f column: 4, 1, 2, 5, 1, 3, 4, 2, 1, 5, 3, 4, 6, 3, 2, 11, 0, 2, 8, 1, 2, 5, 1, 2, 4, 1, 3, 6, 2.

(b) 121, 149, 8.

9. (a) Not of same width (note 128–130 and 131–135), guideline 4 is ignored (most classes are of width 4, an even number).
(b) 126 appears in 2 classes, scores of 146 to 149 are not included, guideline 4 is ignored (120–124, 125–129, . . . would be better.
(c) Fine, except frequency of 120–122 class is 5, not 6.

11. (a) 10, 19, 20–29.
(b) 50, 13, 100–149.
(c) 5, 13, 50–54 (better than 51–55).

13. Fifteen classes of width .25: .26–.50, 51–.75, . . . , 3.76–4.00. Since no GPA is greater than 4.00, you must either have unequal classes (.25–.49, . . . , 3.50–3.74, but the last one, 3.75–4.00), or else end the classes on "nice" numbers instead of beginning them so.

15. (a) .5–3.5, 3, 2; 3.5–6.5, 3, 5; $\Sigma fX = 14 + 25 + 24 + 33 = 96$. (b) $1,499.50–2,499.50, 1,000, 1,999.50; $2,499.50–3,499.50, 1,000, 2,999.50. ΣfX cannot be determined (no midpoint for last interval). (c) 59.5–64.5, 5, 62; 64.5–69.5, 5, 67; $\Sigma fX = 10 * 62 + 150 * 67 + 120 * 72 + 8 * 77 = 19,926$.

17.   (a)   This is categorical data, so a bar chart is called for. (b) Metric data, which implies a histogram or frequency polygon. But two sets of data are difficult on a histogram, so a frequency polygon is called for.

(a)

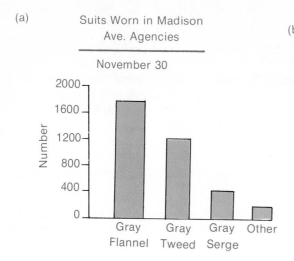

(b)

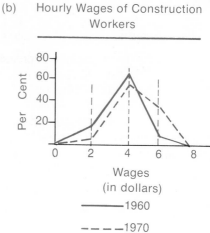

chapter three

measures of center and location

In most of this chapter, you are looking for one number which will best represent the center of scores in a sample. The mode, median and mean are all discussed. The mean especially is essential for further work in statistics. Become very familiar with it. Percentiles, quartiles, and percentile rank are discussed in this chapter also; these are easy once you understand the median.

3.1 INTRODUCTION

We are looking for a single number which will best represent a whole set of scores. There are three ways of doing this in common use: the **mode,** the **median,** and the **mean.** These three terms are lumped together as "measures of center" because each, in its own way, gives an idea of the score about which a whole set centers.

If you look at the coins in your wallet and say, "There are **more** pennies than anything else," you are using the mode. "My verbal aptitude score was well above the **middle** score of 500" uses the median. "The **average** of my quiz grades is 80" almost certainly states that the mean grade is 80, since "average" is most commonly used for "mean" by the general public—although some colleges which claim an average SAT score for entering freshmen of, say, 625 give the median rather than the mean, because their median SAT score is higher and therefore more impressive.

3.2 THE MODE

The **mode** is the score which appears most frequently (abbreviation: *Mo*). The mode is the only measure of center that can be used for categorical data.

Example 1

X: 1, 1, 1, 2, 2, 2, 2, 2, 3, 3, 3, 3, 3, 3, 4, 4, 5, 6, 8.

A score of 3 appears 6 times, more frequently than any other score. *Mo* = 3.

Example 2

Y: 1, 1, 1, 2, 2, 2, 2, 2, 3, 3, 4, 4, 4, 4, 4, 5, 6, 6.

Scores of 2 and 4 each appear 5 times. *Mo* = 2, 4. This set of scores is called **bimodal.** In contrast, the X scores above are unimodal. Occasionally the words **trimodal** (3) and **multimodal** (many) are used. As an adjective, mode becomes "modal": "For the X's, the modal score is 3."

Example 3

　　Z:　1, 7, 8, 10, 12, 14, 17, 23, 24.

　　Each score appears the same number of times (once); the mode doesn't make sense. Often this will be the case when the number of scores is small.
　　For grouped data, the modal class is that class with the greatest frequency; its midpoint is called the **crude mode.**

Example 4

X	f
1–4	2
5–8	7
9–12	5
13–16	3

　　In this example, the modal class is 5–8; the crude mode is 6.5.

　　The mode is the most common score. It is the easiest of the three measures of center to compute, necessitating only a tally of the scores, but no arithmetic beyond adding up the tally marks. It is not affected by extreme scores: the sets 1, 3, 5, 5, 5, 6 and 1, 3, 5, 5, 5, 213 have the same mode. One aberration—for example, a rat, being timed in a maze, who hasn't listened to the explanations and takes a nap on the way—doesn't affect the results. The mode is seldom used, however, except with categorical data. The modal frequency may vary slightly or very much from the frequency of other scores or classes.

3.3　THE MEDIAN

The median is the middle score **after** the scores have been arranged in either increasing or decreasing order (abbreviation: *Md*). The median can be used for ranked or metric data, but not for categorical data.

Example 1

　　X:　8, 1, 10, 7, 17, 23, 12, 15, 24.

　　There are 9 scores here. After ranking them (1, 7, 8, 10, 12, 15, 17, 23, 24), the middle score is 12. *Md* = 12.

Example 2

　　Y:　1, 7, 8, 10, 12, 15, 17, 23, 24, 29.

　　Here there are 10 scores; the middle score is halfway between 12 and 15, so *Md* = 13.5. Note that the median of a set of scores need not be one of the scores.

　　With just a few scores, it is very much better to know what the median means and to use common sense in finding it. Some people like rules, however, and rules are extremely useful with a large number of scores, so here they are: Let *n* stand for the number of scores. If $n/2$ is an integer, the median is halfway between the $(n/2)$th score and the next higher after they have been ranked in increasing order. If $n/2$ is not an integer, then round up to find the number of the median score.

Example 3

n scores have been ranked in increasing order. What is the median if (a) $n = 200$, (b) $n = 201$?

(a) $200/2 = 100$. The median is halfway between the 100th and the 101st scores.

(b) $201/2 = 100.5$. The median is the 101st score.

The median for grouped scores is found by interpolation. The method will not be discussed here since the median is seldom used for grouped data now that computers and calculators have simplified computations for ungrouped data.

The median is about as easy to understand as the mode, but more work to find: you have to arrange the scores in order, and count how many there are. The arithmetic involved is of the simplest sort, however. Like the mode, the median is not affected by extreme scores: the sets 1, 3, 4, 6, 9, 12 and 1, 3, 4, 6, 9, 213 have the same median (5). The median cannot be used for categorical data. It **can** be used for ranked data, and neither the mode nor the mean can. If samples of coffee are tasted and ranked for bitterness, the middle sample can be said to be of "average" bitterness.

3.4   THE MEAN OF UNGROUPED DATA

The mean is the sum of the scores divided by the number of scores. It is used only for metric data. (see p. 15) it un ranked data

A word about abbreviations: *Me* and *Mn* might be confused with median, and *Ma* is phonetically disadvantaged. But an abbreviation is needed: this is the most commonly used measure of center. The **mean of a sample** is called $\overline{X}$ (read this as "X bar") if the scores are called X; the mean of Y scores is $\overline{Y}$ ("Y bar"), and so on. Let n equal the number of scores in the sample.

$$\overline{X} = \frac{\text{sum of scores}}{\text{number of scores}} = \frac{\Sigma X}{n}$$

Example 1

Quiz grades: 70, 75, 80, 80, 85, 90.

$$\text{Mean} = \frac{70 + 75 + 80 + 80 + 85 + 90}{6} = \frac{480}{6} = 80$$

Example 2

Number of inches of rain (X) each day during the week of April 1:   3, 0, 2, 5, 1, 1, 2. Find the mean daily rain.

$$\overline{X} = \frac{3 + 0 + 2 + 5 + 1 + 1 + 2}{7} = \frac{14}{7} = 2.0''$$

To understand the mean better, think of it as the balance point of a balance beam with a weightless board; each weight on the board represents one score. For example, the scores 70, 75, 80, 80, 85, 90 might look like this:

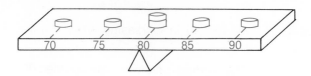

The balance point is at 80, which is the mean of the scores.

What happens when each score in a set is decreased by the same amount? Look at the balance beam and think what happens when each weight is moved a distance of 5 units to the left:

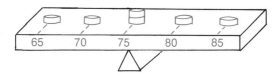

Is it clear that if the beam is to balance, the fulcrum must be moved 5 units to the left? But if you know where the balance point is **after** the weights have been moved 5 units to the left, you can tell where it must have been originally. Similarly, a constant A can be subtracted from each score to give easily handled numbers. After the mean of the new set is found, the constant A is added to it to find the mean of the original set.

Example 3

Find the mean of the X scores in the box below. *of the above wgts.*

X	$A = 5$ $X - 5$	$A = 20$ $X - 20$	$A = 70$ $X - 70$	$A = 80$ $X - 80$
70	$-A =$ 65	50	0	−10
75	$-A =$ 70	55	5	− 5
80	$-A =$ 75	60	10	0
80	$-A =$ 75	60	10	0
85	$-A =$ 80	65	15	5
90	$-A =$ 85	70	20	10
Sum: 480	450	360	60	0
Mean: 80	75	60	10	0
	$= \bar{X}$	$\bar{X} = 75 + 5$	$\bar{X} = 60 + 20$	$\bar{X} = 10 + 70$	$\bar{X} = 80 + 0$

All $\bar{X} = 80$!!

Note that any constant A can be used; choose an A which gives easily handled numbers. Ordinarily, the arithmetic will be simpler if you estimate the mean and let A be some convenient number near this estimate.

Example 4

Find the mean of the Y scores in the box below, letting (a) $A = 40$, (b) $A = 200$, (c) $A = 204$.

	(a)	(b) *Best choice*	(c) $\Sigma(Y - \bar{Y}) = 0$
Y	Y − 40	Y − 200	Y − 204
197	157	− 3	− 7
198	158	− 2	− 6
199	159	− 1	− 5
201	161	1	− 3
203	163	3	− 1
207	167	7	3
212	172	12	8
215	175	15	11
	1312	32	0

$\bar{Y} = 204$

$$\bar{Y} = \frac{1312}{8} + 40$$
$$= 164 + 40 = 204$$

$$\bar{Y} = \frac{32}{8} + 200$$
$$= 204$$

$$\bar{Y} = \frac{0}{8} + 204$$
$$= 204$$

Clearly, 40 is a miserable choice for A (although the correct mean is found). $A = 200$ is probably the obvious choice. (c) was included to show you that if the constant subtracted from each score equals the mean, then $\Sigma(X - \bar{X}) = 0$. This will always be the case. **For any set of scores, the sum of the differences from the mean equals zero. If the sum of the differences from some constant is zero, then that constant is the mean.** See Exercise 23, page 42.

(as -p.5) $\Sigma(X - \bar{X}) = \Sigma X - \Sigma \bar{X} = \Sigma_x - n\bar{X} = \Sigma_x - N\left(\frac{\Sigma_x}{N}\right) = 0$

Rule 1   Rule 3   Def. of $\bar{X}$

If you have only a few scores, the mean is the hardest measure of center to compute. If you have 1,000 scores, it's touch-and-go whether it's more difficult to add them for the mean or to arrange them in increasing order for the median. Be grateful that a computer can do either of these tasks for you, and arrange to use one (or at least a calculator) if you must find the mean of a great many scores.

The mean may not exist. Assume, for example, that a psychologist times 5 rats in a maze, but removes any rat which does not solve the maze in 5 minutes and simply records the time as "over 5." What is the mean time for the rats if their recorded times are 2, 3, 3, 4, over 5 minutes? The first four numbers can be added easily enough, but is "over 5" best represented by 6 or by 17 minutes?

The mean, in contrast to the mode and median, does depend on each of the scores. Sometimes this is not an advantage. For the scores 7, 9, 9, 31, is the median (9) or the mean (14) more representative? If you are an economist reporting on wages (in thousands of dollars) in a small shop, you would probably prefer the median, since the mean is raised so much by one high wage (the owner's).

So far, the mean does not seem to have a clear-cut advantage over the mode and the median, and yet it is used enormously more by statisticians. The first reason is that dependence of the mean on all the scores is usually a tremendous advantage. If your quiz grades in a course are 50, 50, 50, 90, 100, you would certainly prefer the mean (68) to the median (50) — and not just because it is higher, but because it fits your sense of justice. The second reason is that the mean is determined algebraically, and is amenable to algebraic operations. We shall find that when we look at a sample and make inferences about the population from which it is taken, it is the mean that will be of greatest use to us. You will appreciate the mean more fully as you learn to use it—and you will have lots of practice.

A minor algebraic advantage of the mean can be seen immediately. Consider the following yields of bushels of beans in experimental plots:

Field X:　1, 3, 3, 3, 4, 4, 5, 5, 6, 6. $Mo = 3$, $Md = 4$, $\overline{X} = 4$, $n = 10$.

Field Y:　1, 1, 2, 2, 4, 4, 4, 6, 6, 6, 6, 7, 7, 9, 10. $Mo = 6$, $Md = 6$, $\overline{Y} = 5$, $n = 15$.

Now suppose the data from plots in both fields are joined together to give a set of 25 scores. To find the new mode and median it is necessary to rearrange the scores as follows and start over:

Fields X and Y:　1, 1, 1, 2, 2, 3, 3, 3, 4, 4, 4, 4, 4, 5, 5, 6, 6, 6, 6, 6, 6, 7, 7, 9, 10.
$Mo = 6$, $Md = 4$.

$$\overline{X} = \frac{\Sigma X + \Sigma Y}{n} = \frac{40 + 75}{25} = 4.6$$

To find the mean of the joint set, however, it is not necessary to interleave the scores in this fashion; we can reason as follows: $\overline{X} = \dfrac{\Sigma X}{10} = 4$, so $\Sigma X = 40$, and $\overline{Y} = \dfrac{\Sigma Y}{15} = 5$, so $\Sigma Y = 75$. Therefore the sum of all 25 scores is $\Sigma X + \Sigma Y = 40 + 75 = 115$ and the mean of the joint set is $\dfrac{115}{25} = 4.6$. This example may not be persuasive since the number of scores is so small; you can probably count or re-order them mentally to find the mode and median, without writing down the joint set—but imagine doing the same with a hundred or more scores in each set!

3.5　THE MEAN OF GROUPED DATA

To find the mean of grouped scores, treat all the scores in a class as though they were concentrated at the midpoint of the class. Finding the mean of the distribution shown below is, then, equivalent to finding the mean of the set of scores consisting of 15 2's, 10 5's, and 5 8's.

X	f
1–3	15
4–6	10
7–9	5
	30

But adding 15 2's has the same result as 15 * 2, adding 10 5's gives 10 * 5, and the sum of 5 8's = 5 * 8. The mean is $\dfrac{15 * 2 + 10 * 5 + 5 * 8}{30} = \dfrac{30 + 50 + 40}{30} = \dfrac{120}{30} = 4$. It is easier to add a column labelled fX and sum this column. (Note that it is customary to write fX, but we mean by this "multiply the frequency in each class by the midpoint of that class.")

X	f	Midpoint	fX
1–3	15	2	30
4–6	10	5	50
7–9	5	8	40
	30		120

$\frac{\Sigma f(x)}{n}$

$\Sigma f(x)$

$$\overline{X} = \frac{120}{30} = 4.$$

We summed the f and fX columns—that is, we found $\Sigma f = n$ and ΣfX. The formula for the mean of grouped sample data is

$$\overline{X} = \frac{\Sigma fX}{\Sigma f} = \frac{\Sigma fX}{n}$$

Example 1

Find the mean of this distribution:

X	f	Midpoint	fX
10–19	14	14.5	203.0
20–29	14	24.5	343.0
30–39	22	34.5	759.0
	50		1305.0

$$\overline{X} = \frac{1305.0}{50} = 26.1.$$

It will frequently be necessary to find the mean of a grouped distribution; it is worth finding a shortcut. As with ungrouped data, subtracting the same number A from each score (that is, each midpoint) can lead to simpler arithmetic, and then the constant A will be added again to the mean of the new scores to find the mean of the original distribution.

Example 2

Find the mean of the distribution shown in Example 1 by letting $A = 24.5$.

Midpoint	f	X − A = Midpoint − 24.5	f(X − A)
14.5	14	−10	−140
24.5	14	0	0
34.5	22	+10	220
	50		80

$$\overline{X} = \frac{80}{50} + 24.5 = 26.1, \text{ as before.}$$

If A is wisely chosen (if it is a value near the mean), the arithmetic may be considerably reduced. Usually A is chosen to be the midpoint of a class.

Example 3

Find the mean of the distribution in the box below by letting (a) $A = 185$, (b) $A = 175$.

		(a)		(b)	
Midpoint	f	$X - 185$	$f(X - 185)$	$X - 175$	$f(X - 175)$
175	12	−10	−120	0	0
180	12	− 5	− 60	5	60
185	13	0	0	10	130
190	14	5	70	15	210
195	9	10	90	20	180
	60		− 20		580

(a) $\bar{X} = \dfrac{-20}{60} + 185 = 184.7$.

(b) $\bar{X} = \dfrac{580}{60} + 175 = 9.7 + 175 = 184.7$.

Note that in (a) the choice of A (= 185) is quite close to the mean, and the arithmetic is simpler than when 175 is chosen for A.

> ***Round-off:*** Let us agree that computations will, in general, be carried to one more decimal place than in the given scores, and that = rather than ≈ will be used for the answer. Thus $\dfrac{-20}{60} + 185 = 184.7$, rather than $\dfrac{-20}{60} + 185 \approx 184.7$. Remember that 5 is rounded to an even number, so 15.35 becomes 15.4 but 15.25 becomes 15.2.

Whichever method is used for finding the mean, it is well to have an intuitive idea of its value. Always check a computed mean and see if it is reasonable. You will at least discover decimal point errors and other gross arithmetical mistakes.

Example 4

Which of the following **cannot** be the mean of this distribution?
(a) 204, (b) 12, (c) 23, (d) 100, (e) 70.
Why?

X	f
0−49	9
50−99	8
100−149	3
150−199	1

In this distribution, 204 is outside the range of scores; 12 and 23 are both impossible, since we consider all scores in an interval to be at the midpoint of an interval (the smallest midpoint is 24.5); 100 is within the range of midpoints, but only 4 scores are larger and 17 are smaller with equal class widths. Only (e) 70 is reasonable; the true mean is 65.0.

If you have trouble estimating means, visualize balance beams again. Be careful to sketch all scores at the midpoints of class intervals. The balance beam for the above distribution would look like this:

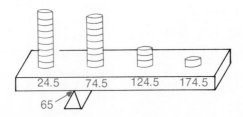

3.6 QUARTILES, PERCENTILES, AND PERCENTILE RANK FOR UNGROUPED DATA

The same reasoning that is used for the median may be used to find quartiles and percentiles. The first quartile (notation: Q_1) is the number that is larger than one quarter of the scores (and smaller than three quarters). The second quartile, Q_2, is the median. The third quartile, Q_3, is larger than three quarters of the scores. The sixtieth percentile (notation: P_{60}) is larger than 60 per cent of the scores (and smaller than 40 per cent). Note that $P_{25} = Q_1$, $P_{50} = Md$, $P_{75} = Q_3$. Quartiles and percentiles are called **measures of location.**

If eight scores are ranked, Q_1 will be larger than two scores and smaller than six; the first quartile will be halfway between the second and third scores. If $\frac{1}{4} * n$ is an integer, then Q_1 is halfway between this score and the next higher. If $\frac{1}{4} * n$ is not an integer, it is customary to round up to find Q_1. Thus if $n = 50$, Q_1 will be the thirteenth score after they are put in increasing order, since 50/4 = 12.5, which rounds up to 13. 12 scores are smaller and 37 are larger. But if you think of the thirteenth score as being half in each group, there are 12.5 below and 37.5 above Q_1.

Q_3 and P_a are determined in a similar fashion: multiply n by $\frac{3}{4}$ and by $\frac{a}{100}$, respectively. If the result is an integer, choose the number halfway between this score and the next; otherwise, round up.

Example 1

Ninety scores are ranked in increasing order. Find (a) Q_1, (b) Q_3, (c) P_{60}, (d) P_{16}.

(a) $\frac{1}{4} * 90 = 22.5$; Q_1 is the 23rd score.

(b) $\frac{3}{4} * 90 = 67.5$; Q_3 is the 68th score.

(c) $\dfrac{60}{100} \star 90 = 54$; P_{60} is halfway between the 54th and 55th scores.

(d) $\dfrac{16}{100} \star 90 = 14.4$; P_{16} is the 15th score.

The **percentile rank** of a score is the percentage of scores which are smaller. If 75 is the 349th score in a list of 400 scores arranged in increasing order, then 348 are smaller. $\dfrac{348}{400} \star 100 = 87$ per cent are smaller than 75; the percentile rank of a score of 75 is 87.

Example 2

A high school student has an average grade of 92.5; he is sixteenth highest in a class of 300. What is his percentile rank?

He does better than $300 - 16 = 284$ in his class. His percentile rank is $\dfrac{284}{300} \star 100 = 94.7$.

3.7 VOCABULARY AND SYMBOLS

mode	*Mo*	mean	$\overline{X}, \overline{Y}$
bimodal		quartiles	Q_1, Q_3
multimodal		percentiles	P_a
crude mode		percentile rank	
median	*Md*		

3.8 EXERCISES

1.   Translate from English into mathematical language:
(a) "To find the mean of X scores in a sample of size n, subtract a constant A from each score, find the mean of the new set, and then add A; label the result $\overline{X}$."
(b) "To find the mean of a grouped frequency distribution of Y scores, subtract a constant A from each midpoint, multiply each of these differences by the corresponding frequency, sum the products, divide by the total number of scores, and add A to the quotient; label the result $\overline{Y}$."

2.   12 students in a statistics class receive the following quiz grades:

Quiz #1:   72, 75, 75, 97, 54, 72, 86, 72, 63, 79, 82, 91
Quiz #2:   78, 42, 72, 88, 86, 97, 91, 79, 82, 86, 91, 74

(a) Find the mode, median, and mean for each quiz.
(b) Find the mode, median, and mean for both quizzes together (24 grades).
(c) Which measure of center best represents the grades on the second quiz?
(d) Find Q_1 and Q_3 for the second quiz.
(e) What is the percentile rank of a score of 88 on the second quiz?

3.   (a) Find the mode, median, and mean for each of the following sets of scores if possible; if not, explain why.
(i) Weekly pay (in dollars) of cafeteria workers: 70, 60, 70, 70, 6, 4, 90, 30, 16, 80, 12, 70.

(ii) Grades of students on a statistics quiz: 50, 100, 90, 100, 50, 60, 80, 70, 40, 100, 90, 95, 75, 50, 74, 81, 76, 69, 75.

(iii) A scale used for weighing letters measures up to 10 ounces. Twenty-four pieces of first class mail have the following weights (in ounces): .7, .9, .8, 1.1, .5, .8, .5, 1.1, .7, .6, .4, .9, 1.3, .4, .6, .5, 1.2, .4, .3, .5, over 10, 3.0, .2, .4.

(iv) Number of cars parked in the Student Center parking lot at noon each day, April 17–23:   212, 650, 435, 672, 510, 621, 287.

(v) Number of each make of car parked in the Student Center parking lot at noon on April 22:   Ford 131, Chevrolet 142, VW 183, other 165.

(b)   For each set of scores in (a), which measure of central tendency seems to be most representative? (That is, if one number is to be chosen to represent the set, should that number be the mode, the median, or the mean?)

4.   The annual salaries of 20 men are given below.

$8000	$ 9800	$11200	$11700	$12400
8500	10000	11400	11800	12800
8700	10400	11600	11900	13500
9200	10800	11700	11900	33000

(a)   Find the median and the mean salary.
(b)   Is the median or the mean more representative? Why?
(c)   Find Q_1, Q_3, and P_{40}.

5.   Which measure of central tendency would each of the following be likely to use to summarize his data?

(a)   The manager of a shoe store contemplating the shoes in his shop listed by size.

(b)   An economist studying the income of heads of families in low-cost housing in Pawtucket, R.I.

(c)   An engineer studying the maximum load supported by each of 100 cables produced by his company.

6.   Find the mean of the following scores; choose A wisely.
(a) X: 15, 17, 18, 19, 26, 22, 24, 16, 21, 17.
(b) Y: 514, 514, 520, 536, 518, 504, 498, 537, 521.
(c) Z: .0021, .0027, .0019, .0018, .0023, .0026, .0029, .0018.

7.   Do **not** compute the mean for the following scores. Instead, decide which of the values given is the best candidate for the mean.

Scores	Candidates for $\overline{X}$
(a) 15, 17, 24, 56, 63, 81	14, 16, 30, 43, 63, 89
(b) 190, 230, 112, 200, 180, 140	100, 120, 170, 200, 291
(c) 24, 12, 18, 36, 10, 12	4, 10, 12, 20, 25

8.   The rainfall each month in Miami, Florida, is shown in the following table. Find (a) the mean monthly rainfall, (b) the median.

Month	Jan	Feb	March	April	May	June	July	Aug	Sept	Oct	Nov	Dec
Inches of rain	2.0	1.8	2.5	3.7	6.8	7.5	6.5	7.1	9.7	8.0	2.6	1.8

9. Thirty families live in an apartment house on Delancey Street in Brooklyn. The number of children in each family is as follows:

| 3 | 1 | 2 | 0 | 4 | 6 | 1 | 0 | 1 | 5 | 2 | 3 | 2 | 0 | 7 |
| 2 | 2 | 0 | 3 | 1 | 3 | 2 | 4 | 2 | 4 | 0 | 4 | 3 | 1 | 4 |

Find (a) the mean, (b) the median, and (c) the modal number of children.

10. The students in a statistics course at Hyde College obtain the following grades on a quiz:

24	30	32	40	48	50	58	60	60	62	63	64	64	67	67	67
70	70	71	72	74	74	75	75	76	77	78	78	78	78	79	80
81	83	83	83	84	84	84	84	85	86	87	88	88	88	89	90
92	92	93	95	95	97	98	99	100	100

    (a) What is the modal grade?
    (b) What is the median grade?
    (c) Find Q_1 and P_{10}.
    (d) What is the percentile rank of a score of 93?

11. Comment on each of the following:
    (a) A high school principal said to one of his teachers, "All the classes in this school took the same test, but your class scored below the average. I don't want any of our classes to be below the average."
    (b) "I am in the first quartile of my class," Ed boasted.
    (c) "My percentile rank in a class of 90 students is 100," Ann boasted.

12. In each of the following, decide whether it is possible for the given measure of center to have the given value if the other requirements are satisfied. If impossible, explain why. If possible, give a set of scores which satisfy all the requirements. (For example: $n = 8$, lowest score = 7, $Mo = 5$. Answer: impossible, since every score is greater than 5.)
    (a) $n = 5$, lowest score = 4, range = 10, $\overline{X} = 14$.
    (b) $n = 5$, lowest score = 4, range = 10, $Md = 14$.
    (c) $n = 5$, lowest score = 50, highest score = 100, $\overline{X} = 55$.
    (d) $n = 8$, range = 6; the mode is meaningless since no score is repeated.

13. In each of the following, give two examples of scores which meet the given requirements. (For example: $n = 6$, range = 10, mode = 4; range = difference between highest and lowest scores. Answers: 3, 4, 4, 4, 7, 13, or 0, 2, 4, 4, 4, 10, for example.)
    (a) $n = 5$, range = 7, $Md = 54$.
    (b) $n = 3$, range = 10, $\overline{X} = 90$.
    (c) $n = 8$, range = 20, $Mo = 42$, $Md = 45$.
    (d) $n = 4$, $\overline{X} = 10$, $Md = 8$.

14. Find the mean of each distribution (a) by using the definition $\left(\overline{X} = \dfrac{\Sigma fX}{n}\right)$; (b) by making the best possible choice of A.

(i)	X	f		(ii)	Y	f		(iii)	Z	f
	1–5	4			30–39	10			0–20	9
	6–10	4			40–49	12			21–40	5
	11–15	2			50–59	14			41–60	6
					60–69	24			61–80	10

15. (a) Which of the following is the best choice for A in finding the mean of the distributions below? 59.5, 34.5, 14, 40, 24.5, 60.

(i) X	f		(ii) Y	f		(iii) Z	f
10–12	4		10–29	2		40–49	2
13–15	7		30–49	3		30–39	5
16–18	3		50–69	8		20–29	4
19–21	2		70–89	7		10–19	3

(b) Use the shortcut method (subtract A from each midpoint) to find the mean for each distribution.

16. Time between needed repairs for a fleet of rental cars were noted as follows:

Time (days)	Number of Cars
0–14	8
15–29	6
30–44	32
45–59	16
60–74	8
75–89	10

Find the mean time between repairs.

17. Two different methods of teaching archery are used for groups of freshmen; their individual scores on rounds at 30 yards are as follows:

Method I	Men	100	95	110	140	80	100		
	Women	110	60	120	100	90	70	130	110
Method II	Men	100	70	110	130	90	100	130	120
	Women	90	100	110	100	150	60	120	120

(a) Compare the mean scores using different methods of instruction of (i) men, (ii) women, (iii) all students.
(b) Find the mean score of (i) all men taking archery, (ii) all women.

18. Students in an Introduction to Music Theory course take part in a course evaluation, using a scale of 1 to 6 where 1 means complete disagreement and 6 means complete agreement. The statement "Your professor makes him/herself available to interested students or those having academic problems" receives the following ratings:

Score: 1 2 3 4 5 6
Frequency: 12 4 8 15 24 17

What is the professor's mean rating on this statement?

19. Find the mean and the crude mode of each of the following:

(a) X	f	(b) Y	f	(c) Z	f	(d) W	f
100–102	30	10–14	2	.5055–.5059	1	10–19	6
103–105	20	15–19	5	.5050–.5054	2	20–29	0
106–108	60	20–24	8	.5045–.5049	8	30–39	8
109–111	40	25–29	0	.5040–.5044	17	40–49	4
112–114	10	30–34	5	.5035–.5039	8	50–59	7
				.5030–.5034	0	60 and over	3
				.5025–.5029	4		

20.   For six years, Everett Whitney invested in ABC stock as follows:

Year	Number of Shares	Total Price Paid
1970	140	$1400
1971	100	1100
1972	300	3800
1973	200	2900
1974	160	1800
1975	700	1500

What is the average cost per share of his ABC stock?

21.   (a) Find the mean of the scores 1, 1, 1, 1, 2, 3, 4, 4, 4, 4, 5, 6.
       (b) Group the scores in two classes, 1–3 and 4–6, and find the (grouped) mean.
       (c) Explain the discrepancy between the answers to (a) and (b).

22.   In a two-week period, a small grocery store sells the following number of quarts of milk: 78, 59, 47, 56, 76, 84, 76, 62, 50, 54, 74, 86.

(a) What is the mean number of quarts of milk sold at this market?
(b) How many quarts of milk should the manager order each day if he wants to be overstocked not more than 10 per cent of the time?

23.   Study this proof that $\Sigma(X - \bar{X}) = 0$:

$$\Sigma(X - \bar{X}) = \Sigma X - \Sigma \bar{X} \qquad \text{(Rule 1 for } \Sigma, \text{ page 5)}$$
$$= \Sigma X - n\bar{X} \qquad \text{(Rule 3 for } \Sigma, \text{ page 5)}$$
$$= \Sigma X - n\left(\frac{\Sigma X}{n}\right) \qquad \left(\bar{X} = \frac{\Sigma X}{n}\right)$$
$$= 0$$

Prove that if $\Sigma(X - A) = 0$, where A is a constant, then $A = \bar{X}$.

24.   The mean grade point average (GPA) of 1000 students in the School of Liberal Arts at Triple University during one semester was 2.85, while it was 2.90 for 200 students in the School of Engineering and 3.10 for 400 students in the School of Education. What was the mean GPA of all the undergraduates in the three schools of Triple University that semester?

25.   TV ratings (per cent of viewers tuned in to that show) of the "Golden Boy" series are as follows over an 8-week period: 12.4, 14.5, 10.6, 18.6, 16.1, 15.2, 13.3, 10.0. What is (a) the median rating? (b) the mean rating? (c) the percentile rank of a 12.4 rating?

26.   Quarts of milk bottled by Freshpack Milk Co. were measured by the Tampa Department of Weights and Measures, with the following results:

Volume	Number of Bottles
At least 30.0 but less than 30.5 oz.	2
At least 30.5 but less than 31.0 oz.	5
At least 31.0 but less than 31.5 oz.	29
At least 31.5 but less than 32.0 oz.	48
At least 32.0 but less than 32.5 oz.	16

Find the mean volume of the bottles tested by the Department of Weights and Measures.

ANSWERS

1. (a) $\overline{X} = \dfrac{\Sigma(X-A)}{n} + A$. (b) $\overline{Y} = \dfrac{\Sigma f(X-A)}{n} + A$.

3. (a) (i) $Mo = \$70$; $Md = \$65$; $\overline{X} = \$48.20$. (ii) $Mo = 50, 100$ (bimodal); $Md = 75$; $\overline{X} = 75.0$. (iii) $Mo = .4$ and $.5$ (bimodal); $Md = .65$; no mean (open-ended). (iv) Mo is meaningless; $Md = 510$; mean $= 483.9$. (v) $Mo = 183$; no median or mean (categorical data).
(b) (i) Probably the median, since part-time workers affect the mean so much. (ii) Mean or median. (iii) Median. (iv) Median. (v) Mode.

5. (a) Mo. (b) $\overline{X}$ (no extraordinarily large income is allowed). (c) $\overline{X}$.

7. (a) 43. (b) 170. (c) 20.

9. (a) 2.4. (b) 2. (c) 2.

11. (a) "Average" probably signifies "mean," but it is not clear. If no class is below the mean, all have exactly the same result; this is highly unlikely.
(b) He could be **at** the first quartile but not **in** it, and he shouldn't boast if he is better than just one quarter of the class. "I am in the top quarter of my class" is probably what he meant.
(c) Even if Ann is first in her class, her percentile rank is $\dfrac{89}{90} * 100 = 99$, not 100.

13. (There are many possible answers to each part.) (a) 48, 50, 54, 54, 55, or 54, 54, 54, 57, 61. (b) 85, 90, 95, or 86, 88, 96. (c) 42, 42, 42, 44, 46, 48, 50, 62, or 40, 42, 42, 42, 48, 50, 55, 60. (d) 7, 8, 8, 17, or 4, 6, 10, 20.

15. (a) (i) 14. (ii) 59.5 (iii) 34.5 or 24.5.
(b) (i) $\overline{X} = 9/16 + 14 = 14.6$. (ii) $\overline{Y} = -0/20 + 59.5 = 59.5$. (iii) $\overline{Z} = -80/14 + 34.5 = 28.8$.

17. (a)

	Men	Women	All students
Method I	$25/6 + 100 = 104.2$	$-10/8 + 100 = 98.8$	$(25 - 10)/14 + 100 = 101.1$
Method II	$50/8 + 100 = 106.2$	$50/8 + 100 = 106.2$	106.2

(b) Men: $(25 + 50)/14 + 100 = 105.4$
Women: $(-10 + 50)/16 + 100 = 102.5$

19. (a) $\overline{X} = -60/160 + 107 = 106.6$; $Mo = 107$. (b) $\overline{Y} = 5/20 + 22 = 22.2$; $Mo = 22$. (c) $\overline{Z} = -.0025/40 + .5042 = .50414$; $Mo = .5042$. (d) No mean (open-ended class); $Mo = 34.5$.

21. (a) $36/12 = 3$. (b) $(2 * 6 + 5 * 6)/12 = 3.5$. (c) They are not equal because the grouped mean treats the scores as though concentrated at the midpoint, but they are not; in each class 2/3 are below the midpoint.

23. $\Sigma(X - A) = 0$ (Given)
$\Sigma X - \Sigma A = 0$ (Rule 1 for Σ, page 5)
$\Sigma X - nA = 0$ (Rule 3 for Σ, page 5)
$A = \dfrac{\Sigma X}{n}$
$A = \overline{X}$

25. (a) 13.9. (b) $-1.3/8 + 14 = 13.84$. (c) $(2/8) * 100 = 25$.

chapter four

measures of variability

To summarize a set of scores, it is customary to give two measures: one (the mean, median, or mode) to describe the center or "average" score, and the other to give some idea of the amount of variability or the spread of the scores. We shall look now at five different measures of variability: the variation ratio, the range, the semi-interquartile range, the variance, and the standard deviation. Of these, the first two are easy to find and to understand; the last three are more difficult to find. Concentrate on variance and standard deviation; these are used so much in the rest of the book that you must have a clear understanding of them. This chapter concludes with a brief section on a measure of skewness.

4.1 INTRODUCTION

"Oklahoma and the Canary Islands have the same mean annual temperature." Would you therefore be equally comfortable throughout the year in both places? Oklahoma may vary from 0° to 110° during the year, the Canary Islands from 60° to 80°, so during most of the year their climates are very different.

The following sets of scores have the same mode, median, mean, and number of scores, but differ markedly:

$$X: \quad 40, 50, 50, 50, 50, \ 60$$

$$Y: \quad 0, 30, 50, 50, 70, 100$$

These sets of scores differ in their spread about the center; we shall look now at five different measures of spread or variability.

4.2 THE VARIATION RATIO

The **variation ratio** is the proportion of nonmodal scores (abbreviation: V). This is the only measure of variability that can be used for categorical data.

$$V = \frac{\text{(number of scores)} - \text{(modal frequency)}}{\text{number of scores}} = \frac{n - f_{Mo}}{n}$$

V is a number between 0 and 1. If V is near 0, most of the scores are at the mode; if V is near 1, the mode is less representative.

Example 1

Compute the variation ratio for each of the three frequencies:

Category	Frequency		
	(a)	*(b)*	*(c)*
A	1	1	2
B	1	3	5
C	9	7	3
D	1	1	2

(a) $V = \dfrac{12 - 9}{12} = .25.$

(b) $V = \dfrac{12 - 7}{12} = .42.$

(e) $V = \dfrac{12 - 5}{12} = .58.$

4.3 THE RANGE

The **range** of a distribution is the difference between the highest and lowest scores.

Example 1

X: 5, 12, 13, 13, 14, 15, 15, 15, 18, 20.

The range is $20 - 5 = 15$.

Example 2

Y: 5, 5, 11, 11, 11, 19, 19, 19, 20, 20.

The range is again $20 - 5 = 15$.

Example 3

Z	f
1–3	4
4–6	7
7–9	3
10–12	2

The range is $12.5 - .5 = 12$. (Note that here we use the class boundaries.)

The range is easy to understand and easy to compute. It depends only on the extreme scores. As a result, two samples of the same size taken from a given population may have quite different ranges. On the other hand, the scores in Examples 1 and 2 have the same mean (14) and range (15) but "look" quite different.

4.4 THE SEMI-INTERQUARTILE RANGE Q

The **semi-interquartile range** (abbreviation: Q) is half the difference between the third and first quartile scores:

$$Q = \frac{Q_3 - Q_1}{2}$$

Example 1

X: 5, 12, 13, 13, 14, 15, 15, 15, 18, 20.

$$Q_3 = 15,\ Q_1 = 13,\ Q = \frac{15 - 13}{2} = 1.0$$

Example 2

Y: 5, 5, 11, 11, 11, 19, 19, 19, 20, 20.

$$Q_3 = 19,\ Q_1 = 11,\ Q = \frac{19 - 11}{2} = 4.0$$

The semi-interquartile range (also called the **quartile deviation**) is harder to find and harder to understand than the range. It is not affected by extreme scores as the range is: the range depends on the largest and smallest scores, while the semi-interquartile range depends on the middle per cent of scores. The next two measures of variability depend on the mean; the semi-interquartile range can be used when the mean cannot be determined. (Do you remember the rat who took over 5 minutes to get through a maze? See Section 3.4, page 33.) When the median is used as measure of center, the semi-interquartile range is often used as measure of variability.

4.5 CRITERIA FOR A MEASURE OF VARIABILITY

The range depends only on the extreme scores, the semi-interquartile range on the middle 50 per cent of the scores. Let us search now for a measure which depends on **all** the scores. It should have other characteristics as well: it should be small when all the scores are close together and large when the scores are widely scattered; it should not get large just because the number of scores is large.

How about taking the sum of the deviations from the mean? The sum $\Sigma(X - \bar{X})$ is always equal to 0, as we proved in Exercise 23, page 42, and as the following example illustrates again:

$$X: 1, 2, 6;\ \bar{X} = 3;\ \Sigma(X - \bar{X}) = (1 - 3) + (2 - 3) + (6 - 3) = 0$$

Next you might decide to ignore the minus signs: add the absolute values† of the deviations from the mean. In the example above, you would get $|1-3| + |2-3| + |6-3| = 2 + 1 + 3 = 6$. If there were 1,000 scores, this sum might be large even if all the scores

---

†The absolute value of a number is its magnitude without regard for sign; the notation is two vertical lines, one on either side of the number. Thus, $|7| = 7$ and $|-7| = 7$.

were between 1 and 6. If the sum of the absolute values of the deviations from the mean is divided by the number of scores, we arrive at a measure of variability called the **mean absolute deviation.** It satisfies the criteria set up in the first paragraph, but is seldom used now. One reason is that absolute values are difficult to work with algebraically; the mean absolute deviation does not enter nicely into formulas indicating its relation to other statistics. Another reason is that, surprisingly, $\dfrac{\Sigma|X-A|}{n}$ is a minimum **not** when A is the mean, but rather when A is the median. (See Exercise 20, page 57.)

4.6 VARIANCE AND STANDARD DEVIATION FOR UNGROUPED DATA

Any number, positive or negative, becomes positive when squared: $(+4)^2 = +16$, $(-2)^2 = +4$, $(-1.2)^2 = +1.44$. Instead of taking the mean of the absolute values of the deviations of scores from the mean, we might take the mean of the squares of these deviations, arriving at the mean square deviation $\dfrac{\Sigma(X-\overline{X})^2}{n}$. This used to be called the variance, and a comparable formula is still used for the variance of a population, as we shall discover in Chapter 7. But recently it has become customary among most statisticians to divide by $n-1$ instead of n to find the **variance** (abbreviation: s^2) of a sample.

$$\text{Variance} = s^2 = \frac{\Sigma(X-\overline{X})^2}{n-1}$$

Why is the denominator changed from n to $n-1$? Later we shall want to estimate the population variance by using the variance of a sample. If n is in the denominator for the sample variance, the estimate will be consistently too low. Note that a fraction is increased if the denominator is decreased (1/2 is greater than 1/3). Changing n to $n-1$ in the denominator increases the fraction just enough so that the sample variance will, on the average, give a good estimate of the population variance. "On the average" means that if the variance of all possible different samples is computed, the average (mean) of all these variances will approximate pretty well the population variances. The variance of any one particular sample may be too small or too large, however. (See Exercise 19, page 57.)

The variance satisfies all the criteria set up in Section 4.5 for a measure of variability. It has one drawback, however: it measures in square units. If X scores are in inches, for example, then so are $\overline{X}$ and the difference $X-\overline{X}$; then $(X-\overline{X})^2$ and the variance are in square inches. To get back to the original unit, the (positive) square root is taken. The square root of the variance is called the **standard deviation.** It is another measure of variability, and it is measured in the same units as the scores. The abbreviation for the standard deviation of a sample is s.

$$s = \sqrt{s^2} = \sqrt{\frac{\Sigma(X-\overline{X})^2}{n-1}}$$

Note that to find the standard deviation using this formula you must first find the mean. Be careful to square first and then add; $\Sigma(X-\overline{X})^2$ is not the same as $[\Sigma(X-\overline{X})]^2$. Use Table 1, Appendix C, to find the square root.

Example 1

Find the variance and standard deviation of the X scores: 1, 2, 6.

$\overline{X} = 9/3 = 3.$

X	$X - \overline{X}$	$(X - \overline{X})^2$
1	−2	4
2	−1	1
6	3	9
9		14

mean = 3

$s^2 = 14/2 = 7.00, s = \sqrt{7.00} = 2.65$

Example 2

Find the variance and standard deviation of Y: 10, 12, 14, 18.

$\overline{Y} = 54/4 = 13.5.$

Y	$Y - \overline{Y}$	$(Y - \overline{Y})^2$
10	−3.5	12.25
12	−1.5	2.25
14	.5	.25
18	4.5	20.25
54		35.00

$s^2 = 35/3 = 11.67, s = \sqrt{11.67} = 3.41$

Be careful not to round off too early. If you are finding s to one decimal place, keep at least two until after you have taken the square root.

This last example involved more arithmetic than Example 1; it is not hard to find examples where the necessary computations are still more complicated. Fortunately, another form of the formula for variance or standard deviation can be found that frequently simplifies computations. These computational formulas are:

$$s^2 = \frac{\Sigma X^2 - \dfrac{(\Sigma X)^2}{n}}{n - 1}$$

$$s = \sqrt{\frac{\Sigma X^2 - \dfrac{(\Sigma X)^2}{n}}{n - 1}}$$

Squares of numbers are given in Table 1, Appendix C; these formulas are also handy for a calculator. Be very careful to distinguish between ΣX^2 (square first and then add) and $(\Sigma X)^2$ (add first and then square the sum).

Where do these computational formulas come from? See if you can follow these steps, at least closely enough to realize the computational formulas are legitimate (remember that $\overline{X}$ is a constant, so $\Sigma \overline{X}^2 = n\overline{X}^2$, by Rule 3 for Σ, page 5):

$$\Sigma(X - \overline{X})^2 = \Sigma(X^2 - 2X\overline{X} + \overline{X}^2)$$

$$= \Sigma X^2 - \Sigma 2X\overline{X} + \Sigma \overline{X}^2 \qquad \text{(Rule 1 for } \Sigma, \text{ page 5)}$$

$$= \Sigma X^2 - 2\overline{X}\Sigma X + n\overline{X}^2 \qquad \text{(Rules 2 and 3 for } \Sigma, \text{ page 5)}$$

$$= \Sigma X^2 - 2\left(\frac{\Sigma X}{n}\right)\Sigma X + n\left(\frac{\Sigma X}{n}\right)^2 \qquad \left(\overline{X} = \frac{\Sigma X}{n}\right)$$

$$= \Sigma X^2 - 2\frac{(\Sigma X)^2}{n} + \frac{(\Sigma X)^2}{n}$$

$$= \Sigma X^2 - \frac{(\Sigma X)^2}{n}$$

Now divide by $n - 1$. This shows that the definition of the variance and the computational formula for it are equivalent.

Example 3

Find the mean and standard deviation of the sample Z: 1, 2, 7.

Z	Z^2
1	1
2	4
7	49
10	54

$$\overline{Z} = 10/3 = 3.3; \; s = \sqrt{\frac{54 - 10^2/3}{2}} = \sqrt{10.33} = 3.2.$$

Example 4

Compare the two methods (definition and computational formula) for finding the variance of the sample X: 1, 2, 3, 5, 10.

Definition:

X	$X - \overline{X}$	$(X - \overline{X})^2$
1	−3.2	10.24
2	−2.2	4.84
3	−1.2	1.44
5	.8	.64
10	5.8	33.64
21		50.80

Shortcut:

X	X^2
1	1
2	4
3	9
5	25
10	100
21	139

$\overline{X} = 21/5 = 4.2$

$s^2 = 50.80/4 = 12.7$

$$s^2 = \frac{139 - 21^2/5}{4} = \frac{139 - 88.2}{4}$$

$$= 12.7$$

In computing the mean, we found that arithmetic can often be simplified by subtracting a constant from each of the scores in a sample. The same device can be used here, and is even more effective because the constant doesn't have to be added again at the end. The "spread" of scores remains the same if a constant A is subtracted from each of them; since the mean is reduced by A also, the square deviations of the new scores about the new mean remain the same. Many times, A can be chosen wisely so arithmetic is much easier.

Example 5

Find the standard deviation of X scores: 15.5, 17.5, 21.5, 23.5 (a) directly, (b) after subtracting 17.5 from each score.

(a)			(b)	
X	X^2		$X - 17.5 = Y$	Y^2
15.5	240.25		−2.0	4.00
17.5	306.25		0.0	0.00
21.5	462.25		4.0	16.00
23.5	552.25		6.0	36.00
78.0	1561.00		8.0	56.00

$$\text{Variance of } X \text{ scores} = s_X^2 = \frac{1561 - (78)^2/4}{3} = \frac{40.00}{3} = 13.33, \ s_X = 3.65$$

$$\text{Variance of } Y \text{ scores} = s_Y^2 = \frac{56 - 8^2/4}{3} = \frac{40.00}{3} = 13.33, \ s_Y = 3.65$$

Note the use of subscripts here to distinguish between the standard deviation of the X scores, s_X, and of the Y scores, s_Y.

Example 6

Find the variance of Z scores: 42, 47, 54, 53, 48, 51, 50, 49, 52.

Z	$Z - 50 = X$	X^2
42	−8	64
47	−3	9
54	+4	16
53	+3	9
48	−2	4
51	1	1
50	0	0
49	−1	1
52	2	4
	−4	108

$$s_Z^2 = s_X^2 = \frac{108 - (-4)^2/9}{8} = \frac{108 - 1.78}{8} = 13.3.$$

4.7 VARIANCE AND STANDARD DEVIATION FOR GROUPED DATA

Treat all the scores in a class as though they were concentrated at the midpoint. (Scores were treated the same way in finding the mean for grouped data.)

Example 1

Find the variance and standard deviation of the distribution in the box:

X	f	Midpoint
4–6	2	5
7–9	5	8
10–12	3	11
	10	

This distribution, then, will have the same variance and standard deviation as the scores 5, 5, 8, 8, 8, 8, 8, 11, 11, 11. The mean is 8.3; we must find the sum of

$$\underbrace{(5-8.3)^2 + (5-8.3)^2}_{2 \text{ terms}} + \underbrace{(8-8.3)^2 + \ldots + (8-8.3)^2}_{5 \text{ terms}}$$

$$\underbrace{+ (11-8.3)^2 + (11-8.3)^2 + (11-8.3)^2}_{3 \text{ terms}}$$

and divide by 9 to arrive at the variance. But again repeated addition is replaced by multiplication, and again extra columns are written to assist the computations:

X	f	Midpoint	fX	$X - \overline{X}$	$(X - \overline{X})^2$	$f(X - \overline{X})^2$
4–6	2	5	10	−3.3	10.89	21.78
7–9	5	8	40	− .3	.09	.45
10–12	3	11	33	2.7	7.29	21.87
	10		83			44.10

$\overline{X} = 83/10 = 8.3$

$s^2 = 44.10/9 = 4.90$

$s = \sqrt{4.90} = 2.2$

Note that the entries in the $f(X - \overline{X})^2$ column are $2(5-8.3)^2$, $5(8-8.3)^2$, and $3(11-8.3)^2$, as expected; these products are added and the sum is divided by 9.

The general formulas for variance and standard deviation for grouped sample data are:

$$s^2 = \frac{\Sigma f(X - \overline{X})^2}{n-1}, \quad s = \sqrt{\frac{\Sigma f(X - \overline{X})^2}{n-1}} \quad \text{degree of freedom}$$

Just as for the mean, X in these formulas represents the midpoint of a class.

Example 2

$\overline{X} = 5$; find s^2 for the distribution in the box below.

X	f	Midpoint	$X - \overline{X}$	$(X - \overline{X})^2$	$f(X - \overline{X})^2$
1-3	3				
4-6	4				
7-9	3				

$n = ?$ $s^2 = \dfrac{\Sigma f(X - \overline{X})^2}{n - 1} = ?$

Answer

$n = 10$; midpoint column: 2, 5, 8; $X - \overline{X}$ column: −3, 0, 3; $f(X - \overline{X})$ column: 9, 0, 9; $f(X - \overline{X})^2$ column: 27, 0, 27; $s^2 = 54/9 = 6.0$.

The computational formula, when used for grouped data, becomes

$$s^2 = \frac{\Sigma fX^2 - \dfrac{(\Sigma fX)^2}{n}}{n - 1}$$

where again X is used for the midpoint of the class. Use the table of squares in Appendix C or use a calculator.

Example 3

Find the standard deviation of the distribution shown in the box below.

X	f	Midpoint	fX	fX²
4-6	2	5	10	50
7-9	5	8	40	320
10-12	3	11	33	363
	10		83	733

$$s^2 = \frac{733 - (83)^2/10}{9} = \frac{733 - 688.9}{9} = 4.90; \; s = 2.2.$$

Now you try one, using the computational formula.

Example 4

Find s. $s = \sqrt{\dfrac{\Sigma fX^2 - (\Sigma fX)^2/n}{n - 1}}$

X	f	Midpoint	fX	fX²
0-4	4			
5-9	10			
10-14	6			

Answer

Midpoints: 2, 7, 12; fX: 8, 70, 72, $\Sigma fX = 150$; fX^2: 16, 490, 864; $\Sigma fX^2 = 1370$;

$$s = \sqrt{\frac{1370 - (150)^2/20}{19}} = \sqrt{\frac{1370 - 1125}{19}} = \sqrt{12.89} = 3.6.$$

As with ungrouped data, if a constant A is subtracted from each midpoint the standard deviation of the new distribution is the same as that of the original set. Arithmetic may be considerably easier if A is carefully chosen. Estimate the mean; a wise choice of A is often the midpoint of the class containing the mean.

Example 5

Find the standard deviation of the distribution in the box below (a) directly, (b) by first subtracting 124.5 from each midpoint.

(a)

X	f	Midpoint	fX	fX²
100 – 109	3	104.5	313.5	32760.75
110 – 119	5	114.5	572.5	65551.25
120 – 129	7	124.5	871.5	108501.75
130 – 139	2	134.5	269.0	36180.50
140 – 149	6	144.5	867.0	125281.50
	23		2893.5	368275.75

$$s_X^2 = \frac{368275.75 - (2893.5)^2/23}{22} = \frac{4261}{22} = 193.68, \; s_X = 13.9.$$

Be careful of rounding off at intermediate steps; if you round off 2893.5 to 2894, you will find $s_X^2 = 4137/22 = 187.95$, $s_X = 13.7$.

(b)

Midpoint – 124.5 = Y	fY	fY²
−20	−60	1200
−10	−50	500
0	0	0
10	20	200
20	120	2400
	30	4300

$$s_Y^2 = \frac{4300 - (30)^2/23}{22} = \frac{4261}{22} = 193.68, \; s_Y = 13.9.$$

Example 6

Try Example 4 again, this time subtracting 7 from each midpoint. Find s.

X	f	Midpoint	Midpoint – 7 = Y	fY	fY²
0 – 4	4				
5 – 9	10				
10 – 14	6			___	___

$s_X = s_Y = ?$

Answer

Y column: −5, 0, 5; fY column: −20, 0, 30, $\Sigma fY = 10$; fY^2 column: 100, 0, 150,

$$s^2 = \frac{250 - (10)^2/20}{19} = \frac{245}{19} = 12.89, s = 3.6.$$

4.8 MEASURE OF SKEWNESS

Sometimes a distribution is very skewed (has a long tail to the left or right). In this case, the mean and median will be quite different because the scores in the tail affect the mean but not the median. This difference is reflected in the following measure of skewness:

$$\text{Skewness} = \frac{3(\bar{X} - Md)}{s} \quad \text{median}$$

Dividing by s scales the skewness measure so that it does not become large (or small) simply because the scores have great (or little) variability.

Example 1

Find the measure of skewness and sketch a histogram for each of the following sets of scores.

X: 1, 1, 1, 1, 2, 3, 5.
Y: 1, 3, 4, 5, 5, 5, 5.

$\bar{X} = 2.0$, $Md = 1$, $s = 1.5$ (check these); skewness $= \dfrac{3(2-1)}{1.5} = 2.0$.

$\bar{Y} = 4.0$, $Md = 5$, $s = 1.5$ (check these, too); skewness $= \dfrac{3(4-5)}{1.5} = -2.0$.

Note that the measure of skewness is greater than 0 for a distribution skewed to the right and less than zero for a distribution skewed to the left.

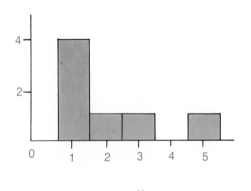

X
Measure of Skewness = 2.0

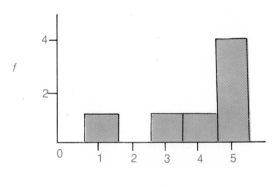

Y
Measure of Skewness = −2.0

4.9 VOCABULARY AND SYMBOLS

| variation ratio | V, f_{Mo} | semi-interquartile range |
| range | Q | mean absolute deviation |

squared deviations from the mean   standard deviation   s
variance s^2 measure of skewness

4.10 EXERCISES

1. Find the variation ratio for the scores in Example 1, page 29.

2. Find the variation ratio for the scores in the Miller Personality Test as grouped in (a) 9 classes, (b) 6 classes on page 17.

3. Find (a) the range, (b) the semi-interquartile range, (c) the mean, (d) the standard deviation using the definition, and (e) the standard deviation using the computational formula and whatever shortcut you can find, for each of these sets of scores:

  (i) *X:*   1, 4, 9, 18
  (ii) *Y:*   51, 54, 59, 62, 63, 63, 67, 69
  (iii) *Z:*   30, 24, 21, 25, 20, 30, 22, 28
  (iv) *W:*   14, 16, 17, over 18

4. Find (a) the range, (b) the semi-interquartile range, (c) the standard deviation, (d) the measure of skewness for the melting points and boiling points of these alcohols:

Alcohol	Melting point, °C	Boiling point, °C
Methyl	−97	64.7
Ethyl	−114	78.3
Isopropyl	−89	82.3
Allyl	−129	97.0
Isobutyl	−108	107.9
Isoamyl	−117	131.5

5. Find (a) the range, (b) the semi-interquartile range, (c) the standard deviation, and (d) the measure of skewness for the grades on a statistics quiz given in Exercise 3 (a) (ii), page 39.

6. Tally and then find s for the number of children in 30 families living in an apartment house on Delancey Street in Brooklyn referred to in Exercise 9, page 40.

7. The systolic and diastolic blood pressures of 10 adults were taken and the difference *d* between the readings for each individual was computed. It was found that $\Sigma d = 300$ and $\Sigma d^2 = 9200$. Find the mean and the variance of these differences.

8. The Upton Calculator Co. received 5 shipments of 1000 batteries. The numbers of defective batteries in each shipment were 4, 12, 6, 4, and 8, respectively. What is the mean number of defective batteries per shipment, and what is the standard deviation?

9. An instructor in a calculus course gave the GLUG algebra test to his students at the beginning of the semester, with the following results:

78	72	84	81	78	76
83	89	81	71	72	78
79	83	81	72	84	68
86	82				

Find the standard deviation of these scores.

10.   In a psychological experiment, 10 subjects were given the same test after 5 of them had been told the test would be very difficult and 5 were told it would be very easy. The scores on the test were as follows:

"It is very difficult."	"It is very easy."
38	30
52	38
46	40
42	24
38	32

Find the standard deviation of each group of 5, and of all 10 subjects together.

11.   Find three scores whose mean is 10 and whose variance is 1.

12.   Find two scores whose mean is 40 and whose variance is 8.

13.   The velocity of sound in meters per second for various metals is as follows:

Metal	Velocity (meters per second)
Aluminum	5000
Cadmium	2300
Copper	3600
Iron	5100
Lead	1200
Nickel	5000
Platinum	2700
Silver	2600
Tin	2500

Find s. Hint: If a set of scores is divided by 100, and s for the new set is computed, it must be multiplied by 100 to find s for the original set.

14.   (a) Find the range, the semi-interquartile range, and the standard deviation for the salaries in Exercise 4, page 39. Hint: To find s, let Y be the scores obtained by dividing the given salaries (X) by 100; let $Z = Y - 80$. $s_Y = s_Z$, $s_X = 100 s_Y$. Do you see why?

(b) Which measure of variability determined in (a) would you use to express the spread of those salaries?

15.   Find the variance for each of these distributions, using $s^2 = \dfrac{\Sigma f(X - \overline{X})^2}{n - 1}$.

(a)	X	f	(b)	Y	f	(c)	Z	f	(d)	W	f
	5–9	2		6–10	2		10, < 20	4		10–19	5
	10–14	5		11–15	4		20, < 30	8		20–29	10
	15–19	2		16–20	3		30, < 40	6		30–39	12
							40, < 50	3		40 and over	4
							50, < 60	4			

16.   Find s for each of the distributions in Exercise 15, using the computational formula. Simplify by subtracting a constant from each midpoint.

17.   Find the variance of each distribution in Exercise 14, page 40; choose A carefully in each case.

18.   Six coins are tossed 100 times; each time the number of coins that land heads up is noted. The results are as follows:

Number of heads up	0	1	2	3	4	5	6
Frequency	3	7	24	34	22	10	1

Find the mean number of heads, and the standard deviation.

19.   Assume a population consists of 5 scores: 1, 2, 4, 6, 7.

     (a)   Compute (to two decimal places) the standard deviation of the population. (Remember that for a population you divide by the number of scores (5) rather than 1 less than this number before taking the square root.)

     (b)   Compute (to two decimal places) the standard deviation of each different sample of size 2, using $s = \sqrt{\dfrac{\Sigma(X - \bar{X})^2}{n - 1}}$ or $s = \sqrt{\dfrac{\Sigma X^2 - (\Sigma X)^2 / n}{n - 1}}$.

     (c)   Compute the standard deviation of each different sample of size 2, using $s = \sqrt{\dfrac{\Sigma(X - \bar{X})^2}{n}}$ or $s = \sqrt{\dfrac{X^2 - (\Sigma X)^2 / n}{n}}$.

Scores in sample	(b) $s = \sqrt{\dfrac{(X - \bar{X})^2}{n - 1}}$	(c) $s = \sqrt{\dfrac{(X - X)^2}{n}}$
1, 2		
1, 4		
1, 6		
1, 7		
2, 4		
2, 6		
2, 7		
4, 6		
4, 7		
6, 7		

     (d)   Compute the means of the standard deviations found in (b) and in (c), and compare them with the standard deviation of the population computed in (a). Which formula gives a better approximation, on the average, to the standard deviation of the population?

20. X: 1, 2, 4, 13. Find (a) Md, (b) $\bar{X}$, (c) $\Sigma|X - Md|$, (d) $\Sigma|X - \bar{X}|$.

ANSWERS

1. $\dfrac{19 - 6}{19} = .68$.

3.   (i) (a) 17. (b) $(13.5 - 2.5)/2 = 5.5$. (c) 8, (d) $\sqrt{166/3} = 7.4$.
(e) $\sqrt{(422 - 32^2/4)/3} = 7.4$.
        (ii) (a) 18. (b) $(65 - 56.5)/2 = 4.2$. (c) $8.0/8 + 60 = 61.0$. (d) $\sqrt{262/7} = 6.1$.
(e) (Subtract 60.) $\sqrt{(270 - 8^2/8)/7} = 6.1$.
        (iii) (a) 10. (b) $(29 - 21.5)/2 = 3.8$. (c) 25.0. (d) $\sqrt{110/7} = 4.0$.
(e) $\sqrt{(110 - 0^2/8)/7} = 4.0$.
        (iv) Cannot be determined (open-ended).

5.   (a) 60. (b) $\dfrac{90-60}{2}=15.0$. (c) 18.5. (d) $\dfrac{3(75-75)}{18.0}=0.0$.

7.   Mean $=300/10=30$, $s^2=(9200-300^2/10)/9=22.2$.

9.   Subtract 80 from each score. $s=\sqrt{(620-(-22)^2/20)/19}=5.60$.

11.   9, 10, 11.

13. $s=1410$ meters per second.

15.   (a) $\overline{X}=12$, $s^2=100/8=12.5$. (b) $\overline{Y}=13.6$, $s^2=122.2/8=15.3$. (c) $\overline{Z}=33.0$, $s^2=4290/25=171.6$. (d) impossible (open-ended).

17.   (i) $A=8$: $s^2=\dfrac{150-(-10)^2/10}{9}=15.6$.

      (ii) $A=54.5$: $s^2=\dfrac{7600-(-80)^2/60}{59}=128.8$.

      (iii) $A=50.5$: $s^2=\dfrac{20762-(-264.5)^2/25}{24}=748.5$.

19.   (a) Standard deviation of the population $=2.28$.
      (b) .71, 2.12, 3.54, 4.24, 1.41, 2.83, 3.54, 1.41, 2.12, .71.
      (c) .50, 1.50, 2.50, 3.00, 1.00, 2.00, 2.50, 1.00, 1.50, .50.
      (d) The mean of the s's in (b) is 2.26, in (c) 1.60. This demonstrates (but does not prove in general!) that $s=\sqrt{\dfrac{\Sigma(X-\overline{X})^2}{n-1}}$ gives a better approximation, on the average, to the standard deviation of the population than does $\sqrt{\dfrac{\Sigma(X-\overline{X})^2}{n}}$, and should help explain why $n-1$ is used instead of n in the denominator of the fraction for a sample.

chapter five

probability

Up to this point you have been studying descriptive statistics, but now you are ready to begin the study of statistical inference: On the basis of data from a sample, you will learn how to draw conclusions about the population from which the sample is drawn. Probability is basic to statistical inference; you will meet the word "probability" dozens of times in every chapter after this.

You could study probability for years and years before getting to the boundaries of what is known about it. Fortunately, for this book you need master only a very few basic facts. Concentrate on (1) the probability that event A or event B will happen, when they cannot occur at the same time, and (2) the probability that event A and event B will both take place when the occurrence of one has no effect on the occurrence of the other.

5.1 WHY STUDY PROBABILITY?

Probability and statistics are very closely related. In statistical inference you may take the list of registered voters in your county, randomly† choose 100 names, and ask these people how they will vote in the election next week; then you draw conclusions about who will be elected mayor next week when everybody votes. In probability, you know how many voted for each of candidates A, B, and C last week; then you randomly choose 100 voters and draw conclusions about how many of these 100 probably voted for candidate A. In statistics, you put your hand into an opaque box of colored and white marbles, examine your handful, and try to answer the question "What is in the box?" In probability, you look into a transparent box, count the colored and white marbles in it,

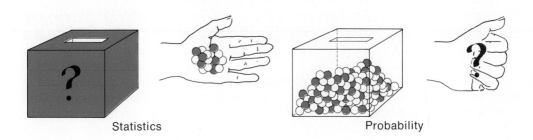

Statistics   Probability

mix them up well, and then blindly take out one handful; without opening your eyes, you predict how many marbles of each kind are in your hand. You will learn how to

---

†How to choose a sample will be discussed in Chapter 8.

forecast the election or answer the question about the black box only after you have studied probability, since you will use this study in choosing the sample, in making your estimate about the population (the contents of the black box), and in deciding what degree of confidence you have in your estimate.

5.2 DEFINITIONS OF PROBABILITY

If an event A can happen in s ways out of a total of n equally likely ways, then the probability that event A will occur is s/n. Let $p(A)$ denote the probability that event A will occur.

$$p(A) = \frac{s}{n}$$

Note that s is a number between 0 and n, so $p(A)$ is a number between 0 and 1. If $s = 0$, then event A never happens and $p(A) = 0$. If A always happens, then $s = n$ and $p(A) = 1$.

> $p(A)$ is always between 0 and 1. If A is certain to happen, $p(A) = 1$; if A is certain **not** to happen, then $p(A) = 0$.

Example 1

If a coin is flipped, it can come up heads (H) or tails (T); there are 2 ways it can land (assuming it doesn't land on edge in the sand), and they are equally likely if the coin is fair. Therefore the probability of heads is $p(H) = 1/2$ and the probability of tails is $p(T) = 1/2$.

Example 2

The population of Jasper County, Iowa, is 35,010, and of Lucas County, Iowa, is 10,031 (1970 census). If one inhabitant of these two counties is chosen at random, what is the probability he lives in Jasper County? ("Chosen at random" means "each person is equally likely to be chosen.")

$$p(\text{from Jasper}) = \frac{\text{number in Jasper}}{\text{total number}} = \frac{35,010}{35,010 + 10,031} = .78$$

> ***Round-off:*** Let's agree that, in general, the answer to a probability problem is rounded to 2 decimal places.

The definition of probability "$p(A) = s/n$ if there are n equally likely outcomes" is the classical definition of probability; it was first developed to help gamblers. The limitations of this definition come from the words "equally likely." If a man on his fortieth birthday asks for one-year term insurance, should a life insurance company say, "One year from now you will be alive or dead; the probability that you will be dead is .50, so you should pay half the insurance as premium"? This is patently absurd, because the two outcomes "the man is alive one year from now" and "the man is dead one year from now" are not equally likely. Of 100,000 men aged 40, past experience indicates that 99,382 will be alive a year later and 618 will be dead. Therefore the insurance company concludes that the probability that a 40-year-old man chosen at random will be alive a year later is $\dfrac{98,312}{100,000} = .98$, not .50 as suggested above. This is a second way of defining probability, then—the probability of an event is the relative frequency with which the event occurs. It is based on past experience (as in this example) or on experiment (as in the next).

If a penny is "fair," the probability that heads will come up when it is tossed is 1/2 (see Example 1 above). But since the designs on the two sides of the penny are not the same, the assumption that heads and tails are equally likely to come up may not be valid.

Example 3

A penny is tossed 200 times; the number of heads every 20 tosses is shown below. What is the probability that this penny will come up heads when tossed?

Number of tosses	Number of heads in last 20 tosses	Cumulative number of heads	Cumulative relative frequency
20	11	11	.55
40	12	23	.58
60	10	33	.55
80	9	42	.53
100	7	49	.49
120	9	58	.48
140	14	72	.51
160	7	79	.49
180	8	87	.48
200	11	98	.49

On the basis of this experiment, the best answer that can be given is that the probability that this particular penny will come up heads when tossed is 98/200 = .49.

The graph below shows the number of tosses and the cumulative relative frequency. Notice how the graph fluctuates around the relative frequency of .5 that is expected if the coin is fair.

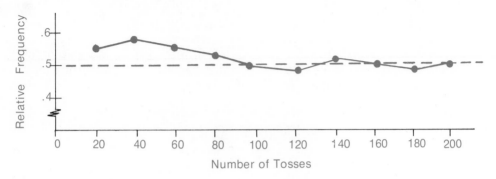

When using the relative frequency definition of probability, you theoretically want an infinite number of trials. Determining the probability after a finite number of trials will cause some inaccuracy perhaps, but this inaccuracy will be less if the number of trials is large. Note that again $p(H)$ will be a number between 0 (if heads never turn up) and 1 (if heads always turn up).

A third definition of probability is **subjective probability**: On the basis of intuition or knowledge of the circumstances of an experiment or sheer guess, a number between 0 (representing strong belief the event will not occur) and 1 (representing a feeling that the event will certainly occur) is assigned to an event. For example, an agricultural chemist who has made a new insecticide may say, "Chemical A kills all flies in 30 seconds. I have added a sulphone group so I think my new chemical acts more quickly, and the probability is .80 that all flies will be killed in 10 seconds or less." If his new insecticide is never tried on flies, no one can argue with him. But if it is tried out, then he may change his choice of probability.

If there are three different definitions of probability, how does one know which one to use? The answer is that, in the modern theory of probability, probabilities are assigned to events in such a way that certain axioms or rules are satisfied, and the same axioms must be satisfied whether probabilities are assigned on the basis of equally likely outcomes or relative frequency or subjectively. Many of the probability problems which you will need to be able to solve in order to understand statistics can be solved using the classical definition $p(A) = s/n$, but the new developments are fairly simple, make some problems simpler, and provide a firmer background for those who will study statistics further. Bon appetit!

5.3   SAMPLE SPACE AND EVENTS

You are familiar with sets of dishes and sets of golf clubs. In mathematics, a **set** is any collection of objects which is distinctly defined: you must be able to tell whether or not a specific object is in the set. Sets may be denoted by listing the items in the set (called its **elements**) or by describing them. Curly brackets {   } are used to denote the contents of a set. Thus, "A is the set of integers between 4 and 8, inclusive" is written either $A = \{$integers between 4 and 8, inclusive$\}$ or $A = \{4, 5, 6, 7, 8\}$. B is a **subset** of A if every element of B is in A. $B = \{4, 5, 7\}$ is a subset of A.

Think of all the possible outcomes of an experiment as elements of a set S; we shall call S a **sample space.**† An **event** is a subset of a sample space. Most people have a splendid geometric sense and find it very helpful to use this geometric sense in picturing relations between the sample space and its subsets. The sample space S is represented by a rectangle or circle, and an event ($=$ subset) by a circle or rectangle inside S; the drawing is called a **Venn diagram.**

Example 1

If a die‡ is tossed once, $S = \{1, 2, 3, 4, 5, 6\}$. $A = \{1, 2, 3\}$ is the event "a 1 or 2 or 3 comes up." $B = \{5\}$ is the event "a 5 comes up." (Note that an event may contain a single element, or more than one.)

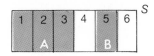

---

†If you studied sets before, you probably used the phrase **universal set.** We shall use **sample space** in this book because it is commonly used in statistics and because it emphasizes a geometrical point of view which will be helpful to you.

‡**Die** is the singular of **dice.**

Example 2

A man is tried for bank robbery. *A* is the event {he robbed the bank and the jury finds him guilty}, *B* is {he robbed the bank and the jury frees him}, *C* is {he did not rob the bank and the jury finds him guilty}, and *D* is {he did not rob the bank and the jury frees him}.

Jury finds him

		Guilty	Not guilty
Man robbed bank?	Yes	A	B
	No	C	D

Example 3

In the Pioneer Club in Mendham, New Jersey, 40 of the 200 members are married college graduates who make over $15,000 a year, 70 are married college graduates, 60 are college graduates who make over $15,000 a year, 100 are college graduates, 45 are married and make over $15,000 a year, 100 are married, and 75 make over $15,000.

(a)   How many members are not married, are not college graduates and make $15,000 or less a year?
(b)   How many members are married but not college graduates?

Draw a Venn diagram. Let *S* = {members of the club}, let *M, CG,* and *15,000+* be the sets of those members who are married, are college graduates, and make over $15,000 a year, respectively, and represent these sets by intersecting circles. There are 40 members in the intersection of the three circles. Since there are 70 married college graduates, of whom 40 have already been accounted for, there are 30 married college graduates who do not make over $15,000 a year. In a similar way, the other portions of each circle are numbered.

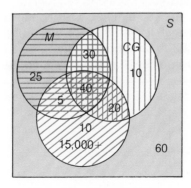

(a) $200 - (25 + 30 + 10 + 5 + 40 + 20 + 10) = 60.$
(b) $25 + 5 = 30.$

If *S* contains only a finite number of elements, then to each element we assign a non-negative number, called a probability, so that the sum of all the numbers assigned to all the elements of *S* is 1. Since the numbers are non-negative and their sum is 1, each of the numbers must be between 0 and 1. If we think an outcome of an experiment is quite likely, we assign to the element consisting of that outcome a number close to 1. If an outcome is quite unlikely to occur, we assign to that element a number close to 0. To outcomes which are totally impossible (including those outside the sample space) we assign the number 0. In many experiments, all the outcomes are equally likely to occur; in this case the same number is given to each element of S.

Example 4

If a fair die is tossed, $S = \{1, 2, 3, 4, 5, 6\}$. "The die is fair" means that it is unchipped and unweighted—that is, the 6 possible outcomes are equally likely and must be assigned equal probabilities. The sum of all 6 probabilities must be 1, so the number 1/6 is assigned to each.

1	2	3	4	5	6	S
1/6	1/6	1/6	1/6	1/6	1/6	

Often the phrase "at random" is used in describing the selection of one object from a group or set of objects. Sometimes this phrase is used to indicate that each member of the group has a known probability of being chosen, but in this book it will be used in a more particular sense: "an object is picked at random" means that **every member of the group has an equal likelihood of being picked.** "A ball is picked at random from a bag containing 4 red and 6 white balls": the balls have the same shape, size, and weight, differ only in color, and have been well mixed; it is not the case that 6 large white balls are sitting on top of 4 tiny red balls. "In a draft lottery, birth dates were chosen at random without replacement": Every birth date was equally likely to be chosen on the first draw. Of those remaining, every date was equally likely to be chosen second and so on.

Example 5

Of every 100,000 American men aged 40, recent studies show that 99,382 will be alive a year later and 618 will die during the year. How can this information be used to assign probabilities to the outcomes **alive, dead** for one American male aged 40 selected at random?

$S = \{alive, dead\}$. Are you tempted to assign the number 99,382 to **alive** and 618 to **dead?** Resist it. The sum of the probabilities must be 1, so 99,382/100,000 = .99382 is assigned to **alive** and 618/100,000 = .00618 to **dead.** (The round-off agreement has been ignored here.)

Alive	Dead	S
.99382	.00618	

Example 6

In Montgomery County Court, 150 defendants are convicted of misdemeanors in October; 90 are fined and put on probation, and 60 receive prison terms of one year or less. If one of those convicted is chosen at random, what is the probability that he is (a) fined and put on probation? (b) sentenced to one year or less? (c) sentenced to more than one year?

Let A, B, and C be, respectively, the events {fined and given probation}, {sentenced to 1 year or less}, and {sentenced to more than 1 year}.

A	B	C
$\dfrac{90}{150}$	$\dfrac{60}{150}$	$\dfrac{0}{150}$

S

(a) $p(A) = 90/150 = .60.$
(b) $p(B) = 60/150 = .40.$
(c) $p(C) = 0/150 = 0.$

> To find the probability of any event $A[= p(A)]$, add all the numbers assigned to the elements of A. If an experiment takes place and S is the set of all the outcomes of the experiment, then S is certain to take place, so $p(S) = 1.$

Example 7

If a die is tossed, what is the probability that a 1, 2, or 3 will come up?
$S = \{1, 2, 3, 4, 5, 6\}$, $A = \{1, 2, 3\}$. 1/6 is assigned to each element of S. The probability that a 1, 2, or 3 will come up $= p(A) = 1/6 + 1/6 + 1/6 = 1/2.$

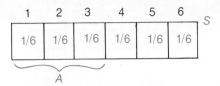

Example 8

In Beach Bluff there are 4,112 Protestants, 1,332 Catholics, and 556 Jews. If one person is picked at random from these, what is the probability that he is (a) Protestant, (b) Catholic or Jewish?
$B = \{\text{Protestant}\}$, $C = \{\text{Catholic or Jewish}\}$.

Protestants .69	Catholics .22	Jews .09
B		C

S

(a) $p(B) = \dfrac{4,112}{4,112 + 1,332 + 556} = \dfrac{4,112}{6,000} = .69.$

(b) $p(C) = \dfrac{1,332 + 556}{6,000} = \dfrac{1,888}{6,000} = .31.$

Notation. In these last two examples, a notation such as $p(A)$ has been used to emphasize that A is an event in the sample space S. But in Example 7, $A = \{1, 2, 3\}$; in Example 8(a), $B = \{\text{Protestant}\}$. By substitution, the solutions would be written $p(\{1, 2, 3\}) = 1/2$, $p(\{\text{Protestant}\}) = .69$. In practice, however, the event whose probability is determined is denoted without curly brackets: $p(1, 2, \text{or } 3) = 1/2$, $p(\text{Protestant}) = .69.$

Example 9

If 2 dice are tossed what is the probability that (a) 11 (total on tops of both dice) will come up? (b) 7 will come up?

(a)   Here is it more difficult to list the elements in the sample space. Each element of S is a pair of numbers such as (4, 3), where the first number is the result on the first die, the second is the toss on the second die. $S = \{(1, 1), (1, 2), \ldots , (1, 6), (2, 1), \ldots , (2, 6), \ldots , (6, 6)\}$. It is easier to see from the Venn diagram that there are 36 elements in S; the probability assigned to each is 1/36 if both dice are "fair." The event "11 comes up" includes the elements indicated by a cross. There are two of them, so $p(11) = p(6$ on the first die and 5 on the second, or 5 on the first and 6 on the second$) = p[(6, 5)$ or $(5, 6)] = 1/36 + 1/36 = 1/18$.

(b)   The elements of the event "the total is 7" are indicated by checks. $p(7) = p[(6, 1)$ or $(5, 2)$ or $(4, 3)$ or $(3, 4)$ or $(2, 5)$ or $(1, 6)] = 6 * (1/36) = 1/6$.

Example 10

One card is drawn at random from an ordinary (bridge) deck of 52 cards. What is the probability that it is (a) a king? (b) a heart? (c) a spade or a heart? What probability is assigned to each element of S?

(a) $p(K) = ?$
(b) $p(\heartsuit) = ?$
(c) $p(\spadesuit$ or $\heartsuit) = ?$

Answer

1/52 is assigned to each element of S; (a) $p(K) = 4/52 = 1/13$; (b) $p(\heartsuit) = 13/52 = 1/4$; $p(\spadesuit$ or $\heartsuit) = 26/52 = 1/2$.

Example 11

A rather extraordinary die is manufactured. It is like the ordinary die only in that it is six-sided, and the sum of the numbers on each pair of "opposite" sides (that is, sides without a common edge) is 7. In the illustration, side 1 is on top and side 6 on the bottom, side 2 is facing you and side 5 is away from you; side 4 is to the left and side 3 is slanting upwards on the right. Because of its unusual construction, this die will not balance on side 3 (it immediately falls on side 6) so side 4 cannot come up, and it will not balance on side 1 (it falls on side 4), so 6 cannot come up on top.

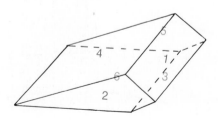

The die is tossed 1,000 times, with the following results:

Side	Number of times on top
1	470
2	110
3	310
4	0
5	110
6	0
	1000

What is the probability that, when it is tossed once, (a) an even number will turn up? (b) a 1 or a 3 will turn up? (c) a 2 or a 5 will come up?

$S = \{1, 2, 3, 4, 5, 6\}$. Probabilities can be assigned here to the elements of the sample space only on the basis of the experimental results. Since 4 and 6 are impossible outcomes in practice, they are each assigned the number 0; 2 occurs 110 times in 1000 trials, and so is assigned the probability $110/1000 = .11$; and so on. (Always check that the sum of the probabilities assigned to all the elements of the sample space is 1.) Now the problem should be simple for you.

(a) $p(2, 4, \text{ or } 6) = .11 + 0 + 0 = .11$.
(b) $p(1 \text{ or } 3) = .47 + .31 = .78$.
(c) $p(2 \text{ or } 5) = .11 + .11 = .22$.

1	2	3	4	5	6	S
.47	.11	.31	0	.11	0	

5.4 MUTUALLY EXCLUSIVE EVENTS

Often it is much easier to determine the probability of an event by first finding the probability of other events. We have already done this by adding the probabilities assigned to elements of the event. If a single die is tossed, we already discovered (Example 7, page 65) that $p(1, 2, \text{ or } 3) = p(1) + p(2) + p(3) = 1/6 + 1/6 + 1/6 = 1/2$. But soon we shall want to tackle problems such as this: What is the probability that heads will come up exactly twice if a fair coin is tossed three times? Here we would need a three dimensional Venn diagram. We could use a "tree" (pushed over by the wind) to show possible outcomes, as follows:

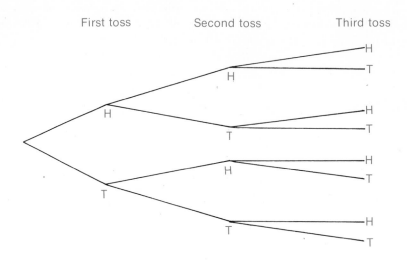

First toss   Second toss   Third toss

Or we could abbreviate "heads on all three tosses" by HHH, "heads on the first two and tails on the third" by HHT, and so on, and then list the elements of *S*: HHH, HHT, HTH, HTT, THH, THT, TTH, TTT. But if the coin is tossed seven times, both these methods become cumbersome too. Some shortcuts should be welcomed.

> ***Definition:***   Two events *A* and *B* are **mutually exclusive** if they have no ele-
> ments in common. If *A* and *B* are thought of as outcomes of an
> experiment, this definition means that *A* and *B* cannot both happen.

Example 1

One die is tossed. Let $S = \{1, 2, 3, 4, 5, 6\}$.
Let $A =$ the event an odd number turns up; $A = \{1, 3, 5\}$.
Let $B =$ the event a 1, 2, or 3 turns up; $B = \{1, 2, 3\}$.
Let $C =$ the event a 2 turns up; $C = \{2\}$.
   A and *B* have the elements 1, 3 in common, so *A* and *B* are **not** mutually exclusive.
   It is impossible to throw an odd number **and** a 2 if one die is tossed, so *A* and *C* have no element in common; they are mutually exclusive. *B* and *C* have the element 2 in common, so *B* and *C* are **not** mutually exclusive.

Example 2

In the 6th Congressional District, the normal election turnout is 180,000 if the weather is good; 100,000 are urban and 80,000 are rural. If the weather is bad, however, 10 per cent of these urban voters and 30 per cent of the rural voters stay home. What is the probability that a voter who normally votes, chosen at random, will stay home when the weather is bad?
   A is the event {urban voters who stay home in bad weather}.
   B is the event {rural voters who stay home in bad weather}.
   A and *B* are mutually exclusive.

If C is the event {urban voters}, then A and C are not mutually exclusive but B and C are.

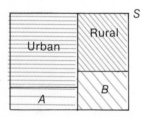

If A and B are mutually exclusive, then $p(A \text{ or } B) = p(A) + p(B)$.

Note: In general, **A or B** means A will occur or B will occur or both will occur. If they are mutually exclusive, **A or B** means A will occur or B will occur (but not both).

Example 3

If a fair coin is tossed twice, what is the probability that heads and tails will each come up once?
$S = \{HT, TH, HH, TT\}$. (Remember that HT means heads on the first toss, tails on the second; TH means tails on the first toss, heads on the second, and so on.) If the coin is fair, each outcome is equally likely and the probability 1/4 is assigned to each. Let $A = \{HT\}$, $B = \{TH\}$. Then A and B are mutually exclusive.

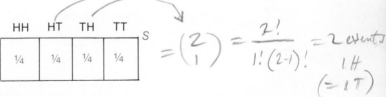

Total # events
$2^2 = 4$

HH	HT	TH	TT
¼	¼	¼	¼

$= \binom{2}{1} = \dfrac{2!}{1!\,(2-1)!} = 2 \text{ events}$

1H
(= 1T)

p(heads and tails each come up once) $= p(HT \text{ or } TH) = p(HT) + p(TH) = 1/4 + 1/4 = 1/2$.

ODDS $= \dfrac{2}{4}$ H. $= \dfrac{1}{2}$

Example 4

If 1 child is chosen at random from a fourth-grade class in which 6 read below grade level, 12 read at grade level, and 7 read above grade level, what is the probability that he reads at or below grade level?
$S = \{$below grade level, at grade level, above grade level$\}$.
p(the chosen child reads below or at grade level) $= p$(he reads below grade level) $+ p$(he reads at grade level) $= 6/25 + 12/25 = 18/25 = .72$.

Below	At	Above
6/25	12/25	7/25

If A, B, and C are all mutually exclusive, then no element in A is in B or C, and none in B is in C, so $p(A \text{ or } B \text{ or } C) = p(A \text{ or } [B \text{ or } C]) = p(A) + p(B \text{ or } C) = p(A) + p(B) + p(C)$. In the same way, probabilities of more than three events which are mutually exclusive may be added.

Example 5

If one score is chosen at random from the following distribution, what is the probability that it is (a) 4 or less? (b) 14 or less?

X	f
0– 4	3
5– 9	4
10–14	7
15–19	2
20–24	9
	25

(a) $p(0-4) = 3/25 = .12$.

(b) $p(0-4 \text{ or } 5-9 \text{ or } 10-14) = p(0-4) + p(5-9) + p(10-14) = 3/25 + 4/25 + 7/25 = 14/25 = .56$.

5.5 AXIOMS OF PROBABILITY

All further results in probability theory can be derived from the following 3 axioms or assumptions:

---

1. $p(S) = 1$

2. $0 \le p(A) \le 1$ (Read $\le$ as "less than or equal to.")

3. If A and B are mutually exclusive events, then $p(A \text{ or } B) = p(A) + p(B)$

If these three are not crystal clear to you, read this chapter again and study the examples carefully. Remember that if an event is certain to happen, then the probability of its happening is 1; and if an event is certain **not** to happen, then the probability of its happening is 0. The second axiom tells you that the answer to every probability problem will be a number between 0 and 1, inclusive. Check your work and find your mistake if you come up with an answer like 2.31 or $-.50$!

If A is an event in the sample space S, let $\overline{A}$ be the event "A does **not** occur." It follows immediately† from the axioms that

$$p(A) = 1 - p(\overline{A})$$

If the probability that George Stone is seasick on a cruise to Bermuda is .05, then the probability that he is not seasick is .95. If the probability of rain today is .40, the probability of no rain today is .60.

---

†Assume, from the definition of $\overline{A}$, that every element of S is in either A or $\overline{A}$, and that A and $\overline{A}$ are mutually exclusive.

$p(S) = 1$ (Axiom 1)

$\therefore p(A \text{ or } \overline{A}) = 1$

$p(A) + p(\overline{A}) = 1$ (Axiom 3)

$\therefore p(A) = 1 - p(\overline{A})$

Example 1

If two dice are tossed, what is the probability that the sum of the two numbers showing is 4 or more? (See Example 9, page 66, for Venn diagram.)

$p(4$ or more$) = 1 - p(2$ or $3) = 1 - [p(2) + p(3)] = 1 - (1/36 + 2/36) = 11/12$. Note that the events "the sum is 2" and "the sum is 3" are mutually exclusive, so $p(2$ or $3) = p(2) + p(3)$.

5.6 *p(A* or *B)* IF *A* AND *B* ARE NOT MUTUALLY EXCLUSIVE

If the events *A* and *B* are mutually exclusive, then $p(A$ or $B) = p(A) + p(B)$.

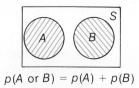

$$p(A \text{ or } B) = p(A) + p(B)$$

What, now, if *A* and *B* are partially overlapping, as shown by the Venn diagram below? The probability of an event is the sum of the probabilities attached to each element in the event.

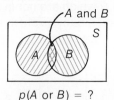

$$p(A \text{ or } B) = ?$$

If we add $p(A)$ and $p(B)$ when *A* and *B* overlap, the elements in the cross-hatched region have been added twice, and hence must be removed once.

$$p(A \text{ or } B) = p(A) + p(B) - p(A \text{ and } B)$$

Example 1

If one card is drawn at random from a standard deck of cards, what is the probability that it is a king or a heart?

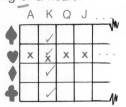

$p(K) = 4/52; \, p(\heartsuit) = 13/52.$

$p(K$ and $\heartsuit) =$ the probability the king of hearts is drawn
$\qquad = 1/52.$

$p(K$ or $\heartsuit) = p(K) + p(\heartsuit) - p(K$ and $\heartsuit)$
$\qquad = 4/52 + 13/52 - 1/52 = 16/52$
$\qquad = 4/13.$

probability

Example 2

Of 200 seniors at Morris College, 98 are women, 34 are majoring in English, and 20 English majors are women. If one student is chosen at random from the senior class, what is the probability that the choice will be either an English major or a woman (or both)?

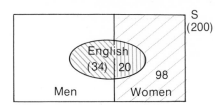

p(English major or woman) $= p$(English major) $+ p$(woman) $- p$(English major and woman)
$= 34/200 + 98/200 - 20/200 = 112/200$
$= .56$.

Before you concentrate your energies on this latest formula, note that you can solve problems without its use. Every element which is in *A* or *B* is in *A* or *C*, where *C* is the set consisting of those elements of *B* which are not in *A*. Furthermore, *A* and *C*, although adjacent in the Venn diagram, are mutually exclusive (they have no elements in common). $p(A \text{ or } B) = p(A \text{ or } C) = p(A) + p(C)$.

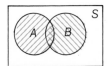 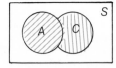

Example 3

$p(K \text{ or } \heartsuit) = p(K \text{ or } [\heartsuit \text{ which is not a } K])$
$= p(K) + p(\heartsuit \text{ which is not a } K)$
$= 4/52 + 12/52 = 16/52$.

Example 4

p(English major or woman) $= p$(English major or [woman not majoring in English])
$= p$(English major) $+ p$(woman not majoring in English)
$= 34/200 + (98 - 20)/200 = .17 + .39$
$= .56$.

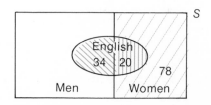

5.7 INDEPENDENT AND DEPENDENT
EVENTS, *p(A* AND *B)*

If a coin is flipped twice, the outcome of the second flip doesn't depend on whether the first one comes up heads or tails. If 2 dice are tossed, the number on the second die is independent of the number which comes up on the first die. Suppose a ball is chosen at random from a bag containing 4 red and 6 white balls, its color is observed, and then the ball is replaced in the bag and the bag is well shaken. Now a second ball is drawn from the bag. What is the probability that the second ball is red? The answer (.4) does not depend on the first choice, since that ball was replaced; the two events are independent: the outcome of one does not affect the outcome of the other.

If, however, the first ball is **not** replaced before the second ball is drawn, what is the probability that the second is red? If the first ball drawn was red, there now remain 9 balls in the bag, of which 3 are red. p(the second is red) = 3/9. But if the first ball was white, then 4 of the 9 balls left in the bag are red and p(the second is red) = 4/9. The result of the second draw depends on what happened on the first; we say the two events are **dependent.**

A and *B* are **independent** events if the occurrence of event *A* has no effect on the probability of occurrence of event *B* and vice versa.

A and *B* are **dependent** events if the occurrence of one event does affect the probability of occurrence of the other.

Example 1

Candidates A and B each have speakers' lists of 10 men and 8 women. One representative is to be chosen at random from each list for a debate on the candidates' policies. The events C = {candidate A's speaker is a woman} and D = {candidate B's speaker is a man} are independent, since a man is neither more nor less likely to be chosen for candidate B if a woman is chosen for candidate A; the selection was specified to be "at random." (Is this example realistic in today's politics?)

Example 2

Two cards are drawn at random, one after the other but without replacement, from a pack of 52 cards. The events A = {the first card is a heart} and B = {the second card is a heart} are dependent; if the first card is a heart, then there are 51 choices for the second card of which 12 are hearts. and the probability that the second card is a heart is 12/51; but if the first card is not a heart, then there are 13 hearts left among the 51 from which the second card is chosen, and the probability of *B* is 13/51.

Notation: The symbol $p(B/A)$ means "the probability of event *B,* given that event *A* occurs."

In Example 1 above (a woman is chosen for candidate A and a man for candidate B from speakers' lists of 10 men and 8 women), $p(D/C) = p$(a man speaking for candidate B given that a woman is to speak for candidate A) = 10/18. Note that in this case $p(D/C) = p(D)$.

In Example 2 above (2 cards drawn from a deck of 52), $p(B/A) = p$(the second card is a heart, given that the first is a heart) = 12/51.

Whenever you wish to find the probability that events A and B both happen, you must first ask (and answer) the question "Are A and B independent events or dependent events?"

If A and B are independent, $p(A \text{ and } B) = p(A)p(B)$.

If A and B are dependent, $p(A \text{ and } B) = p(A)p(B/A)$.

Example 3

Representatives for a debate are chosen at random from separate speakers' list of 10 men and 8 women for candidates A and B. What is the probability that candidate A's speaker is a woman and candidate B's is a man?

$C = \{$Candidate A's speaker is a woman$\}$, $D = \{$Candidate B's speaker is a man$\}$, $p(C) = 8/18$, $p(D) = 10/18$. C and D are independent; $p(C \text{ and } D) = 8/18 * 10/18 = 20/81 = .25$.

Example 4

Two cards are drawn, one after the other but without replacement, from a standard deck of 52 cards. What is the probability that both cards are hearts?

$A = \{$the first card is a heart$\}$, $B = \{$the second card is a heart$\}$. A and B are dependent; $p(A \text{ and } B) = p(A)p(B/A) = 13/52 * 12/51 = 1/17$.

Example 5

A card is drawn from a standard deck of 52 cards, its suit is observed, and the card is then replaced before a second one is drawn. What is the probability that both are hearts?

Here the two events are clearly independent, the probability of each is $13/52 = 1/4$, and p(2 hearts are drawn) = $1/4 * 1/4 = 1/16$. Note: An alternative (and common) way of asking the same question is, "If 2 cards are drawn with replacement from a standard deck of 52 cards, what is the probability that both are hearts?"

Example 6

If two cards are drawn from a standard deck of 52 cards, what is the probability of getting 1 heart and 1 spade?

The small trap: "two cards are drawn" presents to you the same problem as "one card is drawn and not replaced; then a second card is drawn." The larger trap: "getting 1 heart and 1 spade" does not describe order. The first card may be a heart and the second a spade or the first a spade and the second a heart. (If you insist on drawing 2 cards at once, think of both hands reaching out at the same time; you may have a heart in the left and a spade in the right, or vice versa.)

A tree diagram is helpful here:

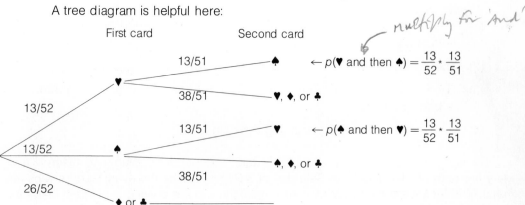

$$p(1 \text{ heart and } 1 \text{ spade}) = \left(\frac{13}{52}\right)\left(\frac{13}{51}\right) + \left(\frac{13}{52}\right)\left(\frac{13}{51}\right) = 2\left(\frac{13}{204}\right) = \frac{13}{102}$$

Sometimes it's easier to find a solution without drawing a tree or a Venn diagram. See if you can follow this reasoning:

$p(1 \text{ heart and } 1 \text{ spade}) = p(\text{[first card is a heart and second is a spade] or}$
$\qquad\qquad\qquad\qquad\qquad \text{[first card is a spade and second a heart]})$
$\qquad\qquad = p(\text{first card is a heart and second is a spade})$
$\qquad\qquad \quad + p(\text{first card is a spade and second a heart})$
$\qquad\qquad = p(\text{first card is a heart}) * p(\text{second is a spade, given}$
$\qquad\qquad \quad \text{that the first is a heart}) + p(\text{first card is a spade})$
$\qquad\qquad \quad * p(\text{second is a heart, given that the first is a}$
$\qquad\qquad \quad \text{spade})$
$\qquad\qquad = (13/52)(13/51) + (13/52)(13/51)$
$\qquad\qquad = 2(13/204)$
$\qquad\qquad = 13/102$

Mutually exclusive events allow use of $p(A \text{ or } B) = p(A) + p(B)$.

Note: Life is too short to write out each problem in such detail. See if you can follow this translation: $p(1\heartsuit \text{ and } 1\spadesuit) = p(\heartsuit_1\spadesuit_2 \text{ or } \spadesuit_1\heartsuit_2) = p(\heartsuit_1\spadesuit_2) + p(\spadesuit_1\heartsuit_2)$
$= p(\heartsuit_1) * p(\spadesuit_2/\heartsuit_1) + p(\spadesuit_1) * p(\heartsuit_2/\spadesuit_1) = 13/52 * 13/51 + 13/52 * 13/51 = 13/102$.

Example 7

What is the probability of heads exactly 3 times if a coin is flipped 4 times?

Here the notation $p(HHTH)$ will be used as shorthand for "the probability of heads on the first, second, and fourth tosses, but tails on the third toss." This is even more abbreviated than the last example, since even subscripts are omitted.

$p(3H \text{ in } 4 \text{ tosses}) = p(HHHT \text{ or } HHTH \text{ or } HTHH \text{ or } THHH)$ (1)
$\qquad\qquad\qquad = p(HHHT) + p(HHTH) + p(HTHH) + p(THHH)$ (2)
$\qquad\qquad\qquad = p(H)p(H)p(H)p(T) + p(H)p(H)p(T)p(H) +$
$\qquad\qquad\qquad \quad p(H)p(T)p(H)p(H) + p(T)p(H)p(H)p(H)$ (3)
$\qquad\qquad\qquad = (1/2)^4 + (1/2)^4 + (1/2)^4 + (1/2)^4$ (4)
$\qquad\qquad\qquad = 4(1/2)^4$
$\qquad\qquad\qquad = 4/16 = 1/4$

Explanations

(1)   These are 4 different events, each consisting of heads 3 times and tails once.

(2)   The 4 events in (1) are mutually exclusive; **or** is translated into +.

(3)   Each of the events in (2) is made up of 4 independent events, so **and** is translated into *.

(4) $p(H) = 1/2$, $p(T) = 1/2$ (assuming the coin is fair).

Do not confuse "mutually exclusive" and "independent." Suppose George and Mary are both playing tennis on Saturday at 3 P.M. Let G be the event George wins his match, and let M be the event Mary wins hers. If the two are playing in different matches, G and M are independent events but are not mutually exclusive; if they are playing against each other, G and M are mutually exclusive events which are not independent.

> If you are finding $p(A$ **or** $B)$, ask "Are A and B mutually exclusive?" If they are, **add** $p(A)$ and $p(B)$.
>
> If you are finding $p(A$ **and** $B)$, ask "Are A and B independent?" If they are, **multiply** $p(A)$ and $p(B)$.

5.8 VOCABULARY AND SYMBOLS

probability
equally likely events
relative frequency
subjective probability
sample space
subset
event
Venn diagram

chosen at random
mutually exclusive events
$p(A$ or $B)$
$\overline{A}$
independent events
dependent events
$p(A$ and $B)$

5.9 EXERCISES

Use a Venn diagram or a tree wherever it is helpful.

1.   Of 400 freshmen, 50 take English, philosophy, and zoology, 160 take both English and philosophy, 100 take both English and zoology, and 60 take both philosophy and zoology; 300 take English, 140 take zoology, and 210 take philosophy.

(a)   How many students take neither English nor philosophy nor zoology?

(b)   How many students take English or philosophy but not zoology?

(c)   How many students take English but do not take either philosophy or zoology?

2.   At an inspection station, 1 per cent of cars tested have bad brakes, bad headlights, and cause too much pollution; 19 per cent cause too much pollution; 14 per cent have bad brakes; 5 per cent have bad brakes and headlights; 3 per cent have bad headlights and cause pollution; 16 per cent have bad headlights, and 3 per cent have bad brakes and cause pollution.

(a)   What percentage have either bad brakes or bad headlights (or both)?

(b)   What percentage have bad brakes or bad headlights but not both?

3.   A market analyst claims that the probabilities that stock in U.S. Rubber will go up more than 5 points, stay the same (within 5 points), or go down more than 5 points this year are .45, .24, and .31, respectively. A second analyst claims these probabilities are .47, .27, and .29; and a third that they are .40, .27, and .35, respectively. Comment on these claims.

4.   Assume that the probabilities that a student's score on an SAT test is over 600, 500–599, 400–499, or under 400 are .16, .34, .34, and .16, respectively. What is the probability that his score is (a) more than 400? (b) either less than 400 or between 500 and 599?

5.   Assume that 70 per cent of crimes occur at night, and that 40 per cent of crimes at night and 20 per cent of daytime crimes are violent. If the record of one crime is chosen at random from the police files on which these percentages are based, what is the probability that it is the record of a violent crime?

6.   If two (fair) dice are tossed, what is the probability that the sum of the two dice is (a) 10? (b) 9? (c) 9 or more? (d) less than 9?

7.   A coin is tossed 1000 times, and comes up heads 440 times. If the coin is tossed again, what is the probability that it will come up (a) heads? (b) tails?

8.   If one card is drawn at random from a well shuffled standard (bridge) deck of 52 cards, what is the probability that it is (a) an ace? (b) an ace, a 4 of diamonds, or a 7 of hearts? (c) a 2 or a 5? (d) a club or a spade? (e) not a 3? (f) a 10 or a 9 or a club?

9.   A ball is drawn at random from a bag containing 5 red, 10 black, and 10 white balls. What is the probability that it is (a) red? (b) red or white? (c) not red? (d) neither red nor white?

10.   Find the probability of a 3 turning up at least once when a die is tossed twice.

11.   If a fair coin is tossed 3 times, find the probability of (a) no heads; (b) exactly 1 head; (c) at least 1 head. Use a tree diagram. $\frac{3 c_0}{2^3} = 1/8$ (b) $\frac{3 c_1}{8} = 3/8$ (c) $1 - \frac{1}{8} = 7/8$

12.   There are three children in a family. Find the probability that (a) all are girls; (b) one is a boy; (c) at least one is a boy. Assume the probability of a boy at any given birth is 1/2.

13.   A friend of yours has just gotten a job as a salesman. You offer to bet him $100 to his $10 that he will make at most $10,000 in commissions during his first year. What can be said of his subjective probability of making $10,000 this year (a) if he accepts the bet, (b) if he does not accept it?

14.   Toss two pennies 48 times, and tally the number of times there are no heads, 1 head, and 2 heads. Compare your tallies with the numbers you expect if the coins are fair.

15.   A man has 6 keys, only one of which fits the door of his house. If he tries the keys one at a time, choosing at random from those that have not yet been tried, what is the probability that he opens the door of his house (a) on the first try? (b) on the third try? (c) on the fifth try?

16.   During the fraternity rush, ΔΦΩ puts a flyer in half the campus mail boxes, choosing them at random, and APO puts a flyer in every third mail box. What is the probability that a randomly selected student will get (a) both flyers in his mail box? (b) neither flyer? (c) at least one flyer? a.) Independent so $\frac{1}{2}(1/3)$ A and B

17.   Acme Computer Co. manufactures 10,000 units each week. Each unit passes

b.) $1 - \left(\frac{1}{2} + \frac{1}{3}\right)$ c.) $\frac{1}{2} + \frac{1}{3}$

through three inspection stations A, B, C before being shipped out. Two per cent are rejected at station A; of those remaining, 5 per cent are rejected at station B, and finally approximately 1 per cent of the rest are rejected at station C. What is the probability that a randomly chosen unit will pass all three inspections?

18.   Fifty-two men and 48 women climb Mt. Washington, and 14 of them get blisters; of those with blisters, 6 are women. If one member of the group is chosen at random, what is the probability that the person chosen either has blisters or is a woman (or both)?

19.   A doctor discovers that the probability is .60 that patients with symptom A have tuberculosis, and that the probability is .50 that those with symptom B have this disease. What is the probability that those with symptom A or B have tuberculosis?

20.   If a coin is flipped 3 times, what is the probability of getting (a) no heads? (b) heads exactly once? (c) heads exactly twice? (d) heads exactly 3 times? (e) heads at least once?

21.   (a)   One of the digits 1, 2, 3 is chosen at random. What is the probability it is an even number?
         (b)   Three of the digits 1, 2, 3 are chosen at random, with replacement. What is the probability that (i) none of the three digits is even? (ii) exactly one is even? (iii) exactly two are even? (iv) all three are even?

22.   A cigarette box contains 10 cigarettes with filters and 5 without filters. If 2 cigarettes are chosen at random from the box, what is the probability that (a) both have filters? (b) at least one has a filter?

23.   Four tennis players are each to play a set against each of the other three. A handicapper assigns probabilities that a player will win at least two sets as follows:

Player	A	B	C	D
Probability	.7	.8	.2	.3

Can the sum of the probabilities be more than 1? Explain.

24.   A child tries to find his way through the maze below. He chooses at random whenever a choice of paths must be made, and never retraces his steps except when returning down a blind alley. If he makes no mistakes, he can find his way to the center in 40 seconds; each mistake causes a delay of 8 seconds. What is the probability that he gets to the center in 70 seconds or less?

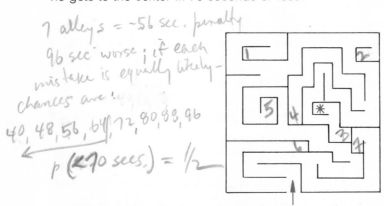

25.   A die is tossed 4 times. What is the probability that the first time a 5 comes up is on the fourth toss?

26. On a multiple choice exam there are 10 questions and 5 choices for each answer. If a student knows the answer to 5 questions and randomly guesses at the answers to the other 5, what is the probability he passes the exam? He needs a grade of 70 or more to pass. *1/25*

27. Mr. and Mrs. Able live in Newark, New Jersey. The probability that Mr. Able will go to New York on December 31 is .2; the probability that Mrs. Able will go is .1; the probability that Mr. Able will go, given that Mrs. Able goes, is .3. What is the probability that

    (a) both Mr. and Mrs. Able will go to New York on December 31? *.3(.1)*

    (b) at least one of them will go to New York on December 31? *.2 + .1 − .3(.1)*

    (c) neither Mr. Able nor Mrs. Able will go to New York on December 31?

28. If 3 pieces of fruit are drawn at random from a bowl containing 4 apples, 5 pears, and 1 peach, what is the probability that

    (a) the peach is selected?

    (b) at least 2 apples are selected?

    (c) a peach and 2 apples are selected?

29. A bag contains 4 red and 6 white balls. Balls are drawn at random; <u>WR</u> means with replacement, <u>WOR</u> means without replacement. Fill in the four columns below.

	2 balls drawn		3 balls drawn	
	WR	WOR	WR	WOR
p(none are white)				
p(exactly one is white)				
p(exactly two are white)				
p(exactly three are white)				

30. A die is rolled twice. What is the probability that the sum of the two throws is (a) 11? (b) 11 or more? (c) 11 if the first die comes up with a 5? (d) 11 if at least one of the two throws comes up 5?

31. If 1 spark plug in your 6-cylinder car is not working, what is the probability that the defective spark plug is replaced (a) if 1 spark plug, chosen at random, is replaced with a new one? (b) if 2 spark plugs, chosen at random, are replaced with new ones? *a.) 1/6* *b.) no luck 1st plug, luck 2nd + luck 1st, no luck 2nd = $\frac{5}{6}(\frac{1}{5}) + \frac{1}{6}\frac{5}{6}$*

32. If a bill must be passed by the House, Senate, and President to become law, and the probabilities of these three actions for a particular bill if presented to that body or person are .40, .30, and .20, respectively, what is the probability of that particular bill becoming law?

★33. In a simplified form of the game of Bingo, balls numbered 0 to 9 are drawn at random and players cover the corresponding numbers on their cards, the winner being the first to cover five numbers in any row or diagonal. If you need the numbers 3, 5, 9, what is the probability that you will complete the row (a) after 3 more calls? (b) after 4 more calls? (compute to 3 decimal places.)

★34. In a hat there are 2 coins, one is "fair" $[p(H) = p(T) = 1/2]$, but the other one has two heads $[p(H) = 1, p(T) = 0]$. One coin is chosen at random from the hat and flipped 6 times. The first 5 times it comes up heads. What is the probability of heads on the sixth toss? Hints: If $p(A) \neq 0$, $p(B/A) = p(A \text{ and } B)/p(A)$; p(heads on first 5 tosses) $= p$([fair coin and heads on first 5 tosses] or [two-headed coin and heads on first 5 tosses]). Let A be the event {heads on the first 5 tosses}, B be the event {heads on the sixth toss}.

ANSWERS

1.   (a)   20. (b) 90 + 110 + 40 = 240. (c) 90.

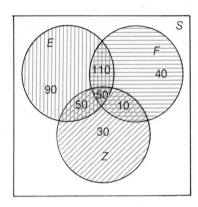

3.   The claims of the second and third analysts are impossible, since their probabilities do not add up to 1.

5.   Of every 1000 crimes, 700 are at night and 280 of these are violent, while 300 are in the daytime and 60 of these are violent; 280 + 60 = 340 crimes are violent. p(the record is of a violent crime) = 340/1000 = .34.

7.   If you are using the relative frequency definition of probability, the answers are (a) 440/1000 = .44, (b) 1 − .44 = .56. If the coin is fair, the answers are (a) 1/2, (b) 1/2. Later (Chapter 7) you will learn how to find the probability that a fair coin will come up heads 440 times in 1,000 tosses; it is very small, but it is not 0.

9.   (a) 5/25 = 1/5. (b) 15/25 = 3/5. (c) 1 − 1/5 = 4/5 **or** 10/25 + 10/25 = 20/25 = 4/5. (d) 10/25 = 2/5.

11.   {HHH, HHT, HTH, THH, HTT, THT, TTH, TTT}. (a) 1/8. (b) 3/8. (c) 1 − 1/8 = 7/8.

13.   (a) His subjective probability is at least .1. (b) It is less than .1.

15.   (a) 1/6. (b) 1/6. (c) 1/6.

17.   Of 10,000 units, 9800 pass inspection at station A; of these, 9310 pass station B, and 9217 pass station C also. p(passing all three) is 9217/10,000 = .92 **or** p(passing station A, and passing station B given that station A has been passed, and passing station C given that stations A and B have been passed) = (1 − .02)(1 − .05)(1 − .01) = .98 * .95 * .99 = .92.

19.   No answer. Did you try $p(A$ or $B) = p(A) + p(B) = .60 + .50 = 1.10$?! Probability can never be >1. Not enough information − such as $p(A$ and $B)$ − is given.

21.   (a) p = 1/3. (b) (i) $(2/3)^3$ = 8/27; (ii) $3(1/3)^1(2/3)^2$ = 4/9; (iii) $3(1/3)^2(2/3)$ = 2/9; (iv) $(1/3)^3$ = 1/27.

23.   Yes. Read "$p(A$ w $B)$" as "the probability that player A wins his set with player B." Then $p(X$ w $Y) + p(Y$ w $X)$ must equal 1 for each pair of players. The answer is easier to understand if players A and B are tennis stars and players C and D are beginners. Then the assigned probabilities of winning at least 2 sets are 1, 1, 0, 0, re-

spectively. The following assignment of probabilities to individual matches may seem unlikely, but would result in the handicapper's predictions:

	A	B	C	D
p(A w −)		.5	.4	1
p(B w −)	.5		1	.6
p(C w −)	.6	0		.3
p(D w −)	0	.4	.7	

25. p(no 5 on the first 3 tosses and then 5 on the fourth toss) $= (5/6)^3(1/6) = .10$.

27. (a) $p(\text{Mr. and Mrs.}) = p(\text{Mrs.})\,p(\text{Mr./Mrs.}) = .1 * .3 = .03$.
   (b) $p(\text{Mr. or Mrs.}) = p(\text{Mr.}) + p(\text{Mrs.}) - p(\text{Mr. and Mrs.}) = .2 + .1 - .03 = .27$.
   (c) $p(\text{neither Mr. nor Mrs.}) = 1 - p(\text{Mr. or Mrs.}) = 1 - .27 = .73$.

29.

Probabilities

Number of white	2 balls drawn WR	2 balls drawn WOR†	3 balls drawn WR	3 balls drawn WOR
0	$(.4)^2 = .16$	$\frac{4}{10} * \frac{3}{9} = .13$	$(.4)^3 = .06$	$\frac{4}{10} * \frac{3}{9} * \frac{2}{8} = .03$
1	$2 * .4 * .6 = .48$	$2 * \frac{4}{10} * \frac{6}{9} = .53$	$3(.4)^2(.6) = .29$	$3 * \frac{4}{10} * \frac{3}{9} * \frac{6}{8} = .30$
2	$(.6)^2 = .36$	$\frac{6}{10} * \frac{5}{9} = .33$	$3(.4)(.6)^2 = .43$	$3 * \frac{4}{10} * \frac{6}{9} * \frac{5}{8} = .50$
3	0	0	$(.6)^3 = .22$	$\frac{6}{10} * \frac{5}{9} * \frac{4}{8} = .17$

Handwritten annotations: Bag has 6w 4r · Denominators are (won) · P # picks · 10 # picks (WR) · Exponent (WR) = # Draws · R W · Notice WR & WOR have close values

31. (a) $1/6 = .17$. (b) $1 - \frac{5}{6} * \frac{4}{5} = \frac{1}{3}$ **or** $\frac{1}{6} * \frac{5}{5} + \frac{5}{6} * \frac{1}{5} = \frac{1}{3}$.

33. $.3 * .2 * .1 = .006$, (b) To win on 4 calls, the last one must give you bingo. A further complication is that if you get a 3, say, on the first call, you can get 3 again on the second or third calls and still win in 4; but if you do not get a 3, 7, or 9 on the first call then there can be no repetitions in the 4 calls.

Handwritten: no luck 1st ball

$p(3, 7, 9\ [\text{in any order}]\ \text{on calls } 2, 3, 4) = .7 * .3 * .2 * .1 = .0042$
$p(3, 7, 9\ [\ ''\ \ ''\ \ '']\ ''\ \ ''\ 1, 3, 4) = .3 * .8 * .2 * .1 = .0048$
$p(3, 7, 9\ [\ ''\ \ ''\ \ '']\ ''\ \ ''\ 1, 2, 4) = .3 * .2 * .9 * .1 = \underline{.0054}$
$.0144$

$p(\text{winning on 4 calls}) = .014$.

†Because of round-off, the sum of these probabilities does not equal 1.00.

chapter six

discrete probability distributions

In this chapter you are introduced to probability distributions. You are going to face hundreds of them in the future. You will find that they are special kinds of frequency distributions, and that the probability of a score between a and b can be found by determining the area of the bars of the histogram between a and b. The binomial and normal distributions are particular kinds of probability distributions. The binomial distribution, taken up in this chapter, is important in itself—you will see it applied several times in later chapters—but it also heads you toward the normal distribution, the subject of the following chapter. And an understanding of the normal distribution is basic to much of the rest of the book.

6.1 PROBABILITY DISTRIBUTIONS

You have long since (Chapter 2) become familiar with frequency distributions and with histograms to illustrate them:

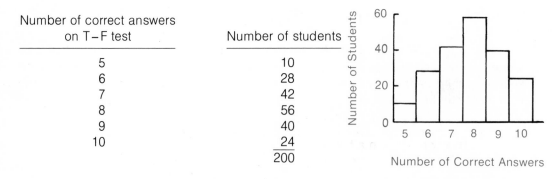

Number of correct answers on T–F test	Number of students
5	10
6	28
7	42
8	56
9	40
10	24
	200

Now if one student is chosen at random from this class, the probability that he answered 5 questions correctly is 10/200 = .05. In a similar fashion, the probability that his number of correct answers is 6, 7, 8, 9, or 10 may be determined, and the results summarized in the following table:

Number of correct answers on T–F test	Probability
5	.05
6	.14
7	.21
8	.28
9	.20
10	.12
	1.00

This table presents a **probability distribution.** Note that the events in the "scores" column are mutually exclusive—the student cannot get 5 and 8 correct answers on the test at the same time—but that a particular student is certain to answer 5, 6, . . . , or 10 questions correctly, so the sum of the probability column is 1. This probability distribution is a **relative frequency distribution** [p(5 correct answers) = .05 since 10 students out of 200 answered 5 questions correctly]; since it is a particular kind of frequency distribution, it may be represented by a histogram. The area under the histogram is 1.

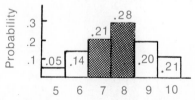

Number of Correct Responses

Further, the probability that he answered 7 or 8 questions correctly is .21 + .28 = .49; this can be determined by adding the areas of the bars over the responses 7 and 8, indicated by shading in the histogram.

Since a probability distribution is a special kind of (relative) frequency distribution, it too will have a mean and standard deviation. Ordinarily we shall be interested in the probability distribution of a population, not of a sample, and it is therefore time to introduce two more Greek letters: μ (mu) is the Greek (lower case) m, and σ (sigma) is the Greek (lower case) s. (You have already met capital sigma, Σ, used to denote a sum.)

μ = mean of a population

σ = standard deviation of a population

The formulas for the mean and standard deviation of a discrete probability distribution are:

$$\mu = \Sigma X p(X)$$
$$\sigma = \sqrt{\Sigma (X - \mu)^2 p(X)} \;\dagger$$

Example 1.

A fair coin is flipped twice. What is the probability of heads 0 times? 1 time? twice? Make up a probability distribution, find its mean and standard deviation, and sketch a histogram.

Let X equal the number of heads.

X	$p(X)$	$Xp(X)$	$X - \mu$	$(X - \mu)^2$	$(X - \mu)^2 p(X)$
0	1/4	0	−1	1	1/4
1	1/2	1/2	0	0	0
2	1/4	1/2	1	1	1/4
		1			1/2

$p(X)$ 1/2, 1/4

$$\mu = \Sigma X p(X) = 1.$$
$$\sigma = \sqrt{\Sigma (X - \mu)^2 p(X)} = \sqrt{1/2} = .7.$$

---

†These are your old friends the mean and standard deviation of a grouped frequency distribution in different clothing: μ and σ instead of $\overline{X}$ and s, $p(X)$ instead of f, and no denominators because $\Sigma p(X) = 1$. More details will be given on page 109.

Example 2

After much experimentation, the distribution shown in the box below has been determined for the probability that a rat will solve a certain maze within 60 seconds after X trials. (a) Find the mean and standard deviation of this distribution. (b) What is the probability that a rat will need 3 trials or less before solving the maze in the allotted time? (c) Sketch a histogram.

X	$p(X)$	$Xp(X)$	$X - \mu$	$(X - \mu)^2$	$(X - \mu)^2 p(X)$
1	.20	.20	−1.5	2.25	.450
2	.40	.80	− .5	.25	.100
3	.20	.60	.5	.25	.050
4	.10	.40	1.5	2.25	.225
5	.10	.50	2.5	6.25	.625
		2.50			1.450

(a) $\mu = \Sigma Xp(X) = 2.5$
     $\sigma^2 = \Sigma(\dot{X} - \mu)^2 p(X) = 1.450$
     $\sigma = \sqrt{1.450} = 1.20$

(b) $p(X = 1, 2, \text{ or } 3)$
       $= .20 + .40 + .20 = .80$

(c)

If you have a frequency distribution, you can find a probability distribution by dividing each entry in the f column by the total number of scores. If you have a probability distribution, you can immediately find the percentage of scores in any class (multiply by 100), and can find the number of scores in any class if you know the total number of scores:

X	f	$p(X)$		Y	$p(Y)$	Per cent	If $n = 20$: f
10–19	6	6/50 = .12		1–3	.3	30	6
20–29	26	26/50 = .52		4–6	.6	60	12
30–39	18	18/50 = .36		7–9	.1	10	2
	50	1.00			1.00	100	20

Remember that a probability distribution is a special kind of frequency distribution in which (a) the sum of the probability column is 1, and (b) the probability that a score is between a and b equals the area under the histogram between a and b.

6.2   PRE-TEST ON PERMUTATIONS AND COMBINATIONS

Before continuing with probability distributions, there will be a brief side-trip into permutations and combinations. If you can answer the following questions correctly, skip immediately to Section 6.6, page 89, as soon as you have absorbed the following notations that will be used in this book:

         $_nP_X$ = number of permutations of n things taken X at a time   w/ "Replacement"

         $_nC_X$ = number of combinations of n things taken X at a time

1. In how many ways can a committee of 2 members be chosen from a club with 10 members?

2. In how many ways can a president and vice-president be chosen from a club of 10 members?

3. $5! = ?$

4. In how many ways can 3-letter "words" be made from the letters a, e, i, o, u if no letter is repeated in the same word?

6.3 PERMUTATIONS

A **permutation** is an ordered arrangement without repetitions of the objects in a set. The question "How many 2-letter permutations are there of the letters in the set $\{a, b, c\}$?" implies that ab is a different permutation from ba; they consist of the same 2 letters, but the order is different.

Soon enough we shall have formulas for permutations, but let us start by constructing tree diagrams. Remember that the wind blows so the tree falls to the right. Consider the following examples.

Example 1

How many 2-letter permutations are there of the letters in the set $\{a, b, c\}$?

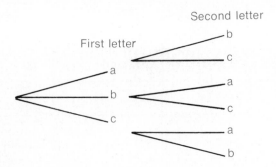

This "tree" lets us now list the six different permutations: ab, ac, ba, bc, ca, cb. Of course you could have made this listing without making a tree first, but since it has already been blown down, let's take another look at it. Notice that there are 3 branches of the tree on the lowest level, representing the 3 choices possible for the first letter—a, b, or c. Each of these branches leads to 2 more at the top of the tree. (If the first letter is a, then the second can be b or c.) There are 3 * 2 permutations.

The notation for "the number of permutations of 3 things taken 2 at a time" is $_3P_2$. In this example, the number of 2-letter permutations of 3 letters is $_3P_2$, where here the "things" are letters. We have discovered, then, that $_3P_2 = 3 * 2 = 6$.

It is fair to give warning that the question may be re-phrased in many different ways. The essential ingredients are order and lack of repetition. Here are some other questions to which the answer is $_3P_2 = 3 * 2 = 6$:

1. How many different 2-digit numbers can be formed from the digits 1, 3, 4 if no digit is repeated?

2. In how many different ways can a president and vice-president be chosen from 3 members of a club?

Answers: 1. $_{10}C_2 = 45$. 2. $_{10}P_2 = 90$. 3. 120. 4. $_5P_3 = 60$.

3. If a student takes chemistry, physics, or zoology his freshman year and 1 of the remaining 2 in his sophomore year, in how many ways can his choices be made?

[handwritten: $P = 4! = 24$ w/ Replacement]

Example 2

[handwritten margin notes: $\frac{P}{4}$ Note duplication of values! Sets (total values) would merge the dups... Combination... order not important... $C_3 = \frac{4}{4}$ w/ Replacement]

How many permutations are there of 4 objects if they are taken 3 at a time? Suppose the objects are labeled *a, b, c, d*. We might list them:

[handwritten: combined 1st] ... [combined 2nd] ... [combined 3rd]

abc	acb	bac	bca	cab	cba	dab	dba
abd	adb	bad	bda	cad	cda	dac	dca
acd	adc	bcd	bdc	cbd	cdb	dbc	dcb

Answer: 24

[handwritten: combined 4th time]

Or we can make a tree diagram and count the branches in the tree:

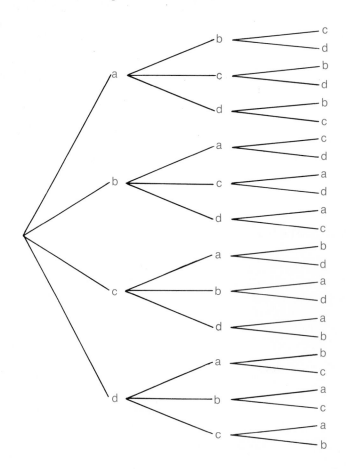

Again the answer is 24. There are 4 branches at the first level (that is, 4 choices are possible for the first letter); each of these leads to 3 choices for the second letter, and, finally, each of these to 2 choices for the last letter; there are altogether 4 * 3 * 2 = 24 permutations. $_4P_3 = 4 * 3 * 2 = 24$.

Before tackling a tougher problem, it will be well to solve this problem in yet another way. Three letters are to be in a given arrangement, so start with 3 blank spaces: __ __ __. There are 4 possible choices for the first letter (*a, b, c,* or *d*), so the first blank is filled in with the number 4: 4 __ __. Once the first letter has been chosen, there are 3 choices left for the second; we arrive at 4 3 __. Then finally there are 2 choices left for the third blank (see the tree diagram); here we are at $\underline{4}$ $\underline{3}$ $\underline{2}$. Of course we are just abbreviating the reasoning behind the tree diagram: 4 branches at the first level, each of which leads to 3 at the second level, means $4 * 3 = 12$ branches so far. Then each of these 12 leads to 2 more at the third level, so there are $4 * 3 * 2 = 24$ permutations of 4 things taken 3 at a time.

Example 3

$_{10}P_4 = ?$ Here listing is a tasteless kind of affair (and if you are tempted to try it, then work instead on $_{1000}P_4$). A tree diagram is not appetizing either. So try filling in blanks: $\underline{10}$ $\underline{9}$ $\underline{8}$ $\underline{7} = {}_{10}P_4 = 10 * 9 * 8 * 7 = 5040$. Now go back and visualize the tree in your mind without actually sketching it: how many branches are there at the first level above the root? Each of these leads to how many at the second level? What is the total number of branches at the second level, then? And so on.

We have now discovered that
$$_3P_2 = 3 * 2$$
$$_4P_3 = 4 * 3 * 2$$
$$_{10}P_4 = 10 * 9 * 8 * 7.$$

Can you solve: $_{12}P_5 = ?$ (The answer is at the bottom of the page.) If n and X are positive integers with $X \le n$, $_nP_X = ?$ It should be clear now that $_nP_X$ will have X factors (that is, X numbers multiplied together); the first is n, the second $n - 1$, the third $n - 2$, ..., the Xth number will be $n - (X - 1)$. Thus $_{10}P_8 = 10 * 9 * 8 * 7 * 6 * 5 * 4 * 3$, $_3P_1 = 3$, $_5P_2 = 5 * 4 = 20$.

6.4 FACTORIAL NOTATION

It is useful sometimes to abbreviate products of the form $3 * 2 * 1$ or $5 * 4 * 3 * 2 * 1$. Such products are written with an exclamation mark after the highest integer in the sequence. Thus $3! = 3 * 2 * 1 = 6$; $5! = 5 * 4 * 3 * 2 * 1$.

> **Definition:** If n is a positive integer, $n! = n(n - 1)(n - 2) \ldots * 2 * 1$.

$3!$ is read as "3 factorial" (not "3 exclamation mark"), $n!$ as n factorial.

Example 1

$$_5P_2 = 5 * 4 = \frac{5 * 4 * 3 * 2 * 1}{3 * 2 * 1} = \frac{5!}{3!} = \frac{5!}{(5 - 2)!}$$

Example 2

$$_{10}P_4 = 10 * 9 * 8 * 7 = \frac{10 * 9 * 8 * 7 * 6 * 5 * 4 * 3 * 2 * 1}{6 * 5 * 4 * 3 * 2 * 1} = \frac{10!}{6!} = \frac{10!}{(10 - 4)!}$$

Example 3

$$_{27}P_{12} = \frac{27!}{(27 - 12)!} = \frac{27!}{15!}$$

Answer: $_{12}P_5 = 12 * 11 * 10 * 9 * 8$.

Example 4

$$_nP_k = ?$$

By analogy with the first 3 examples, this should and does equal $\dfrac{n!}{(n-k)!}$. But what about $_nP_n$ (the permutations of n things taken n at a time)? $_nP_n = \dfrac{n!}{(n-n)!} = \dfrac{n!}{0!} = ?$ We already discovered that $_3P_3 = 3*2*1 = 3!$ (3 decreasing factors, starting with 3). But by formula $_3P_3 = \dfrac{3!}{(3-3)!} = \dfrac{3!}{0!}$. Therefore $3! = \dfrac{3!}{0!}$, and this makes sense only if $0! = 1$. So that's how we define it: $0! = 1$.

Then the formula for $_nP_k$ holds when $k = n$; $_nP_n = \dfrac{n!}{0!} = n!$

6.5 COMBINATIONS

If A is a set with n elements, then a subset of A containing X elements is called a **combination** of n things taken X at a time. The notation for the total number of combinations (the number of different subsets of A with X elements in each subset) is $_nC_X$. As with permutations, repetitions are not allowed: the set containing elephants, tigers, and elephants is the same as the set containing elephants and tigers. The essential difference between combinations and permutations is that order is not important in combinations: $\{a, b\} = \{b, a\}$.

Example 1

How many combinations are there of the 3 letters a, b, c taken 2 at a time? (If $A = \{a, b, c\}$, how many 2-element subsets of A are there?)

1. We can list them: $\{a, b\}, \{a, c\}, \{b, c\}$. $_3C_2 = 3$.

2. Just as the letters a, b give 2 permutations (ab and ba) but 1 combination $\{a, b\}$, so do any 2 letters in A give $_2P_2 = 2$ permutations but only 1 combination. There are, altogether, $_3P_2 = 6$ permutations. So to find the number of combinations, we must divide $_3P_2$ by $_2P_2$:

$$_3C_2 = \frac{_3P_2}{_2P_2} = \frac{6}{2} = 3$$

Example 2

In how many ways can a committee of 3 members be chosen from a club with 10 members?

A committee consisting of Arthur, Bill, and Cindy is the same as a committee of Cindy, Bill, and Arthur; order is not important, so this is a problem in combinations. $_{10}C_3 = ?$ Here listing is a lengthier business, so we try the method used in the second solution of Example 1. There are $_{10}P_3$ permutations of 10 members taken 3 at a time. But any 3 members of the club have $_3P_3 = 3!/0!$ permutations but form only 1 combination. (ABC, ACB, BAC, BCA,

CAB, and *CBA* are all permutations of Arthur, Bill, and Cindy, but {*A, B, C*} is the only combination of these three.) So

$$_{10}C_3 = \frac{_{10}P_3}{_3P_3} = \frac{10!/7!}{3!/0!} = \frac{10!}{3!7!} \ (= 120).$$

Generalizing, the number of combinations of *n* things *X* at a time is given by:

$$_nC_X = \frac{_nP_X}{_XP_X} = \frac{n!}{X!(n-X)!}$$

Once you understand factorials and the formula $_nC_X = \frac{n!}{X!(n-X)!}$, you will realize that, for example,

$$_5C_3 = \frac{5!}{3!2!} = \frac{5*4*3*2*1}{(3*2*1)(2*1)} = 5*2 = 10$$

Be sure to "cancel" factors in the numerator and the denominator; don't dream of multiplying first to get $\frac{120}{12} = 10$.

You can also evaluate $_nC_X$ by starting with *n* in the numerator and writing *X* factors, each decreasing by 1; and starting with *X* in the denominator and writing decreasing factors until you get to 1:

$$_5C_3 = \frac{5*4*3}{3*2*1} = 10$$

Study these two methods until you realize they are essentially the same.

Your greatest difficulty in solving problems involving permutations or combinations is to decide which is to be used. Always decide whether or not order is important. Here are four questions; the answer to each is $_4C_3 = 4$:

1. In how many ways can 3 of the 4 numbers 1, 3, 5, 7 be chosen?

2. If the same coin is tossed 4 times, how many different sequences of heads and tails are there which include heads 3 times?

3. A hat contains 4 balls numbered 1 through 4. If 3 balls are drawn from the hat, how many different combinations of numbers are possible?

4. A university is to give tenure to 3 faculty members in a department of 4 people. In how many different ways can this be done?

6.6 ASSUMPTIONS FOR A BINOMIAL DISTRIBUTION

We shall continue with the study of probability distributions and the corresponding histograms, but shall limit our study for the moment to problems which satisfy the following assumptions:

1. The same experiment is carried out *n* times (*n* trials are made).

2. Each trial has 2 possible outcomes. Let us call these 2 outcomes "success" and "failure." (A successful outcome doesn't imply a good one, nor failure a bad outcome. If a trial consists in throwing one die, success may mean "a 2 comes up"; then failure would mean "1, 3, 4, 5, or 6 comes up.") Let p be the probability of success in one trial, and let q be the probability of failure. Since one outcome or the other must happen, $p + q = 1$ or $q = 1 - p$.

3. The result of each trial is independent of the result of any other trial.

We shall call these the binomial assumptions; the reason for the word **binomial** will be clear shortly (p. 93).

If a fair coin is tossed three times, the binomial assumptions are satisfied. $n = 3$; "success" is interpreted, say, as getting a head on any one toss, so $p = 1/2$, $q = 1/2$, and whether you get heads or tails on one toss doesn't affect the outcome of any other toss.

Now consider a doctor testing a new drug on 35 patients with a certain disease. He wants to compare the new drug with one now used, and he knows that this relieves the symptoms of 70 per cent of patients who use it. Then $p = .70$, $q = .30$, $n = 35$.

On the other hand, suppose that 4 cards are drawn, without replacement, from a deck of 52 cards, and suppose "success" means drawing a heart. Here the binomial assumptions are **not** satisfied: the probability of a heart on the first draw is 13/52, but on the second draw it is 12/51 if a heart was chosen on the first draw, and 13/51 if a club, diamond, or spade was drawn first; the result of the second trial is not independent of the result of the first trial.

6.7 $p(X$ SUCCESSES IN n TRIALS)

Here is a summary of results of two problems you have already worked on or should be able to solve easily.

Example 1

A fair coin is flipped n times. What is the probability of heads exactly 0, 1, . . . , n times if (a) $n = 1$, (b) $n = 2$, (c) $n = 3$? $p = 1/2$, $q = 1/2$; let X equal the number of heads.

(a) $n = 1$

X	$p(X)$
0	1/2
1	1/2
	1

(b) $n = 2$

X	$p(X)$
0	1/4
1	1/2
2	1/4
	1

(c) $n = 3$

X	$p(X)$
0	1/8
1	3/8
2	3/8
3	1/8
	1

Example 2

A bag contains 4 red and 6 green balls. If n balls are drawn at random, with replacement, what is the probability of getting exactly 0, 1, . . . , n red balls if (a) $n = 1$, (b) $n = 2$, (c) $n = 3$? $p = .4$, $q = .6$; let X equal the number of red balls.

(a)	
$n = 1$	
X	p(X)
0	.6
1	.4
	1

(b)	
$n = 2$	
X	p(X)
0	$.6^2 = .36$
1	$2 * .4 * .6 = .48$
2	$(.4)^2 = .16$
	1

(c)	
$n = 3$	
X	p(X)
0	$(.6)^3 = .216$
1	$3 * .4 * (.6)^2 = .432$
2	$3 * (.4)^2(.6)^1 = .288$
3	$(.4)^3 = .064$
	1

A quick review of how two of these probabilities were determined: p(exactly 1 head in 3 tosses of a coin) $= p$(HTT or THT or TTH) $= p$(HTT) $+$ p(THT) $+ p$(TTH) $= 1/2 * (1/2)^2 + 1/2 * 1/2 * 1/2 + (1/2)^2 * 1/2 = 3 * 1/8$. And p(exactly 2 red balls when 3 are drawn from the bag, with replacement) $= p$(RRG or RGR or GRR) $= p$(RRG) $+ p$(RGR) $+ p$(GRR) $= .4^2 * .6 + .4 * .6 * .4 +$ $.6 * .4^2 = 3 * .4^2 * .6 = .288$.

We shall need soon to solve such problems for larger values of n. Do you relish the prospect of finding the probability of exactly 17 heads coming up when a coin is tossed 20 times? It is worth finding a shortcut.

Take another look at the probability of exactly 1 head coming up in 3 flips of a coin. We have, up until now, listed the different possibilities that will give 1 head: p(HTT or THT or TTH). But instead of listing, let the 3 flips be represented by 3 blanks __ __ __, each to be filled in with H or T and with only 1 H being used. But this takes us back to combinations: the number of different ways 1 H can be filled in is $_3C_1$; then the remaining blanks are filled in with T. Then any combination consisting of 1 H and 2 T's will have a probability of $1/2 * (1/2)^2$, since p(H) $= 1/2$ and p(T) $= 1/2$. Thus p(exactly 1 H in 3 times) $= _3C_1 * 1/2 * (1/2)^2$.

Another example: p(exactly 2 red balls when 3 balls are drawn with replacement from a bag with 4 red, 6 green). Three balls drawn again gives 3 blanks: __ __ __; this time 2 are to be filled with R. This can be done in $_3C_2$ ways. Once 2 blanks are filled in with R's, then the third must be filled with G. These $_3C_2 = \dfrac{3 * 2}{1 * 2} = 3$ ways (RRG, RGR, GRR) all have the same probability of occurring. $p = p$(R) $= .4$, $q = p$(G) $= .6$, so the probability of any one is $.4^2 * .6$. But these are mutually exclusive outcomes, so p(exactly 2 R in 3 draws) $= _3C_2 * .4^2 * .6$.

Now it is time to generalize. If an experiment consists of n trials and the binomial assumptions are satisfied, what is the probability of exactly X successes? Here there are n blanks, of which X are to be filled with S (for success) and the rest with F (for failure). The X blanks for S can be chosen in $_nC_X$ ways; once the X positions for S's are chosen, the remaining $n - X$ places are filled in with F's. But we now have $_nC_X$ outcomes, each consisting of X S's and $n - X$ F's. Because of the binomial assumptions, these outcomes are independent. The probability of each of them is the same, $p^X q^{n-X}$, where $p = p$(S), $q = p$(F). We have arrived at the following important result:

> If the binomial assumptions are satisfied, the probability of exactly X successes in n trials is $_nC_X p^X q^{n-X}$.

Of course, X must be less than or equal to n. Notice that the second subscript of C is the same as the exponent of p, and that the sum of the exponents of p and q is n. Remember that, by definition, $p^0 = q^0 = 1$, $p = p^1$, $q = q^1$. Use this delightful formula to check the results of Examples 1 and 2, and then try some new ones.

Table 2, Appendix C, gives values of $_nC_Xp^Xq^{n-X}$ for n from 2 to 20 for various values of p.

$$C_n^{''} = \frac{n!}{x!(n-x)!}$$

Example 3

Find $_{15}C_4p^4q^{11}$ if $p = .60$.

Here $n = 15$, $X = 4$. Find 15 in the n column of Table 2, Appendix C, and then go down that section in the table until you find 4 in the X column. Now look across the row to the $p = .60$ column; you will find the number 007. The decimal point has been omitted from the beginning of each table entry to save space. Therefore $_{15}C_4p^4q^{11} = .007$.

Example 4

A fair coin is flipped 3 times. (a) What is the probability of heads exactly 0, 1, 2, 3, times? (Compare with Example 1, page 90.) (b) What is the probability of at most 1 head in 3 flips? (c) What is the probability of at least 1 head?

(a) $n = 3$, $p = .50$; let $X = $ number of heads.

X	$_nC_Xp^Xq^{n-X} = p(X)$
0	.125
1	.375
2	.375
3	.125

(b) $p(X = 0 \text{ or } 1) = {_3C_0}p^0q^3 + {_3C_1}p^1q^2 = .125 + .375 = .50$.

(c) $p(X = 1, 2, \text{ or } 3) = 1 - p(X = 0) = 1 - {_3C_0}p^0q^3 = 1 - .125 = .875$.

Example 5

A family has 5 children. What is the probability (a) that at least 1 child is a boy? (b) that at least 2 children are boys? Assume the probability of the birth of a boy is 1/2 at each birth.

$p = .5$, $q = .5$; let X equal the number of boys, and use Table 2, Appendix C.

Prob No Boys.

(a) $p(X = 0) = .031$ $= {_5C_0}\left(\frac{1}{2}\right)^0\left(\frac{1}{2}\right)^{5-0} = 1 \cdot \frac{1}{2}^5 = .031$

$p(X = 1 \text{ or more}) = 1 - p(X = 0) = 1 - .03 = .97$.

(b) $p(X = 1) = .156$; $p(X = 2 \text{ or more}) = 1 - p(X = 0 \text{ or } 1) = 1 - p(X = 0) - p(X = 1) = 1 - .031 - .156 = .81$; or $p(X = 2 \text{ or more}) = .312 + .312 + .156 + .031 = .81$. But don't just blast your way through to an answer; see if you can find the most graceful way.

$P(1) = {_5C_1}\left(\frac{1}{2}\right)^1\left(\frac{1}{2}\right)^{5-1} = .156$

6.8   WHY IS IT CALLED A *BINOMIAL* DISTRIBUTION?

The probability distributions we have been developing are called binomial distributions; they are based on the binomial assumptions. In high school you probably studied the binomial theorem for the expansion of $(x + y)^n$:

$$(x + y)^n = x^n + {}_nC_{n-1}x^{n-1}y + {}_nC_{n-2}x^{n-2}y^2 + \ldots + {}_nC_1xy^{n-1} + y^n.$$

For example, $(x + y)^3 = x^3 + {}_3C_2x^2y + {}_3C_1xy^2 + y^3$

$$= x^3 + 3x^2y + 3xy^2 + y^3.$$

Note that the probabilities in the binomial distribution are terms in the binomial theorem if we replace x by p, y by q:

$$(p + q)^n = p^n + {}_nC_{n-1}p^{n-1}q + {}_nC_{n-2}p^{n-2}q^2 + \ldots + {}_nC_1pq^{n-1} + q^n$$

This last equation also shows why the sum of the binomial probabilities is 1: $q = 1 - p$, so $p + q = 1$ and $(p + q)^n = 1$.

6.9   THE MEAN AND STANDARD DEVIATION OF THE BINOMIAL DISTRIBUTION

A binomial distribution is a special kind of probability distribution and therefore a very special kind of frequency distribution — and so we can expect to find its mean and standard deviation as was done on page 83. It is particularly easy to find these for a binomial distribution, however. The mean and standard deviation of the binomial distribution are given by the following formulas:

$$\mu = np, \sigma = \sqrt{npq}$$

$$q = (1 - p)$$

The formula for the mean can be intuitively found: if a fair coin is flipped 100 times, you would expect heads to come up 50 times on the average. But here $n = 100, p = 1/2$, $np = 100 * 1/2 = 50$. If a die is tossed 60 times, you would expect, on the average, 3 would come up 10 times. $n = 60, p = 1/6, np = 60 * 1/6 = 10.$

Example 1

Find the mean and standard deviation of a binomial distribution with $n = 3, p = .5$.

$$\mu = np = 3 * .5 = 1.5$$
$$\sigma = \sqrt{npq} = \sqrt{3 * .5 * .5} = .87$$

Example 2

Balls are taken twice, with replacement, from a bag containing 4 red and 6 green balls. Find the mean and standard deviation for the probability distribution of the number of red balls which are chosen.

$$p = .4, q = .6, n = 2$$
$$\mu = np = 2 * .4 = .8$$
$$\sigma = \sqrt{npq} = \sqrt{2 * .4 * .6} = \sqrt{.48} = .7$$

6.10 APPLICATIONS OF THE
BINOMIAL DISTRIBUTION

Knowledge of the binomial distribution will help you understand the normal distribution to be taken up in the next chapter, but some applications are of immediate interest.

Example 1

A doctor studying cures for leukemia knows that 30 per cent of patients will be alive eight years after onset of the disease with the present treatment. He tries a new drug on 10 leukemia patients, chosen at random, and finds that 7 are still alive eight years after onset. Does it seem likely that the new treatment is much preferable to the present treatment?

The question is first re-phrased in terms of probabilities: What is the probability that at least 7 out of 10 leukemia patients will be alive after eight years of the present treatment? $p = .3$, $q = .7$; p(at least 7 are alive) $= p$(7 or 8 or 9 or 10 are alive) $= .009 + .001 + .000^+ + .000^+$, $\approx .01$. There is approximately a 1 per cent probability that at least 7 out of 10 patients will live eight years under the present treatment, so the new drug looks pretty promising. The doctor should certainly continue his work with it, and try the new treatment on more patients.

Example 2

A machine shop foreman has found that, in the past, 1 per cent of the bolts produced by a machine are defective. He tests a sample of 600 and finds that 8 are defective. What is the probability that 8 or more will be defective in a sample of 600 if the 1 per cent figure still holds?

$p = .01$, $q = .99$; p(8 or more defective out of 600) $= 1 - p$(7 or less are defective). But $n = 600$, and Table 2, Appendix C, does not contain binomial probabilities for such large n. Computing the answer does not seem to be a lovely prospect. But in the next chapter we shall learn how to approximate it easily using the normal distribution, so don't stay up all night carrying out such computations.

6.11 VOCABULARY AND SYMBOLS

probability distribution	factorial	$n!$
relative frequency distribution	combination	$_nC_k$
permutation $\quad _nP_k$	binomial distribution	$_nC_x p^x q^{n-x}$
ordered arrangement	binomial coefficients	
tree diagram	binomial theorem	

6.12 EXERCISES

1.   (a) Scores in a population are distributed as shown in the table below. If one score is chosen at random, what is the probability it equals 2? 5? 8? 5 or 8? that it is less than 11?

X	f
2	6
5	12
8	8
11	14
	40

(b) Form the corresponding probability distribution and illustrate it with a histogram.

(c) Find the mean and standard deviation of the probability distribution you found in (b).

2. (a) Three coins are to be tossed 800 times. Use the accompanying table to determine the number of times you expect (i) none of the coins to come up heads, (ii) exactly 1, (iii) exactly 2, (iv) all three coins to come up heads.

X = number of heads	p(X)
0	1/8
1	3/8
2	3/8
3	1/8

(b) Find the mean and standard deviation of the probability distribution shown in the table.

(c) On what per cent of tosses do you expect two of the coins will come up heads?

3. (a) A bag contains 4 red and 6 green balls. If 3 balls are drawn at random, with replacement, the probability distribution for drawing X red balls is shown in Example 2, Section 6.7. Illustrate the probability distribution with a histogram.

(b) Find the mean and standard deviation of the probability distribution.

4. The following probability distribution is based on the grade-point averages of 2000 students:

GPA	Probability
3.00–4.00	.24
2.00–2.99	.62
1.00–1.99	.12
.00– .99	.02

(a) Find the mean and standard deviation of this probability distribution.
(b) Write the corresponding frequency distribution.

5. (a) $_9P_3 = ?$ (b) $_9C_3 = ?$ (c) $_9C_6 = ?$

6. If a license plate contains 3 different numbers followed by 2 different letters, how many different plates can be issued?

In Exercises 7–18, decide first whether permutations or combinations are asked for.

7. In how many ways can a president and vice-president be chosen from a club of 60 members? $_{60}P_2$

8. A child has a penny, a nickel, a dime, a quarter, a half-dollar, and a silver dollar. If she puts 3 coins in her piggy bank, how many different amounts could be in the piggy bank? *They are replaced each time so and order isn't important.* $_6C_3 = \frac{6!}{3!\,(6-3)!} = 20$

9. In how many ways can a committee of 4 members be chosen from a club of 20 members?

10. In how many ways can 5 people be seated in a row? *order imp. so* $_5P_5 = 5!$

11. In how many ways can a Committee on Tenure and Promotion of 5 members be chosen from a faculty of 100?

12. In how many ways can 3-letter "words" be formed from the letters of the word **magazine** if no letter is repeated in the same "words"?

13. (a) In how many ways can a teacher choose a committee of 3 from a class of 4 boys and 3 girls? $_7C_3 = 35$
    (b) How many of the possible committees contain exactly one girl? $_3C_1 \times _4C_2$
   ★(c) How many of the possible committees contain at least one girl?

14. An urn contains 8 balls, numbered 1 through 8. In how many ways can 3 balls be chosen from the urn, if the balls are chosen without replacement?

15. A set U contains 4 elements. Find the number of subsets of U.

★16. In how many ways can 4 men and 3 women be seated in a row, if men and women alternate?

★17. From 5 economists and 6 sociologists, a committee of 3 economists and 3 sociologists is to be selected. In how many ways can the committee be chosen if (a) all 11 are willing to serve on it? (b) one economist refuses to serve on it? (c) 2 particular sociologists must serve on it?

★18. How many different 3-digit numbers are there (a) if repetitions are not allowed, 0 cannot be the first digit, and the number is even? (b) if each digit is 2 less than its predecessor (864 is allowed, but not 863 or 368)? (c) if 1 repetition is allowed, but the first digit cannot be 0?

19. What is the probability of getting 1 (a) exactly 4 times if a die is tossed 5 times? (b) at least 4 times? (c) at most 4 times?

20. If 4 cards are dealt from a standard deck of 52 cards, what is the probability that they are all hearts?

21. There are 10 questions in a multiple-choice quiz, with 5 answers to each. If answers are chosen at random, what is the probability of 8 correct answers?

22. A bag contains 10 red, 50 green, and 40 black balls. What is the probability of getting a red ball 3 times in 4 draws if each ball is replaced before the next one is drawn?

23. What is the mean number of successes expected in 200 trials if the binomial assumptions are satisfied and the probability of success in any one trial is .4? What is the standard deviation of the probability distribution?

24. Assume that, when a child is born, the probability it is a girl is 1/2 and that the sex of the child does not depend on the sex of an older sibling. (a) Find the prob-

ability distribution for the number of girls in a family with 4 children. (b) Find the mean and standard deviation of this distribution.

25. In the past, 10 per cent of the students taking statistics have been absent from the first class in March because of illness. What is the probability that not more than 2 of a class of 30 will be absent from that class this year on account of illness?

26. After extensive tests, a seed company determines that 90 per cent of its petunia seeds will germinate. It sells the seeds in packets of 100 and guarantees 85 per cent germination. If the binomial assumptions are met, what is the probability that a particular packet will not meet the guarantee?

27. A pair of dice is tossed 200 times. (a) What is the mean number of times the sum of the 2 dice is 7? (b) What is the most probable number of times the sum of the 2 dice is 7?

28. If 10 per cent of Americans have an IQ over 125, what is the probability that at least 1 of 4 Americans chosen by lot will have an IQ over 125?

ANSWERS

1. (a) $6/40 = .15, .30, .20, .50, 1 - .35 = .65$.

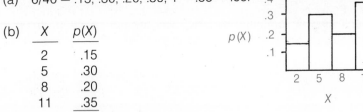

(b)

X	p(X)
2	.15
5	.30
8	.20
11	.35
	1.00

(c) $\mu = 2 * .15 + 5 * .30 + 8 * .20 + 11 * .35 = 7.25$
$\sigma^2 = (-5.25)^2 * .15 + (-2.25)^2 * .30 + .75^2 * .20 + 3.75^2 * 35$
$= 10.69$
$\sigma = \sqrt{10.69} = 3.3$

3. (a)

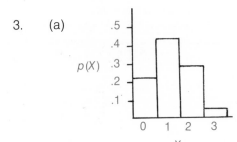

(b) $\mu = 1.6, \sigma = \sqrt{3} \ .4 \ .6 = 85$

5. (a) $9 * 8 * 7 = 504$. (b) $\dfrac{9 * 8 * 7}{1 * 2 * 3} = 84$. (c) $\dfrac{9!}{6!3!} = {}_9C_3 = 84$.

7. Permutations, since A for president and B for vice-president is quite different from B for president and A for vice-president. ${}_{60}P_2 = 60 * 59 = 3540$.

9.   Combinations. $_{20}C_4 = \dfrac{20 * 19 * 18 * 17}{4 * 3 * 2 * 1} = 4845.$

11.   Combinations. $_{100}C_5 = 75{,}287{,}520.$

13.   (a) Combinations. $_7C_3 = 35.$ (b) Choose 1 of 3 girls, 2 of 4 boys. $_3C_1 * {_4C_2}$ $= 3 * 6 = 18.$ (c) Exactly 1 or exactly 2 or exactly 3 girls. $_3C_1 * {_4C_2} + {_3C_2} * {_4C_1} + {_3C_3} * {_4C_0}$ $= 18 + 12 + 1 = 31$ **or** $35 -$ (number of committees with no girls) $= 35 - 4 = 31.$

15.   Solution 1: A subset has size 0, 1, 2, 3, or 4. $_4C_0 + {_4C_1} + {_4C_2} + {_4C_3} + {_4C_4}$ $= 1 + 4 + 6 + 4 + 1 = 16.$ Solution 2: For each element of $U,$ there are two possibilities for each subset: that element is either in or not in the subset. There are, then, $2 * 2 * 2 * 2$ $= 16$ subsets.

17.   (a) $_5C_3 * {_6C_3} = 10 * 20 = 200.$ (b) $_4C_3 * {_6C_3} = 4 * 20 = 80$ (choose 3 of the 4 economists willing to serve, and 3 of the 6 sociologists). (c) $_5C_3 * {_4C_1} = 10 * 4 = 40$ (choose 3 of the 5 economists and 1 of the 4 other sociologists).

19.   (a) $_5C_4 \, (1/6)^4 (5/6)^1 = 25/6^5 \approx .003.$ (b) $25/6^5 + {_5C_5} \left(\dfrac{1}{6}\right)^5 = \dfrac{26}{6^5} \approx .003.$

     (c) $1 - .003 = .997.$

21. $_{10}C_8 \, (.2)^8 (.8)^2 = .00007.$

23. $\mu = 200 * .4 = 80.$
     $\sigma = \sqrt{200 * .4 * .6} = \sqrt{48} = 6.9.$

25.   Can't answer; binomial assumptions are not satisfied, since illness of one student (flu or cold, say) may cause illness in another.

27.   (a) $200 * \dfrac{1}{6} = 33.3.$ (b) 33.

chapter seven

normal distributions and standard scores

At the end of Chapter 2 continuous distributions were briefly mentioned, but in most of the frequency distributions with which you have worked, X could take on only a finite number of values. Now you will get better acquainted with continuous population distributions, in which an infinite number of different scores are possible. The frequency polygon used to illustrate a distribution with discrete values of X will become a smooth curve.

By far the most important of the continuous distributions are the normal distributions. There are many of these; you will first become familiar with one of them, called the standard normal distribution, whose mean is 0 and whose standard deviation is 1. Then standard scores are introduced in which the unit of measurement is one standard deviation. Standard scores have many uses, one of which is to make familiarity with other normal distributions easy once you know the standard normal distribution.

In the remainder of the book you will face a great many formulas, but most of them will involve simply changing scores in a normal distribution into standard scores. If you thoroughly master the standard normal curve, use of the z table for probabilities in any normal distribution won't cause any trouble. Use of the normal distribution to approximate a binomial distribution will need some careful attention. Remember that the mean of a binomial distribution = np, and its standard deviation is $\sqrt{npq}$.

7.1 HISTOGRAMS OF BINOMIAL DISTRIBUTIONS

Consider some binomial distributions, all with $p = 1/2$ but various values of n. Here are the histograms for $n = 4$, 10, and 50 (remember that these are relative frequency or probability distributions; the labels on the vertical axes have been omitted so that you can concentrate on the shapes of the histograms):

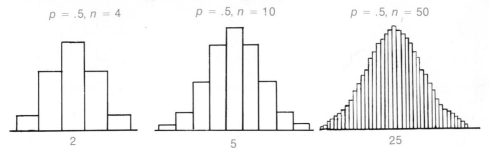

$p = .5, n = 4$ $p = .5, n = 10$ $p = .5, n = 50$

2   5   25

Observe, in each case, that the histogram is symmetrical about its mode, that the area under each histogram is 1, and that if frequency polygons are drawn instead of histograms, these polygons get closer to a bell-shaped curve as n gets larger.

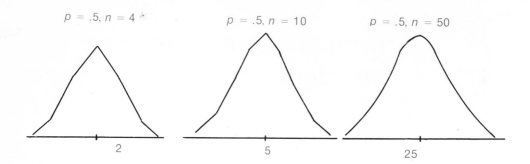

If the height of the bar is known, then its area is known (its width is 1); the probability of getting a value between two numbers may be determined by finding the area under the histogram corresponding to those numbers.

Example 1

If a fair coin is flipped 10 times, what is the probability that heads will turn up exactly 3, 4, or 5 times?

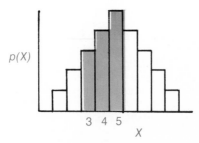

If 3 is thought of as the midpoint of a class whose class limits are 2.5 and 3.5, and correspondingly for 4 and 5, then the probability of a number between 2.5 and 5.5 when $n = 10$, $p = .5$ can be found in the same way.

Now look at the histograms for $p = .3$, $n = 4$, 10, or 50:

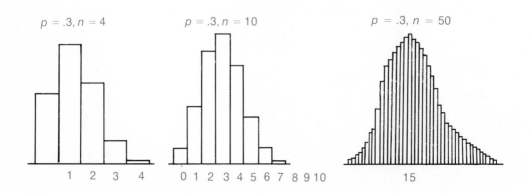

Here the areas under the histograms still equal 1, but the corresponding frequency polygons are not symmetrical about the mode for small values of *n*.

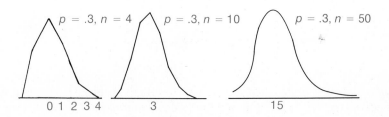

$p = .3, n = 4$ $p = .3, n = 10$ $p = .3, n = 50$

As *n* gets larger, however, the polygon becomes more bell-shaped (and more symmetrical about the mode).

7.2 NORMAL CURVES

The area under the histogram of a probability distribution is 1, and the probability of a score between *a* and *b* is represented by the area of the histogram between *a* and *b*. If $p = .5$, $n = 5$, for example, the probability of success exactly 3 or 4 times (between 2.5 and 4.5 times) in 5 trials equals the area of the bar over 3 plus that of the bar over 4 (= the area of the bars between 2.5 and 4.5).

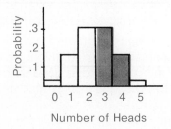

In the same way, a continuous curve represents a probability distribution if

(a) The area under the whole curve is 1. "Under the whole curve" means between the curve and the horizontal axis; there is no ambiguity since the whole curve is above this axis.

(b) The probability of a score between *a* and *b* equals the area under the curve between *a* and *b*.

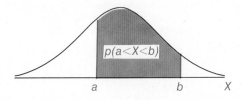

One kind of probability distribution, called the **normal distribution,** is fundamental in modern statistics. It is a continuous distribution; the curve representing it is called the **normal curve.** Historically, it was discovered in the eighteenth century in the study of errors (if a gun is shot at a target, and the horizontal distance from the point where each bullet hits the target to its center is measured, these distances are normally distributed). It is easier to understand, however, as the symmetrical bell-shaped curve

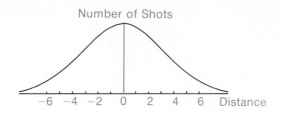

which a binomial distribution approaches as *n* gets very large. There is a whole family of normal curves, differing in mean or standard deviation. First we shall look only at one member of the family — the **standard normal curve** with $\mu = 0, \sigma = 1$ — and shall study the rest of the family in the following section.

7.3 THE STANDARD NORMAL CURVE

You certainly met simple equations such as $y = x + 1$ and $y = x^2$ in high school. The equation of the standard normal curve is rather more complicated; it is

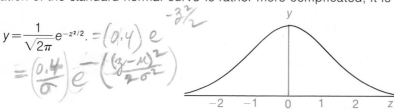

$$y = \frac{1}{\sqrt{2\pi}} e^{-z^2/2}.$$

There is no need whatever to memorize this equation! It is included simply to reassure you that the standard normal curve is the graph of a particular equation. Here z is used for the independent variable instead of the *x* to which you are accustomed. This is done not willfully or maliciously but to distinguish the standard normal curve from other normal curves. *e* is an irrational number which is usually first met in studying calculus; its approximate value is 2.7. But even if you don't need to know the equation, you should become very familiar with certain characteristics of the standard normal curve:

1. It extends indefinitely in both directions. The curve gets very close to the horizontal axis without actually touching it as z gets either very large or very small. Fortunately, most of the area under the curve occurs between −3 and +3 and it won't be necessary to extend the tails very far.

2. The highest point of the curve is at 0; the mode of the standard normal distribution is 0.

3. The curve is symmetrical about 0: the height of the curve above the z axis is the same for 3 and −3, for 1.4 and −1.4, for z and −z. Therefore **the mean of the standard normal distribution is 0.**

4. A curve is "concave up" if it is part of a bowl that will hold water; ⌣ or ⌣ or ⌣ ; a curve is "concave down" if it is part of a bowl that will spill water: ⌢ or ⌢ or ⌢ . The standard normal curve changes from concave up to concave down at +1 and −1: This knowledge will help you in sketching the curve.

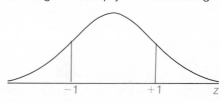

5. **The standard deviation of the standard normal distribution is 1.**
   A z score of 2.3 is, then, 2.3 standard deviations above the mean. For this reason, z scores are sometimes called **scores in standard deviation units.**

6. The standard normal distribution is a probability distribution: the area under the whole curve is 1, and the probability that a score is between a and b equals the area under the normal curve between a and b.

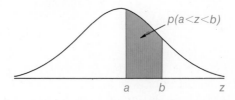

7.4 AREAS UNDER THE NORMAL CURVE

Areas under the normal curve are given in Table 4, Appendix C. The entries in this table give the area under the normal curve between the mean ($z = 0$) and the given value of z. The integral value and the first decimal place of z are given in the vertical column at the left; the second decimal place of z is given in the horizontal row at the top. **Always** sketch the area needed.

Example 1

What is the probability that a z score lies between 0 and 1.32?
Look down the left-hand column of Table 4 to 1.3 and across the top row to .02. The intersection of this row and column contains the number .4066. Rounding off to two decimal places, $p(0 < z < 1.32) = .41.$

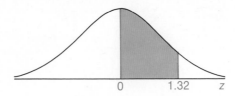

Example 2

$p(-2.03 < z < 0) = ?$
Negative values of z are not given in the table. However, the standard normal curve is symmetrical about $z = 0$, so $p(-2.03 < z < 0) = p(0 < z < 2.03) = .48.$

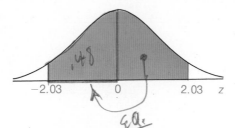

Example 3

If a z score is chosen at random, what is the probability that it is at least 1.25 standard deviation units above the mean?

Since the mean is 0 and the standard deviation is 1, this question is a more interesting way of asking "$p(z > 1.25) = ?$"

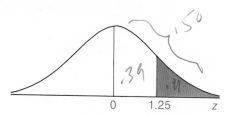

Table 4 gives areas starting at $z = 0$. Because of symmetry, and since the total area under the curve is 1, the area to the right of $z = 0$ is .50; the area between 0 and 1.25 can be found in the table. So $p(z > 1.25) = .50 - p(0 < z < 1.25) = .50 - .39 = .11$. See if you understand this graphical "arithmetic":

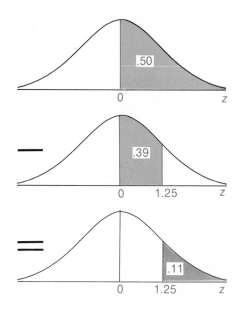

Example 4

$p(z > -0.34) = ?$

By symmetry, the area between $z = -.34$ and $z = 0$ is the same as that between $z = 0$ and $z = +.34$. To this must be added the area of the right half of the curve.

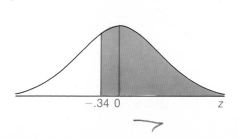

Symbolically,

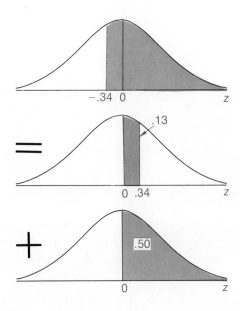

$$p(z > -0.34) = p(0 < z < .34) + p(0 < z) = .13 + .50 = .63$$

Example 5

If one score is chosen at random from a population which is normally distributed with mean 0 and standard deviation 1, what is the probability the score is between 0.34 and 2.30 standard deviation units above the mean?
$p(0.34 < z < 2.30) = ?$

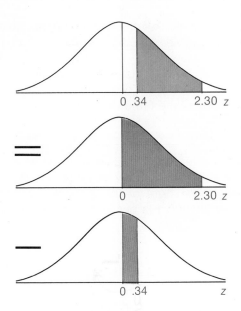

$$p(0.34 < z < 2.30) = p(0 < z < 2.30) - p(0 < z < 0.34) = .489 - .133 = .36.$$

Example 6

What proportion of z scores is between -1.96 and 2.30?
This is equivalent to the question "$p(-1.96 < z < 2.30) = ?$"

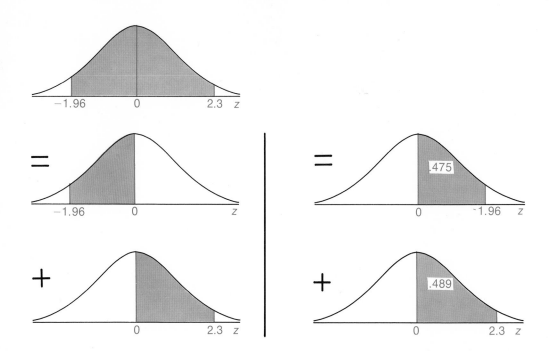

$$p(-1.96 < z < 2.3) = p(0 < z < 1.96) + p(0 < z < 2.3) = .475 + .489 = .96.$$

Compare Examples 5 and 6 carefully; note that probabilities are subtracted when boundary values of z have the same sign, but are added when the boundary values have opposite signs.

Always rely on drawing quick sketches of the normal curve and the area involved until you have done enough problems yourself so that you can see the sketch in your mind.

Sometimes you will be faced with the reverse problem: an area under the normal curve is given, and a value of z is to be determined. This still involves use of Table 4, but here a number in the body of the table determines a value of z on the edges.

Example 7

Find z if the area under the normal curve between 0 and z is .36.
In Table 4, the number in the body of the table which is closest to .36 is .3599; it is located to the right of the z value 1.0 and below the value .08; $z = 1.08$. If $z = -1.08$, the area between z and 0 is also .36, by symmetry.

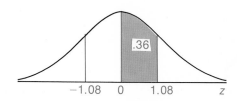

Example 8

Find the 98th percentile of z scores.

This is equivalent to "find z if the area to the left of z is .98." The area below z = 0 equals .50, so the area bounded by 0 and z is .98 − .50 = .48; using Table 4, z = 2.05.

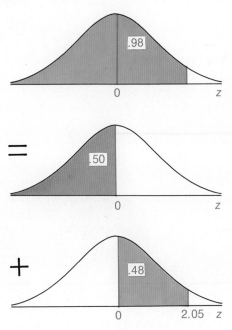

Example 9

What is z_0 if the probability is .60 that a randomly chosen z score is greater than z_0?

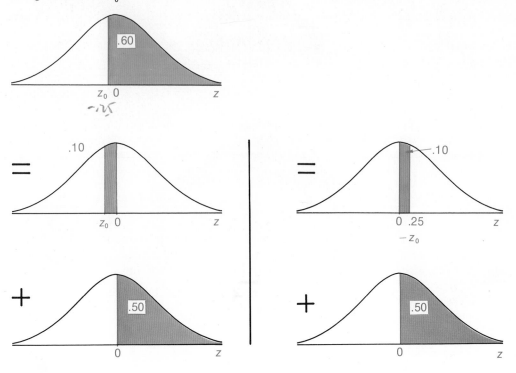

z_0 must be negative, since the area sought is greater than .50. An area of .10 corresponds to a z value of 0.25; the value sought, then, must be $z_0 = -0.25$.

Always remember that a normal distribution is a probability distribution, and therefore there is a direct correspondence between probabilities and areas under the curve.

7.5   PERCENTAGES OF SCORES WITHIN 1, 2, AND 3 STANDARD DEVIATIONS OF THE MEAN

For the standard normal curve, $\sigma = 1$. For this reason, z scores are often called **scores in standard deviation units.** For example, a z score of 2.56 is 2.56 standard deviation units away from the mean ($\mu = 0$).

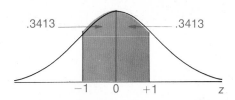

$p(0 < z < 1) = .3413$, so $2 * .3413 = .6826$; approximately 68 per cent of scores **in a normal distribution** fall within 1 standard deviation of the mean.

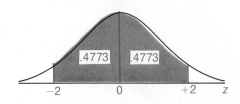

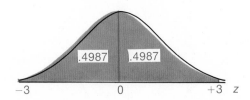

In like fashion it is determined that, **in a normal distribution,** approximately 95.5 per cent of scores fall within 2 standard deviation units of the mean, and approximately 99.7 per cent of scores fall within 3 standard deviations of the mean.

The words "in a normal distribution" are in bold type above to emphasize that the statements made are true only for this case. Many novices have the odd belief that, for any distribution, roughly 2/3, 95 per cent and 99.7 per cent of the scores fall within 1, 2, and 3 standard deviations, respectively, of the mean. This is errant nonsense, of course, as the following example shows.

Example 1   *use calc.* $x_1 = 1$ *freq. 9*
                              $x_2 = 9$ *freq. 1*

X: 1, 1, 1, 1, 1, 1, 1, 1, 1, 9; mean = 1.8, standard deviation = 2.4.† What percentage of scores falls within (a) 1, (b) 2, and (c) 3 standard deviations of the mean?

The answer to all these questions is 90 per cent!

i.e only 1 score out of 10 is > 3σ

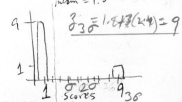

mean = 1.8
$3σ = 1.8 + 3(2.4) = 9$

7.6 MEAN AND STANDARD DEVIATION OF A POPULATION

It is time to look at the mean μ and standard deviation σ of a (discrete) population. The formulas will cause little surprise. Let N be the size of the population.

$$\text{Mean:} \quad \mu = \frac{\Sigma X}{N}$$

$$\text{Standard deviation:} \quad \sigma = \sqrt{\frac{\Sigma(X - \mu)^2}{N}}^{‡}$$

(Look again at page 47 to see why the sample standard deviation has $n - 1$ instead of n in the denominator.) *for better results*

Example 1

A population consists of X scores: 10, 30, 40, 40. Find μ and σ.

or check, 1 Var.

X	$X - \mu$	$(X - \mu)^2$
10	−20	400
30	0	0
40	10	100
40	10	100
120		600

$\mu = 120/4 = 30.$
$\sigma = \sqrt{600/4} = \sqrt{150} = 12.2.$

The usual shortcuts still work. Subtract a constant A from each score, find the mean of the new set, and then add A to it to get the mean of the original set. The computational formula for σ becomes

$$\sigma = \sqrt{\frac{\Sigma X^2 - (\Sigma X)^2/N}{N}}.$$

Example 2

Y scores in a population are 1, 2, 4, 4. Use the computational formula to find σ.

---

†Check these values of μ and σ after you have read the next section.

‡The corresponding formulas for grouped population scores are $\mu = \dfrac{\Sigma fX}{N}$ and $\sigma = \sqrt{\dfrac{\Sigma f(X - \mu)^2}{N}}$.

We shall have little use for these in this form, but you have already used them for probability distributions with $f = p(X)$ and $N = \Sigma p(X) = 1$ (see page 83).

$\Sigma Y = 11$, $N = 4$; Y^2: 1, 4, 16, 16, $\Sigma Y^2 = 37$.

$$\sigma = \sqrt{\frac{37 - 11^2/4}{4}} = \sqrt{1.69} = 1.3.$$

What about a continuous population? The individual scores can't even be listed, since there are an infinite number of them. In general, one has to know the equation of the distribution and use calculus. The exception is for a symmetrical distribution: the mean is the point of symmetry.

Example 3

Find the mean for each of these distributions:

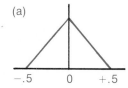

(a)

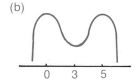

(b)

(a) $\mu = 0$. (b) $\mu = 3$.

7.7 STANDARD SCORES, OR SCORES IN STANDARD DEVIATION UNITS

In general, a normal distribution will not have a mean of 0 and a standard deviation of 1. It will be bell-shaped and symmetrical as is the standard normal distribution.

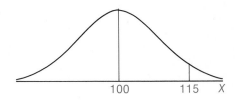

But if we try to answer the question "Scores are normally distributed with $\mu = 100$, $\sigma = 10$; what percentage of the scores are between 100 and 115?" we would seem to need another table—similar to that we used for the standard normal curve, but adjusted for the shift in μ and σ. But since there are infinitely many possible values for both μ and σ, and since each new combination would require a new page, no book would be fat enough to include all the necessary tables. Instead, a way must be found for comparing any normal distribution with the standard normal distribution, and our study of the former must be delayed until we take a detour and look at standard scores.

Standard scores are handy in comparing 2 scores from 2 different distributions. (This type of problem is typical: Tommy gets 78 in a spelling test; the mean for his class is 75 with a standard deviation of 2, while Billy gets 81 in a different spelling test; the mean for his class is 80 with a standard deviation of 5. Did Tommy or Billy spell better?) Standard scores will be immediately useful in studying normal distributions whose mean is not 0 or whose standard deviation is not 1, but you should realize that standard scores are of use in many other contexts.

The important question in this section is this one: If you have a set of X scores and know its mean μ and its standard deviation σ, how can you find a related set (of z scores or standard scores) whose mean is 0 and whose standard deviation is 1? The answer is this:

To find z scores: Subtract μ from each of the X scores, and then divide each of the new set by σ, thus arriving at the set of z or standard scores.

$$z = \frac{X - \mu}{\sigma}$$

Example 1

Given the X scores in the box below, find the corresponding z scores.

$$z = \frac{X - 8}{3}$$

$$\frac{X - \mu}{\sigma}$$

X	$X - \mu$	$(X - \mu)^2$	$z = \dfrac{X-8}{3}$
5	−3	9	−1
7	−1	1	−1/3
7	−1	1	−1/3
13	5	25	5/3
32		36	

$\mu = 32/4 = 8.$
$\sigma = \sqrt{36/4} = 3.$

The corresponding z scores are $-1, -1/3, -1/3, 5/3$. Their mean is 0 and their standard deviation is 1 (check these!).

Why does this work? Let's look at the same example a little more slowly.

Example 2

(a) For the X scores in the box below, find μ_X and σ_X.
(b) Let $Y = X - \mu_X$, find μ_Y and σ_Y.
(c) Let $z = \dfrac{X - \mu_X}{\sigma_X} = \dfrac{Y}{\sigma_Y}$; find μ_z and σ_z.

X	$Y = X - \mu_X$	$(X - \mu_X)^2 = (Y - \mu_Y)^2$	$z = \dfrac{X - \mu_X}{\sigma_X} = \dfrac{Y}{\sigma_Y}$	$(z - \mu_z)^2$
5	−3	9	−1	1
7	−1	1	−1/3	1/9
7	−1	1	−1/3	1/9
13	5	25	5/3	25/9
32	0	36	0	4

(a) $\mu_X = 32/4 = 8$; $\sigma_X = \sqrt{36/4} = 3$.

(b) $\mu_Y = 0/4 = 0$; $\sigma_Y = \sqrt{36/4} = 3$. Subtracting from each of the X scores their mean, 8, gives a set of Y scores whose mean is 0, but the standard deviation is unchanged. (Recall that if a constant A is subtracted from each score in a distribution, the mean is decreased by A [see p. 32]; if the mean is subtracted from each score, the new mean is 0. Subtracting a constant doesn't affect the standard deviation since the spread of the original and new sets about their means is the same.)

(c) $\mu_z = 0/4 = 0$; $\sigma_z = \sqrt{4/4} = 1$. Dividing each of the Y scores by their standard deviation, 3, gives a set of z scores whose mean is still 0 but whose standard deviation is only 1/3 as much. (If $\mu_Y = \dfrac{\Sigma Y}{N} = 0$, then $\dfrac{\Sigma(Y/\sigma)}{N} = \dfrac{1}{\sigma} \cdot \dfrac{\Sigma Y}{N}$

$= \dfrac{1}{\sigma} \cdot 0 = 0$. But the standard deviation of the quotients is 1: if $\dfrac{\Sigma(Y-0)^2}{N} = \sigma^2$,

then $\dfrac{\Sigma(Y/\sigma - 0)^2}{N} = \dfrac{1}{\sigma^2} \dfrac{\Sigma Y^2}{N} = \dfrac{1}{\sigma^2} \sigma^2 = 1$.)

Example 3

(a) Find μ and σ for the X scores in the box below; (b) find the corresponding z scores; (c) find the mean and standard deviation of the z scores.

X	$X-3$	$(X-3)^2$	$z = \dfrac{X-3}{\sqrt{6}}$	z^2
1	−2	4	$-2/\sqrt{6}$	4/6
1	−2	4	$-2/\sqrt{6}$	4/6
3	0	0	0	0
7	4	16	$4/\sqrt{6}$	16/6
12		24	0	4

(a) $\mu = 12/4 = 3$; $\sigma = \sqrt{24/4} = \sqrt{6}$.

(b) $z = \dfrac{X - \mu}{\sigma} = \dfrac{X-3}{\sqrt{6}}$; the corresponding z scores are $-2/\sqrt{6}, -2/\sqrt{6}$, 0, and $4/\sqrt{6}$.

(c) Mean of z scores $= 0/4 = 0$; standard deviation of z scores

$= \sqrt{\dfrac{\Sigma(z-0)^2}{N}} = \sqrt{\dfrac{4}{4}} = 1$.

In previous problems X was given and z was determined; sometimes z is known and the equation must be solved for X.

Example 4

In a population with mean 100 and standard deviation 10, what score is 1.3 standard deviation units below the mean?

"1.3 standard deviation units below the mean" implies that $z = -1.3$. $\mu = 100$, $\sigma = 10$, $z = (X - \mu)/\sigma$, so

$$-1.3 = \dfrac{X - 100}{10}$$

$$-13 = X - 100$$

$$X = 87.$$

Example 5

In a distribution with $\mu = 40$, $\sigma = 3.1$, what score corresponds to a z score of 2.1?

$$z = (X - \mu)/\sigma$$
$$2.1 = (X - 40)/3.1$$
$$2.1 * 3.1 = 6.5 = X - 40$$
$$X = 46.5$$

We shall use z scores many times in succeeding chapters. The formula will always give directions for subtracting the mean and then dividing by the standard deviation, but will have many varied forms. If Y scores in a sample have a mean $\bar{Y}$, the formula will be $z = (Y - \bar{Y})/s$. There will also be much more astonishing forms, however. Soon we shall look at a distribution made up of sample means $\bar{X}$ (that is, every score is the mean of some sample). If the symbols $\mu_{\bar{X}}$ (read "mu sub X bar") and $\sigma_{\bar{X}}$ (read "sigma sub X bar") are used for the mean and standard deviation of this peculiar distribution, then the z formula will be $z = (\bar{X} - \mu_{\bar{X}})/\sigma_{\bar{X}}$. Here are some other examples:

Scores	Mean	Standard deviation	$z =$
p	P	σ_p	$(p - P)/\sigma_p$
$\bar{X} - \bar{Y}$	$\mu_X - \mu_Y$	$\sigma_{(\bar{X}-\bar{Y})}$	$[\bar{X} - \bar{Y} - (\mu_X - \mu_Y)]/\sigma_{(\bar{X}-\bar{Y})}$
$\bar{X}$	μ	$\sigma_{\bar{X}}$	$(\bar{X} - \mu)/\sigma_{\bar{X}}$
X	$\bar{X}$	s	$(X - \bar{X})/s$
$p_X - p_Y$	$P_X - P_Y$	$\sigma_{(p_X-p_Y)}$	$\dfrac{(p_X - p_Y) - (P_X - P_Y)}{\sigma_{(p_X-p_Y)}}$

If these formulas worry you, pass them by. But if you understand the principle you will have an easier time later. Try working out this example.

Example 6

The proportion P of a population which smokes more than a pack a day is .40. A p distribution has mean P and standard deviation $\sigma_p = .06$. What are the standard scores corresponding to p scores of (a) .46, (b) .28?

Answer

(a) $z = (p - P)/\sigma_p = (.46 - .40)/.06 = 1$.
(b) $z = (.28 - .40)/.06 = -2$.

Note that a p score of .46 is 1 standard deviation (.06) above the mean (.40); its z score is 1. A p score of .28 is 2 standard deviations (2 * .06) below the mean (.40); its z score is -2. A z score tells how many standard deviations above or below the mean the corresponding score in the original distribution is; this again demonstrates why z scores or standard scores are also called scores in standard deviation units.

7.8 USES OF STANDARD SCORES

Example 1

A student gets 82 in a final examination in mathematics; the mean grade is 75 with a standard deviation of 10. In economics he gets 86 in a final examination on which the mean grade is 80 with a standard deviation of 14. Is his relative standing better in mathematics or in economics?

In mathematics, $z = \dfrac{82-75}{10} = .70$; in economics $z = \dfrac{86-80}{14} = .43$. His grade in mathematics is .70 standard deviation units above the mean, while in economics his grade is .43 standard deviation units above the mean. His relative standing is higher in mathematics than in economics.

     Note: No assumption is made here that the grades in either mathematics or economics are normally distributed; z scores have many other uses besides comparing any normal distribution to a standard normal distribution.

Example 2

     On a mathematical aptitude test, Philip Nolan gets a raw score of 88.3. The mean of all scores on that test is 75 with a standard deviation of 8. Each year scores are standardized with a mean of 500 and a standard deviation of 100 before reports are sent to those taking the test. What score is reported to Philip Nolan?

     Philip's z score is $\dfrac{88.3-75}{8} = 1.66$. In a distribution with $\mu = 500$, $\sigma = 100$, a score of 666 corresponds to a z score of 1.66: solve $1.66 = \dfrac{X-500}{100}$.

The score reported to Philip Nolan is 666.

     We shall find standard scores most useful, however, in showing relationships between a general normal distribution and the standard normal distribution with which you are already familiar.

7.9 NORMAL DISTRIBUTIONS IN GENERAL

     There is a (double) family of normal distributions. Particular values of μ and σ give particular members of the family. Here are some examples:

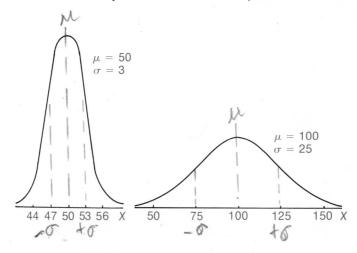

The standard normal curve is one member of the family (with $\mu = 0$, $\sigma = 1$).

     All members of the family have certain things in common; if you are friendly with the standard normal curve none of these should surprise you:

1.   They are all probability distributions—that is, the area between the curve and the horizontal axis is always equal to 1, and the area between the curve, the X axis, and

<u>the lines $X = a$, $X = b$</u> represents the <u>probability that an X score will fall between</u> a and b.

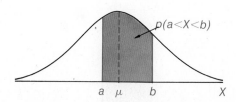

$p(a<X<b)$

$a \quad \mu \quad b \qquad X$

2. All normal curves extend indefinitely both to the right and to the left. Only a small portion about the mode is sketched, since the curve gets (and remains) very close to the horizontal axis on both sides of the mode.

3. Each normal curve is bell-shaped and is symmetrical about the mode, and the mode, median, and mean coincide.

4. Each normal curve changes from concave down ⌢ to concave up ⌣ <u>at a point 1 standard deviation away from the mean.</u> Thus if $\mu = 50$, $\sigma = 20$, these "points of inflection" will be at 30 and 70. Also check the figures at the beginning of this section.

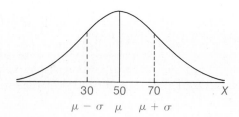

$$30 \quad 50 \quad 70 \qquad X$$
$$\mu - \sigma \quad \mu \quad \mu + \sigma$$

7.10 AREA UNDER ANY NORMAL CURVE

To find the area under a normal curve (with mean μ and standard deviation σ) between $X = a$ and $X = b$, find the z scores corresponding to a and b (call them z_1 and z_2), and then find the area under the standard normal curve between z_1 and z_2, using Table 4, Appendix C. Some examples will demonstrate how easily this is done.

Example 1

X scores are normally distributed with $\mu = 100$, $\sigma = 10$. What percentage of X scores will fall between (a) 100 and 110? (b) between 80 and 105?

(a) $z = \dfrac{X - 100}{10}$. $z = \dfrac{X - \mu}{\sigma}$

$$100 \quad 110 \qquad X$$
$$0 \quad\quad 1 \qquad\quad z$$

When $X = 100$, $z = \dfrac{100 - 100}{10} = 0$; when $X = 110$, $z = 1$. $p(100 < X < 110)$

$= p(0 < z < 1) = .341$; 34 per cent of X scores will fall between 100 and 110.

    (b)   When $X = 80$, $z = \dfrac{80 - 100}{10} = -2$; when $X = 105$, $z = \dfrac{105 - 100}{10} = .5$.

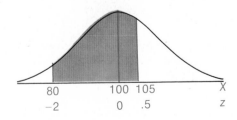

	80	100 105		X
	−2	0 .5		z

$p(80 < X < 105) = p(-2.0 < z < .5) = .4773 + .1915 = .669$; 66.9 per cent of the scores are between 80 and 105.

   It will frequently be helpful to show on the horizontal axis both X scores and their corresponding z scores, as was done in the sketches above.

Example 2

    X scores are normally distributed with $\mu = 50$, $\sigma = 5$. What is the probability that a score is (a) between 45 and 48? (b) over 58?

    Here $z = \dfrac{X - \mu}{\sigma}$ becomes $z = \dfrac{X - 50}{5}$. It is often convenient to arrange corresponding X and z scores in parallel columns:

X	z
45	−1
48	− .4
58	1.6

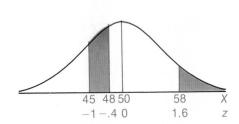

	45 48 50		58	X
	−1 −.4 0		1.6	z

    (a) $p(45 < X < 48) = p(-1 < z < -.4) = .341 - .155 = .196$.
    (b) $p(X > 58) = p(z > 1.6) = .5 - .445 = .055$.

    In Examples 1 and 2 it was assumed that X scores formed a continuous distribution. It is time to meet a complication which arises when they form a discrete distribution.

Example 3

    The weights of 1000 male students are found to be normally distributed with mean of 175 lb. and standard deviation of 12 lb. How many of them weigh 180 lb. or less? (All measurements were made to the nearest pound.)

$z = \dfrac{X - \mu}{\sigma}$ becomes $z = \dfrac{X - 175}{12}$. Since measurements are made to the nearest pound, a man who weighs 180 lb. weighs less than 180.5, and it is this latter figure which is used in computing the z score.

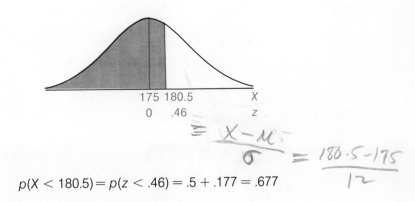

175 180.5   X
0   .46   z

$$\frac{X - M}{\sigma} = \frac{180.5 - 175}{12}$$

$p(X < 180.5) = p(z < .46) = .5 + .177 = .677$

.677 * 1000 = 677 of the students weigh 180 lb. or less.

If the number of X scores is finite, then they must come from a discrete distribution. Before reading on, read again the first sentence of Example 3 and see if you still walk into a trap.

How can 1000 scores (a discrete distribution) be normally distributed, since a normal distribution is continuous? This "mistake" will be made frequently and casually in future pages. Of course an approximation is implied: It is assumed that if X scores are changed to standard scores, the distribution is sufficiently close to the standard normal distribution so that Table 4, Appendix C, may be used to determine probabilities.

Example 4

The diameters of ball bearings are normally distributed with mean .5000 inch and standard deviation .0010 inch. Of 10,000 ball bearings, measured to .0001 in., how many will have diameters (a) between .5005 and .5020 in.? (b) equal to .4992 in.?

$$z = \frac{X - .5000}{.0010}$$

	X	z
(a)	.50045	.45
	.50205	2.05
(b)	.49915	− .85
	.49925	− .75

.49925 .49915 .5 .50045   .50205 X
−.85   −.75   0 .45   2.05   z

(a) $p(.50045 < X < .50205) = p(.45 < z < 2.05) = .480 - .174 = .306$; 3060 bearings will have diameters between .5005 and .5020 in.

(b) $p(X = .4992 \text{ in.}) = p(.49915 \text{ in.} < X < .49925 \text{ in.}) = p(-.85 < z < -.75)$
$= .302 - .273 = .029$; 290 bearings will have diameters equal to .4992 in.

Sometimes it is necessary to look up areas and find the corresponding z value as the following example demonstrates.

Example 5

On a statistics quiz the grades are normally distributed with a mean of 77 and a standard deviation of 7. If the top 10 per cent of students get A's, what is the lowest grade to which an A will be assigned?

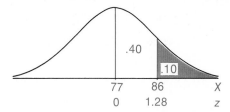

$.50 - .10 = .40 = p(0 < z < 1.28)$.

$$z = \frac{X - 77}{7}.$$

If $z = 1.28$, $X = 8.96 + 77 = 85.96$.
The lowest grade for an A is 86.

The notation **A(z)** will be used for the area under the standard normal curve between 0 and z. In Example 5, $A(z) = .40$, so $z = 1.28$.

Binomial Assumptions:
1. n trials are made (for table 2, n≤20)
2. Each trial two outcomes
3. Each trial result independent other trials

7.11   NORMAL APPROXIMATION TO THE BINOMIAL DISTRIBUTION

It was mentioned earlier (page 102) that, as the number of trials n gets larger and larger, the binomial distribution gets closer and closer to a normal distribution. A normal distribution with mean $\mu = np$ and standard deviation $\sigma = \sqrt{npq}$ can be used to estimate binomial probabilities **if both np and nq are greater than 5.**

Example 1

Find the probability of getting <u>at most 3 heads</u> in 12 flips of a fair coin by using (a) the binomial distribution and Table 2, Appendix C; (b) the normal approximation to the binomial distribution.   *p. 345*

(a) $n = 12$, $p = .5$

X	$p(X)$
0	.000+
1	.003
2	.016
3	.054
	.073+

ODDS 1 H
$12^{c_1}(\frac{1}{2})^1(\frac{1}{2})^{11} = .003$

ODDS 2 H
$12^{c_2}(\frac{1}{2})^2(\frac{1}{2})^{10} = .016$ .073

ODDS 3 H
$12^{c_3}(\frac{1}{2})^3(\frac{1}{2})^9 = .053$

$p(\text{at most 3 heads}) = .073$.

(b) To apply the normal distribution to discrete data, it is necessary to treat the scores as though they were continuous. It follows that "at most 3 heads" should be treated as "less than 3.5 heads" for the normal approximation.

$$\mu = np = 12 * 1/2 = 6$$

$$\sigma = \sqrt{npq} = \sqrt{12 * 1/2 * 1/2} = 1.732$$

$$z = \frac{X-6}{1.732}$$

If $X = 3.5$, $z = -2.50/1.732 = -1.45$.

$p(X < 3.5) = p(z < -1.45) = .5 - .427 = .073$.

Example 2

A fair coin is tossed 900 times. Find the probability that heads will turn up between 400 and 500 times.

$n = 900$, $p = 1/2$, $q = 1/2$, $\mu = np = 900 * 1/2 = 450$, $\sigma = \sqrt{npq} = \sqrt{900 * 1/2 * 1/2} = 15$, $z = \dfrac{X - 450}{15}$. 400 to 500 heads become 399.5 to 500.5 when considered as continuous data.

X	z
399.5	−3.37
500.5	+3.37

$p(399.5 < X < 500.5) = p(-3.37 < z < +3.37) = 2 * .4996 = .9992$. We are almost certain that, if a fair coin is tossed 900 times, heads will turn up between 400 and 500 times. If heads turn up only 300 times in 900 tosses, we would have a very strong feeling that the coin is not fair.

7.12 VOCABULARY AND SYMBOLS

normal distribution	standard score
standard normal curve	z score
concave up (down)	score in standard deviation units
μ	$A(z)$
σ	

7.13 EXERCISES

In each exercise, make a quick sketch of the normal curve and the area involved.

1. Find the area between
   (a) $z = 0$ and $z = 1.41$.
   (b) $z = -.6$ and $z = 0$.
   (c) $z = -1.23$ and $z = 0.53$.
   (d) $z = -1.23$ and $z = -0.53$.
   (e) $z = .46$ and $z = 2.31$.
   (f) $z = -\infty$ and $z = 1.28$.

2. What is the probability that a z score chosen at random from a normal distribution is
   (a) greater than 1.65?
   (b) between 0.46 and 2.59?

(c) between −0.46 and 2.59?
(d) less than 0.46?
(e) greater than −1.3?
(f) less than −1.17?
(g) within 1.96 standard deviation units of the mean?

3.   Find z if the area under the normal curve between 0 and z is .4236.

4.   Find z if the probability that a score falls between z and 3.42 is .2058.

5.   In a normal distribution, what percentage of the z scores falls between −.16 and +1.97?

6.   If a set of measurements are normally distributed with $\mu = 0$ and $\sigma = 1$, what per cent of the measurements are
(a) inside the range $\mu \pm 1.2\sigma$?
(b) outside the range $\mu \pm 2\sigma$?

7.   Find z if (a) 95 per cent (b) 98 per cent of normally distributed scores fall between −z and +z.

8.   Find (a) the ninety-ninth percentile, (b) the seventy-fifth percentile of normally distributed z scores.

9.   Find the semi-interquartile range Q of normally distributed z scores.

10.   20,000 scores are normally distributed with $\mu = 0$, $\sigma = 1$. How many of them are (a) greater than 1? (b) between .50 and 1.53? What assumptions have you made?

11.   A psychologist gives a personality test to a sample of 6 students, who obtain the following scores: 28, 34, 32, 31, 30, 31. In writing a paper he wishes to report these as "standardized" (not **standard**) scores with a mean of 50 and a standard deviation of 10. What scores does he report?

12.   If the scores in Exercise 11 were those of a population, what would their mean μ and standard deviation σ be?

13.   X scores are normally distributed with $\mu = 30$, $\sigma = 20$. If one score is chosen at random, what is the probability that it is (a) between 60 and 80? (b) less than 60? (c) greater than 40?

14.   Three hundred women students have mean height of 65.0 inches and standard deviation 2.0 inches. The 300 heights are normally distributed and are measured to the nearest inch.
(a) How many of these students are between 66 and 70 in.?
(b) How many of them are 64 in. or less?
(c) Thirty per cent of the students are below what height?
(d) How many of the students have heights which differ from the mean by more than one standard deviation?

15.   If a fair coin is tossed 100 times, what is the probability that the number of heads is
(a) at least 40 and not more than 60?
(b) at least 41 and not more than 59?
(c) at least 43 and not more than 57?

16.   On a 20-question True−False exam, what is the probability that a know-nothing student who answers each question by sheer guess will get 15 or more correct answers?

17.   John Bradford's grades on hour exams in physics during one semester are as follows:

Exam	John's Grade	Class Mean	Class Standard Deviation
1	86	85	5
2	90	85	10
3	60	55	5
4	74	72	5

(a) What is John's average grade?

(b) What is John's grade in the course if the teacher averages z scores for each exam, and assigns final grades with a mean of 78 and a standard deviation of 6?

(c) Does (a) or (b) give a more just way of determining John's grade in the course?

18.   Five hundred students taking Psychology I have a mean grade for the course on all exams of 77 with a standard deviation of 10. Final grades are to be assigned "on the curve" so that 10 per cent of the class gets A, 30 per cent gets B, 50 per cent gets C, and 10 per cent gets No Credit. Find the student averages for each of these grades.

19.   A machine makes bolts whose mean diameter is .240 in. with standard deviation .005 in. (a) If the diameters of 6000 bolts produced by the machine are normally distributed, how many are more than .239 in.? (b) If the machine is adjusted so the mean diameter of the bolts is increased by .006 in. but the standard deviation is unchanged, what per cent of bolts it produces are over .255 in. in diameter?

20.   A department store uses fluorescent lights which have a mean life of 3500 hours with a standard deviation of 600 hours; the lives are normally distributed. (a) If the lights are on 10 hours a day, 6 days a week, for 52 weeks, what proportion of the lights would need replacement? (b) After how many weeks would 10 per cent of the lights need to be replaced? $\mu = 30 \times \frac{1}{4} = 7.5 ; \sigma = \sqrt{npq} = \sqrt{30(\frac{1}{8})\frac{3}{4}} = 2.37 ; z = \frac{14-7.5}{2.37} = 2.74$

21.   Richard Smartt claims that he can distinguish wine A from three other brands $prob = 99.67.$ of the same type. Over a period of a month he is given 30 trials, and is able to distinguish wine A 14 times. What is the probability that he would identify wine A 14 or more times if he were just guessing? Is his claim justified? $(q = 1-n)$

22.   On the basis of tests and past experience with Model 101XE, a washing machine manufacturer decides its mean life in normal family use is 5.75 years with a standard deviation of 2 years. If the life of this model is normally distributed, (a) what guarantee should he offer if he is willing to repair only 1 per cent of the machines sold during the guarantee period? (b) If he gives a 2-year guarantee, what percentage of machines will need repair before the guarantee period runs out?

$z = (72-75)/4.33 = .693 ; .5 - .2549 = .245$

23.   Mendel discovered that when the seeds produced by tall F_1 peas were planted, 75 per cent of the resulting plants were tall and 25 per cent were dwarf. If 100 F_1 seeds are planted, what is the probability that at least 28 plants are dwarf?

$p.43 \; Binomial ; M = np = 100(\frac{1}{4}) = 25 ; \sigma = \sqrt{npq} = \sqrt{100(.25).75} = 4.33 ;$

24.   A pair of fair dice is rolled 180 times. What is the probability that the same number comes up on both dice

(a) at least 40 times?

(b) between 30 and 40 times?

(c) exactly 35 times?

25.   A coffee machine is set to fill cups with 6 oz. of coffee, with a standard deviation of .40 oz. If 7 oz. cups are used, what proportion of them will overflow?

26.   The mean IQ of students admitted to St. James University is 118 with a standard deviation of 5. If 40 students admitted have an IQ between 120 and 125, how many students were admitted? Assume their IQ's are normally distributed.

ANSWERS

1.   (a) .4207.   (d) $.3907 - .2019 = .1887$.
      (b) .2257.   (e) $.4896 - .1772 = .3124$.
      (c) $.3907 + .2019 = .5926$.   (f) $.5 + .3997 = .8997$.

3.   1.43 (or -1.43).

5. $.0636 + .4756 = .5392$; 53.9 per cent.

7.   (a) $.95/2 = .4750$; $z = 1.96$. (b) $.98/2 = .49$; $z = 2.33$.

9. $A(z) = .25$ so $z = .67 = Q_3$, $Q_1 = -.67$, $Q = \dfrac{.67 - (-.67)}{2} = .67$.

11. $\overline{X} = 31.0$, $s = 2.0$.

      z scores: -1.5, 1.5, $.5$, 0, $-.5$, 0.

      Standardized scores: 35, 65, 55, 50, 45, 50.

13. $z = \dfrac{X - 30}{20}$.   (a) $p(60 < X < 80) = p(1.5 < z < 2.5) = .4938 - .4332 = .06$.

X	z
60	1.5
80	2.5
40	.5

(b) $p(X < 60) = p(z < 1.5) = .5 + p(0 < z < 1.5) = .5 + .4332 = .93$.

(c) $p(X > 40) = p(z > .5) = .5 - p(0 < z < .5) = .5 - .1915 = .31$.

15. $n = 100$, $p = .5$, $\mu = np = 50$, $\sigma = \sqrt{100 * .5 * .5} = 5$; $z = \dfrac{X - 50}{5}$.

X	z
39.5	−2.10
60.5	2.10
40.5	−1.90
59.5	1.90
42.5	−1.50
47.5	1.50

(a) $p(39.5 < X < 60.5) = p(-2.10 < z < 2.10) = 2 * .482 = .96$.

(b) $p(40.5 < X < 59.5) = p(-1.90 < z < 1.90) = 2 * .471 = .94$.

(c) $p(42.5 < X < 47.5) = p(-1.50 < z < 1.50) = 2 * .433 = .87$.

17.   (a) 77.5. (b) Average of z scores $= (+.2 + .5 + 1 + .4)/4 = .52$: $.52 = \dfrac{X - 78}{6}$, so $X = 81$. (c) Using standard scores gives a more accurate reflection of his relative ability — but whether he thinks (b) is more just may depend on the teacher's choice of mean and standard deviation for the class in assigning final grades.

19. (a) $z = \dfrac{.2395 - .240}{.005} = -.10$; $p(z > -.10) = .5 + .0398 = .5398$; $.5398 * 6000$

= 3239 bolts.

(b) $z = \dfrac{.2555 - .246}{.005} = 1.90$; $p(z > 1.90) = .5 - .4713 = .029$; 2.9 per cent are

over .255 in. $= 216 \; bolts$

21. $n = 30$, $p = 1/4$, $\mu = np = 7.5$, $\sigma = \sqrt{npq} = 2.37$, $z = \dfrac{13.5 - 7.5}{2.37} = 2.53$.

p(identify 14 or more) $= .5 - .4943 = .0057$. The chance he is guessing is less than 6 out of 1000; his claim is probably justified.

23. p(dwarf) $= p = 1/4$, $n = 100$, $\mu = 25.0$, $\sigma = \sqrt{100 * 1/4 * 3/4} = 4.33$. If $X = 27.5$,

$z = \dfrac{27.5 - 25.0}{4.33} = .58$; $p(X > 27.5) = p(z > .58) = .5 - .219 = .28$.

$25 \; \sigma=1 = 29.33$

25. $z = \dfrac{X - 6.0}{.40}$; $p(X > 7.0) = p(z > 2.5) = .5 - .4938 = .006$; .6 per cent will overflow.

chapter eight

how to choose a sample

You have the background in probability and probability distributions necessary for an understanding of statistical inference. In this chapter there is a very brief overview of how to set up an experiment using statistics. In most such experiments, one does not take a census (that is, get data on the whole population), but rather one takes a sample and collects data on just this portion of the population. Most of this chapter is concerned with how this sample is chosen. Concentrate on simple random samples.

Be of good cheer: there is almost no mathematics in this chapter.

8.1 FORMULATING A STATISTICAL EXPERIMENT

Five steps are necessary:

1. You must decide very carefully what question you want answered. In particular, you must identify the population about which the question is asked.

If you plan a study of the relationship between smoking and lung cancer, for example, one of the very many questions you will have to answer is "who is a smoker?" You might consider only cigarette smokers; are cigar and pipe smokers classified with non-smokers or not included in the study at all? You would probably need to classify cigarette smokers by number of cigarettes per day and the number of years of smoking. Is a smoker a person who smokes over two packs a day? over a pack a day? or anyone who has smoked this year? Do you want to compare the incidence of lung cancer in all smokers and all non-smokers, or only between those in certain occupations or in certain areas or in particular kinds of environment? The questions are endless; the answers depend on the time and money you have available as well as on the aims of your study.

Usually you must restrict your aims because it takes so much effort to collect and analyze data, but then you must also restrict the generality of your conclusions. Counting hawks as they fly over Hook Mt. in New York will not tell you anything about the hawk population of Minnesota, for example. But a restricted experiment with results that are meaningful is worth much more than an experiment with scope but meaningless results. Plan carefully before you start collecting data.

2. You must decide whether the experiment is to be conducted on the whole population or on only a part of it.

Can you take a census, or must you deal with a sample? If you can take a census, the statistical techniques you will need are mostly those of descriptive statistics with which you are already familiar: you must present the data in tables and graphs, and you may summarize it by finding the mean and standard deviation, for example.

If the question you are asking is simple enough ("What is the average height of third graders in Meadowlands Elementary School?") and the population small enough, a census may be easily carried out. But for a large population, a census is ordinarily so costly and time-consuming that it is carried out only for the most important questions ("What is the population by age of each town in the United States?"). To answer some questions ("What percentage of patients who have this disease now **or in the future** will be cured with this drug?") a census of the whole population is impossible regardless of money and effort now available.

For most experiments you will use a sample. From the sample data you will make inferences about the population; most of the rest of this book will deal with the techniques by which this is done. How should the sample be chosen? This question will be answered in the rest of this chapter. How large should the sample be? Some tentative answers will be given in Chapter 10, and there will be further discussion in later chapters. For the moment, assume that you have advice on sample size from your friendly neighborhood statistician. One point should be emphasized here, however: the whole experiment, including the sample size, should be planned before data is collected.

3. You must determine what data are to be collected, how they are to be collected, and what criteria are to be used to interpret the results.

Suppose, for example, you are determining the IQ of 40 pairs of identical twins separated soon after birth and raised in different environments. You will need to decide not just which IQ test is to be given, but also whether the same person or two different people should give it to each pair (will the use of one person assure uniformity, or will his knowledge of the result of testing one twin unconsciously affect the results of the second test?); if you are interested in the effect of different environments on identical twins, you will have to consider very carefully what factors in their environment are to be measured, how to get the necessary information, and how comparisons are to be made.

It is essential to decide before an experiment is carried out what criteria will be used to interpret the results. If you are trying to decide whether a coin is fair, for example, you might decide to experiment by tossing it 100 times. If it comes up heads 50 times you will certainly decide the coin is fair; if it comes up heads only 3 times, you will be quite certain it is not a fair coin. But what if it comes up heads 20 times? 40 times? 48 times? If the coin is fair, there is a probability of .95 that heads will come up 41 to 59 times (see Exercise 15, page 120). So you might decide to toss the coin 100 times, and accept the hypothesis that the coin is fair if heads turn up between 41 and 59 times. (Of course, the probability is .05 that heads will come up less than 41 or more than 59 times even if the coin is fair; pure chance may still give the unusual result. You will have to learn not only what criteria to use to test your results but also how to determine the probability of error in using those criteria.) Much more will be said about this later, but again you should realize at the beginning of your study of statistical inference that criteria for interpreting the results should be set up **before** data are collected.

4. You must select the sample, if one is to be chosen, and collect the data.

This process will take much of your time and energy, and possibly much money. It will be easier and certainly your results will be more meaningful if you have laid careful plans first.

5. You must analyze the results, draw conclusions, and estimate the precision of the results.

The last requirement is related, of course, to the criteria you set up for interpreting the results before data were collected. This estimate is sometimes not included in reports of experiments that seem otherwise respectable; such reports are usually meaningless.

8.2   RANDOM SAMPLES

Suppose that you plan to measure the heights of 30 people, a sample of all the inhabitants of a small town. Clearly you would not stop the first 30 children entering an elementary school, nor would you stand outside the high school gym and measure 30 candidates for the basketball team. Our aim very shortly will be to take the results of a sample and generalize them to statements about the whole population from which the sample has been taken. Obviously, the two samples suggested above would not be representative of the heights of residents of the town.

A famous example of an unrepresentative or biased sample is a poll of voting opinion conducted by the *Literary Digest* magazine in 1936 before the presidential election. The Digest poll showed Alfred Landon winning by a wide margin, but on election day Franklin Roosevelt won by a landslide. The *Literary Digest* soon folded, partly because readers lost confidence in it after this fiasco. The sample was taken from those who had a telephone or owned a car. Why were the conclusions so erroneous? The poll was taken at the height (or should I say "bottom"?) of the depression; only relatively well-to-do people had a telephone or owned a car, and most of these supported Landon. But the sample was very unrepresentative of all voters, most of whom supported Roosevelt.

How should a sample be chosen? Intuitively, one wants the sample to be "representative." But how is this to be put into practice? If the population is completely uniform, you can choose any sample of it. This assumption is made, for example, when a sample of your blood is tested. But suppose you want to choose 30 residents of the small town and measure their heights. Can you decide to include a quota of 2 tall men, 3 infants under two years old, and allot the other choices in similar ways? No matter how the quotas for each group are allotted, it is done well only if you already know well the characteristics of the population—and, in that case, why bother getting data on a sample? The sample must not be chosen because of a subjective decision about what is typical of the population, since this causes the results of the experiment to be biased by the prejudices of the selector.

Similarly, a sample should not be chosen by convenience. The interviewer who stops people in front of a department store to ask their preference for governor may avoid hippies or blacks, or may get too many suburban housewives and too few farmers or miners. The *Literary Digest* had convenient telephone directories and lists of car owners, but failed miserably in forecasting the election. If you are experimenting on 12 white mice, you may find it necessary to use mice ordered at the same time from one supplier—but you must be aware that you may have litter-mates in your sample, and must ask yourself whether this may affect the outcome of your experiment.

A sample should be chosen objectively, and should not be determined by the convenience of the experimenter. What method of selection should be used? As you will soon see, valid conclusions about the population can be made if the sample taken is a **simple random sample;** this is a sample chosen so that each sample of that size in the population has an equal probability of being chosen. Such a selection is said to be made **at random.** The techniques of statistical inference that you will study in this book are based on the assumption that your sample is a simple random one. There are other kinds of random samples, but we shall work always with simple random samples and therefore the word "simple" will not be used in future.

A sample can, theoretically, be chosen either with replacement or without replacement. In practice, however, it is almost always taken without replacement. If you are interviewing students, for example, you would feel foolish indeed—and would probably not get a warm reception—if you presented yourself again to repeat the same interview. If you are a doctor testing a drug, you cannot try it on the same patient twice and hope to get the same results as when it is tried independently on two different patients.

If a sample is chosen without replacement, it isn't necessary to list all possible samples and then choose one of them at random from the list. Instead, the sample is a random one if each additional member added to the sample is chosen so that all remaining members of the population are equally likely to be chosen. For example, the possible samples of size 2 without replacement from the population A, B, C are $\{A, B\}$, $\{A, C\}$, $\{B, C\}$; if these three samples are written on identical slips of paper and mixed up in a hat, the probability of drawing the sample $\{A, B\}$ is 1/3. But if one chooses the first member of the sample at random and the second member from those remaining after the first has been drawn, the probability of choosing the sample $\{A, B\} = p$(both A and B) $= p(A_1B_2$ or $B_1A_2) = p(A_1B_2) + p(B_1A_2) = 2 * (1/3) * (1/2) = 1/3$, as before.

How can a random sample of 30 students be chosen from a freshman class of 400 students? One method is to write the name of each freshman on a slip of paper, put the identical slips of paper in a hat, mix well, and blindly draw 30 names. Another method is to assign each student a number and then choose 30 numbers from a table of random numbers. Such a table can be made up by writing the 10 digits 0 through 9 on a separate cards and mixing them in a box. One card is drawn and the digit noted, then that card is replaced and the cards mixed again. A space is left after the process has been carried out five times for ease in reading the table. Table 3, Appendix C, is such a table.

Selection of a sample of 30 from 400 students is straightforward. Each student is assigned a number from 001 to 400. A starting point is chosen at random in Table 3 and any direction you wish: left, right, up, down, or move like a knight on a chess board if you prefer. You may take only the first 3 digits of each 5-digit group, or the middle 3, or ignore the spaces in the table and take each successive set of 3 numbers (3 because in this example 400 has 3 digits; 2 if you were choosing from a class of 99 or less, 4 if from a class of 1,000 to 9,999, etc.). If the three digits form a number over 400 or a number previously chosen, ignore it; otherwise, assign the student with that number to the sample.

Example 1

Select a sample of size 12 from a population of size 80, starting with row 18, column 5 of Table 3 and reading horizontally, continuing with row 19. Part of this table is reproduced here:

```
18 |
19 | 70791 39030 . . .              . . . 62542 30536 14777 72360
```

Assume the population is assigned numbers 01, 02, 03, . . . , 80. Starting with 62 in the random number table, the numbers marked A are chosen for the sample:

```
                    A  A  A  A  A  A  A  B  C  A
                    62 54 23 05 36 14 77 77 23 60
```

```
A  A  A  D  A
70 79 13 90 30
```

The second 77 (marked B) is omitted because it has appeared before; the second 23 (marked C) is also omitted. The number 90 (marked D) is omitted because the population is of size 80, and no member of it was assigned the number 90.

Choosing a random sample either by writing names on slips of paper or by assigning numbers and using a random number table may be too time-consuming, too expensive, or simply impossible. Would you look forward to choosing a sample of 8000 voters from all the registered voters in the United States by either of these methods? So other methods have been devised.

8.3   OTHER METHODS OF SAMPLING

Stratified Sampling.   If your population is divided into strata having common characteristics, it is sometimes useful to choose a random sample from each stratum. You might do this because data on separate strata are easier to assemble, or because you want information about separate strata as well as about the combined population, but the most important reason is that there is less variation in the means of samples.

Suppose, for example, that a population has four scores: 0, 2, 6, 8. If samples of size 2 are taken (without replacement), the means of the samples are as follows:

Sample	Mean	Sample	Mean
0, 2	1	2, 6	4
0, 6	3	2, 8	5
0, 8	4	6, 8	7

The six sample means have a mean of 4, a range of 6, and a standard deviation of 2.0. But if the scores 0 and 6 form one stratum and 2 and 8 another, the possible samples of size 2 from each stratum have means of 3 and 5; these means have a mean of 4 as before, but now the range is 2 and the standard deviation = 1.4.

Stratified sampling is indeed important in statistics, but sufficiently complicated so that it will not be considered in this book.

Cluster Sampling.   To choose a sample of 30 students from 400 freshmen living in four dormitories, you might pick at random one floor in each dormitory, and then choose randomly from the students on the selected floors. Cluster sampling is relatively inexpensive, since there is less travel time between interviews, but cluster sampling gives less accurate predictions about the population if individuals in one cluster are alike. If you wish to predict how New York City voters will vote for President, you would hardly be wise to pick all your sample from one block, since rich and poor generally live in separate neighborhoods and family income may affect voting patterns. If, on the other hand, your clusters are very variegated, predictions about the population may be more precise than when a simple random sample is used.

Systematic Sampling.   To choose your sample of 30 students from the freshman class of 400, you might pick a starting place at random in the freshman directory and then list every thirteenth name after that until you have 30 names, continuing with the A's after you finish the Z's (thirteenth because 400/30 ≈ 13). This method of sampling is simple and frequently used. Systematic sampling is a particular kind of cluster sampling.

8.4 VOCABULARY

census
representative sample
random sample

stratified sampling
cluster sampling
systematic sampling

8.5 EXERCISES

1.   Which of the following are random samples from the suggested population? Explain the reason if not a random sample.

(a) A chemical society is interested in the number of technical papers written by those who have earned a Ph.D in chemistry in the past 5 years. It sends a questionnaire to all those awarded such a degree in the past 5 years, and bases its estimate on the replies received.

(b) A magazine takes a presidential preference poll by sending a questionnaire to one subscriber from each zip code district in its mailing list.

(c) A manufacturer of automobile horns sends every hundredth horn it manufactures to quality control for testing.

(d) A psychologist chooses 30 students from a large class by assigning each member of the class a number 001, 002, . . . , 485. He uses the middle three digits in a group of 5 numbers in a random number table, but skips every other group of 5. He skips every number which repeats one previously chosen, and every number over 485.

(e) A sample of 50 is to be taken from telephone subscribers in Buffalo. You choose numbers at random from the Buffalo telephone directory and spend every evening calling selected numbers on the telephone until you have 50 answers.

(f) You are a doctor carrying out research at a famous cancer hospital in New York. You have developed a new chemical for treatment of lung cancer. You give the chemical to every third patient who enters with lung cancer if he is willing, leaving the others for controls for comparison of treatment.

2.   In a table of random numbers, what proportion of numbers do you expect to be 4's?

3.   What is the probability that 2 adjacent digits in a table of random numbers are identical?

4.   Choose a sample of 10, without replacement, from a population numbered 0001 to 8432, starting with the third row, second column of Table 3, Appendix C, and reading horizontally from left to right with each succeeding line above the previous one.

5.   Choose (without replacement) a sample of size 4 from a population of size 10, starting with the tenth row, fifth column of Table 7 and reading the first digit in each group of 5 vertically.

6.   Criticize the following:

(a) A market researcher asks people entering a supermarket, "Do you use Kleenex?"

(b) A sociologist asks a random sample of 75 students in his Introduction to Sociology class, "How many times have you stolen something valued over $10?"

7.   Choose 10 numbers (between 0 and 9) at random, using Table 3, Appendix C. What do you expect the mean and standard deviation of 10 random numbers to be?

What is the mean and what is the standard deviation of the 10 random numbers you chose?

    8.   Is a random sample a representative sample?

ANSWERS

    1.   (a) Not random. Those who have written no papers are less likely to answer.

    (b) Not random. Rural areas are represented more than big-city areas. Nor does the description claim that the one subscriber is chosen at random from his district. Also, a voluntary response is needed; this may further bias the results (towards people with a better education?).

    (c) This is systematic, not random, sampling. It may not be objectionable, but what if a machine that makes part of the horn has a cyclic defect that causes trouble on every hundredth or thousandth horn?

    (d) This is a random sample. He will probably need a longer table of random numbers than that given in Table 7.

    (e) Not random. You will almost certainly get few business telephones, and will get answers from individuals who stay at home and not those who go out in the evening.

    (f) It is a very poor sample. Those who are willing to try the new drug may be hopeless of cure by any other means. It is not even a systematic sample, much less a random sample, of those entering that hospital, and certainly not a random sample of all patients with lung cancer—more "late" cases would probably be sent on to this hospital. And you are about to lose your medical license; a doctor can't just try out new drugs freely and ignore established methods of treatment.

    3.   1/10. There are ten pairs of digits (00, 11, . . . , 99) and 100 2-digit numbers (00, 01, 02, . . . , 99).

    5.   Number the population 0 to 9 rather than 1 to 10, so you need choose only 1-digit numbers. The numbers that appear vertically in the random number table starting at the designated spot are 8, 9, 8, 9, 0, 3, 1, ... The numbers chosen for the sample are 8, 9, 0, 3.

    7.   This is a uniform probability distribution. $p(0) = p(1) = \ldots = p(9) = .1$. $\overline{X} = 4.5$, $s = \sqrt{9.17} = 3.0$.

chapter nine

confidence intervals for means

Here at last you begin to do one of the most important things a statistician needs to do: to take a sample from the population, study the sample, and then make inferences (draw conclusions) about the population from which the sample was taken.

Part of this chapter is highly theoretical, but the sampling distribution of means must be thoroughly understood if you are to understand how inferences about the population mean are made. Then you finally get down to the nitty-gritty of estimating the population mean, knowing the mean of one random sample.

*Two cases are discussed: (1) if σ is known, and (2) if σ is unknown but n is large (at least 30). If σ is unknown **and** n is smaller than 30, you will have to wait until Chapter 12 before estimating the population mean.*

9.1 SAMPLING WITH AND WITHOUT REPLACEMENT

Suppose numbered balls are placed in a fishbowl. If a ball is drawn from the fishbowl, we have the choice of replacing or not replacing the ball before a second one is drawn. If each ball is replaced (and well mixed) before the next is drawn, then each may appear again; we have sampling with replacement. If a ball is not replaced before the next one is drawn, then that ball can be drawn only once.

Example 1

Four students (A, B, C, and D) take a test with 20 questions, and the number of questions each answers correctly is

Student:	A	B	C	D
Number of correct answers:	4	8	12	20

List all possible samples of size 2 that can be taken with replacement from this population.

There will be $4 * 4 = 16$ samples, since each student chosen as the first element of the sample may be associated with any of the 4 students as the second element. The scores in the 16 samples are:

4, 4	8, 4	12, 4	20, 4
4, 8	8, 8	12, 8	20, 8
4, 12	8, 12	12, 12	20, 12
4, 20	8, 20	12, 20	20, 20

Example 2

Five salesmen (M, N, O, P, and Q) sell the following number of cars on June 15:

Salesman:	M	N	O	P	Q
Number of cars:	1	2	3	4	4

List all possible different samples of size 2 that can be taken **without** replacement from this population.

There are $_5C_2 = 10$ samples that can be drawn from this population.

(SP$_x^2$ = 20)

	Samples	Scores
1	M, N	1, 2
2	M, O	1, 3
3	M, P	1, 4
4	M, Q	1, 4
5	N, O	2, 3
6	N, P	2, 4
7	N, Q	2, 4
8	O, P	3, 4
9	O, Q	3, 4
10	P, Q	4, 4

9.2 SAMPLING DISTRIBUTION OF MEANS

In Example 1 of the preceding section, think of the 4 students as a population ($N = 4$). All possible samples of size 2 were taken, with replacement, from this population ($n = 2$ for each sample). Each of these samples has, of course, its own mean; let us list all of them:

Mean

taken w/ Replacement
various x 4 choices = 16

$\overline{X}$:	(4,4)	4		6	8	12
	(4,8)	6		8	10	14
	(4,12)	8		10	12	16
	(4,20)	12		14	16	20

These 16 means are now to be thought of as scores in a new distribution. We can tally them as we would any distribution:

"sampling distribution of the means"

freq. of $\overline{x}$

$\overline{X}$	f
4	1
6	2
8	3
10	2
12	3
14	2
16	2
20	1

This new distribution, made up of the means of all possible samples of size 2 taken from the given population, is called the **sampling distribution of means.** It is one of the most fundamental concepts of statistical inference, and we shall soon discover some of its remarkable properties.

Since the sampling distribution of means is a frequency distribution, it has its own mean and standard deviation. Since the scores ($\overline{X}$) in the sampling distribution of means are themselves means (of individual samples), we shall use the notation $\mu_{\overline{X}}$ (read "mu sub X bar") for the mean of the distribution: μ because it is a mean, and the subscript $\overline{X}$ to distinguish this mean from that of the population. Similarly, the notation $\sigma_{\overline{X}}$ (read "sigma sub X bar") is used for the standard deviation of the sampling distribution of means.

One further complication: the standard deviation of the sampling distribution of means is called the **standard error of the mean.** = Std. deviation of the sampling dist for

Let us compute the mean $\mu_{\overline{X}}$ and the standard error of the mean $\sigma_{\overline{X}}$ for the sampling distribution of means written above. (To simplify arithmetic, we will subtract 11 from each value of $\overline{X}$ before multiplying by f—see page 35 if you don't recall this shortcut; for $\sigma_{\overline{X}}$ use the population rather than the sample formula for standard deviation—that is, divide by the number of scores, not 1 less than this number).

*(handwritten annotations: freq.; col'n f * col'n($\overline{x}$-11); Col'n f * (Col'n $\overline{x}$-11)²)*

$\overline{X}$	f		$\overline{X} - 11$		$f(\overline{X} - 11)$	$f(\overline{X} - 11)^2$
4	1	⊥	−7	≡	− 7	49
6	2	⊬	−5	≡	−10	50
8	3	⊬	−3	≡	− 9	27
10	2	⊬	−1	≡	− 2	2
12	3	⊬	1	≡	3	3
14	2	⊬	3	≡	6	18
16	2	⊬	5	≡	10	50
20	1	⊬	9	≡	9	81
	16				0	280

(handwritten: use calc, one variable w/ freq.)

$$\mu_{\overline{X}} = 11 + \frac{0}{16} = 11 \qquad \sigma_{\overline{X}} = \sqrt{280/16} = \sqrt{35/2} = 4.18$$

(handwritten: if you subtract 11, you have to add 11)

$\sqrt{35/2}$ is left in this form rather than written as 4.2 since its relation to the standard deviation of the population will be easier to see.

Now compute μ and σ for the original population:

X	$X - \mu$	$(X - \mu)^2$
4	−7	49
8	−3	9
12	1	1
20	9	81
44		140

(handwritten: Same!) $\quad \mu = 44/4 = 11 \qquad \sigma = \sqrt{140/4} = \sqrt{35} = 4.18$

Comparing these results with the sampling distribution of means, two astonishing relations are noted:

1. $\mu_{\overline{X}} = \mu$. (11 = 11).
2. $\sigma_{\overline{X}} = \sigma/\sqrt{n}$. ($\sqrt{35/2} = \sqrt{35}/\sqrt{2}$.)

Example 1

Consider the 5 salesmen in Example 2, page 132, to be a population. (a) Find the mean of each of the 10 samples of size 2 drawn without replacement, and then form the sampling distribution of means for samples of size 2 from this population; (b) find $\mu_{\bar{X}}$ and $\sigma_{\bar{X}}$; (c) compute μ and σ; (d) compare the results in (b) and (c).

mean of the sampling distrib.

(a) $\bar{X}$: 1.5, 2, 2.5, 2.5, 2.5, 3, 3, 3.5, 3.5, 4; see the box below.

Samples MN

use calc

$\bar{X}$	f
1.5	1
2	1
2.5	3
3	2
3.5	2
4	1
	10

$\sum f \cdot col \cdot (\bar{X} - 2.5)$ $\sum f \cdot col \cdot (\bar{X} - 2.5)^2$

(b) $\bar{X} - 2.5$

$\bar{X}-2.5$		$f(\bar{X}-2.5)$	$f(\bar{X}-2.5)^2$
−1.	=	−1.	1.
−0.5	=	−.5	.25
0	>	0	0
.5	=	1.	.5
1.	=	2.	2.
1.5	=	1.5	2.25
		3.0	6.0

use calc, one variable w/ freq.

(b.) $A = 2.5;\ \mu_{\bar{X}} = 2.5 + \dfrac{3}{10} = 2.8.$

$$\sigma_{\bar{X}} = \sqrt{\frac{6 - 3^2/10}{10}} = \sqrt{\frac{51}{100}} = .714$$

(c) X: 1, 2, 3, 4, 4; $\mu = 14/5 = 2.8.$

actual #'s

$$\sum X^2 = 1 + 4 + 9 + 16 + 16 = 46,\ \sigma = \sqrt{\frac{46 - 14^2/5}{25}} = \sqrt{\frac{34}{25}} = 1.17$$

(needs corr. factor - see below)

(d) $\mu_{\bar{X}} = \mu$, as before.

This time it is not true that $\sigma_{\bar{X}} = \sigma/\sqrt{n}$; rather, $\sigma_{\bar{X}} = \dfrac{\sigma}{\sqrt{n}}\sqrt{\dfrac{N-n}{N-1}}$. (This rather

messy correction factor $\sqrt{\dfrac{N-n}{N-1}}$ is necessary in theory when sampling is with-

out replacement; in practice it is seldom necessary to use it. If N is greater than $20n$, the correction factor is almost 1. If $N = 10{,}000$ and $n = 100$, for ex-

ample, $\sqrt{\dfrac{N-n}{N-1}} = \sqrt{\dfrac{9900}{9999}} = .995.$ The correction factor is never needed for an

infinite population.) In this example, $\sigma_{\bar{X}} = \sqrt{51/100},\ \sigma = \sqrt{34/25},\ N = 5, n = 2,$

and $\dfrac{\sigma}{\sqrt{n}}\sqrt{\dfrac{N-n}{N-1}} = \dfrac{\sqrt{34/25}}{\sqrt{2}}\sqrt{\dfrac{5-2}{5-1}} = \sqrt{\dfrac{34 \cdot 3}{25 \cdot 2 \cdot 4}} = \sqrt{\dfrac{51}{100}} = \sigma_{\bar{X}}.$

It is time to generalize what has been done in these two cases, and to describe a procedure and results for samples of any size from any population. The procedure we have used involves 3 steps:

1.) All possible samples of a given size n are taken from a given population whose mean is μ and whose standard deviation is σ. Let the samples be $S_1, S_2, S_3, \ldots$

2.) The mean of each of these samples is determined. These means, $\bar{X}_1, \bar{X}_2, \bar{X}_3, \ldots,$ are the scores in the **sampling distribution of means.**

3.) The mean $\mu_{\bar{x}}$ and standard deviation $\sigma_{\bar{x}}$ of the sampling distribution of means are computed. If this process is carried out, then the following relations hold:

(a) $\mu_{\bar{x}} = \mu$.

(b) $\sigma_{\bar{x}} = \dfrac{\sigma}{\sqrt{n}}$ if the population is infinite **or** if the samples are taken with replacement **or** if $N > 20n$.

$\sigma_{\bar{x}} = \dfrac{\sigma}{\sqrt{n}} \sqrt{\dfrac{N-n}{N-1}}$ if the population is finite **and** the samples are taken without replacement **and** $N \le 20n$.

A general proof that these conclusions are valid will not be given here. Two more examples will help persuade you of their truth, perhaps. First, let $n = 1$. Each sample consists of 1 score from the population, and the mean of 1 score is the score itself; therefore the sampling distribution of means consists of the same scores as the original population. Of course, then, $\mu_{\bar{x}} = \mu$ and $\sigma_{\bar{x}} = \sigma$. But since $n = 1$, $\sigma_{\bar{x}} = \dfrac{\sigma}{\sqrt{1}} \sqrt{\dfrac{N-1}{N-1}}$, and the second conclusion above follows.

Secondly, let $n = N$. Then there is only 1 sample, identical to the original population. The sampling distribution of means consists of one score, μ. The absurd problem of finding the mean and standard deviation of a distribution consisting of a single score is one you have not had to face before, but a little thought or use of the usual formulas will tell you that the mean is the score itself and the standard deviation is 0. So $\mu_{\bar{x}} = \mu$ and $\sigma_{\bar{x}} = 0$, as should be expected when $n = N$ since $\dfrac{\sigma}{\sqrt{N}} \sqrt{\dfrac{N-N}{N-1}} = 0$.

Ordinarily we shall be interested only in samples of size greater than 1, so $\dfrac{\sigma}{\sqrt{n}} < \sigma$.

The correction factor $\sqrt{\dfrac{N-n}{N-1}}$ is also less than 1, so $\sigma_{\bar{x}}$ is less than σ. As a result, the means of all the samples cluster about **their** mean more closely than do the individual scores in the population from which the samples are taken. Suppose the weights of 1000 women students range from 85 to 165 with a mean of 125. A sample of 10 students may include overweight and underweight girls, but they tend to "cancel each other out" so that the sample mean ends up somewhere near 125. It would be most extraordinary for a random sample to have a mean of 90 or of 160. Later (in Chapter 11) you will learn how to calculate just how extraordinary this would be.

9.3   CENTRAL LIMIT THEOREM

So far we have discussed the mean and standard error of the sampling distribution of means, but not the shape of the distribution. The incredible power of the sampling distribution of means depends upon the following theorem:

Central Limit Theorem:

1. The sampling distribution of means is a normal distribution if the population is normally distributed.

2. Even if the population is not normally distributed, the sampling distribution of means is approximated by a normal distribution for large n. The more the population distribution differs from a normal distribution, the larger n must be for the sampling distribution to approximate a normal one. In practice, we shall assume the approximation is a workable one if n is 30 or more.

The proof will not be given, since it requires advanced mathematical techniques quite a bit beyond the prerequisites for this course. It is difficult to give examples that are persuasive. Would you happily find the means of all samples of size 30 from a population of 1,000, say? Even without replacement there would be $_{1000}C_{30}$ of them, a huge number ($\approx 2.5 \times 10^{54}$). The only simple example is the trivial case $n = 1$ when the population is normally distributed; clearly the sampling distribution of means is normal too, since it is identical with the population. The two examples we took up earlier are hardly persuasive; neither sampling distribution is normal, since neither population is normal and $n(=2)$ is too small. You will, alas, have to accept the theorem on faith.

9.4 USES OF THE SAMPLING DISTRIBUTION OF MEANS

A psychologist may wish to measure the reaction time of 30 students to an electric shock and estimate the reaction time of all individuals to this shock. A manufacturer may find the mean life of 100 light bulbs and estimate the mean life of all light bulbs manufactured by a certain process. These are two examples of statistical inference which can be answered with the help of the sampling distribution of means and the Central Limit Theorem. For the moment, however, let us look at simpler problems about random samples when information about the population is given.

Example 1

A random sample of size 64 is chosen from a population of size 100,000 whose mean is 10 and whose standard deviation is 3. What is the probability that the mean of the sample is (a) between 9 and 11? (b) greater than 11?

The sampling distribution is approximately normal, since $n > 30$. Its mean $\mu_{\bar{x}} = \mu = 10$. Since $N(= 100,000)$ is greater than $20n = 20 * 64$,

find Std. Deviation of Mean: $\sigma_{\bar{x}} = \dfrac{\sigma}{\sqrt{n}} = \dfrac{3}{\sqrt{64}} = \dfrac{3}{8} = .375.$

$z = \dfrac{X - \mu \bar{x}}{\sigma \bar{x}}$

$= \dfrac{11 - 10}{.375} = 2.67$

Sketch the sampling distribution of means.

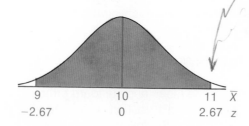

9	10	11 X
−2.67	0	2.67 z

Note that the horizontal axis is labelled $\bar{X}$, since the distribution is made up of sample means.

(a) $p(9 < \bar{X} < 11) = ?$ This is a familiar problem of finding areas under a normal curve. To transform to z scores, we always subtract the mean of the distribution and then divide by its standard deviation. In this case, the appropriate equation will be

$$z = \frac{\bar{X} - \mu_{\bar{x}}}{\sigma_{\bar{x}}} = \frac{\bar{X} - \mu_{\bar{x}}}{\sigma/\sqrt{n}}.$$

When $\overline{X} = 11, z = \dfrac{11-10}{3/\sqrt{64}} = 2.67$; when $\overline{X} = 9, z = \dfrac{9-10}{3/\sqrt{64}} = -2.67$. $p(9 < \overline{X} < 11)$

$= p(-2.67 < z < 2.67) = 2 * .496 = .997$

$\qquad$ *same as $(1 - .992)/2$*

(b) $p(\overline{X} > 11) = p(z > 2.67) = .5 - .496 = .004$, rounding to three decimal places.

Example 2

All third-graders in New York City take the Peterson Interval Test; the mean score is 200 with a standard deviation of 20. What is the probability that a sample of 16 children chosen at random from the third grade in Browning School will have a mean score greater than 205?

There are two traps here; see if you can find both of them before you read on.

1) w/o replacement
2) N < 30 n

First, the sample is not a random sample from the specified population (all third-graders in New York City). Perhaps Browning School is a school for retarded children, and its third-graders do badly on all tests. All deductions made in this section about the probability of sample means, and all inferences that we shall make about the population mean after studying a sample in the next sections, are based on the premise that the sample is a random one. If this requirement is not satisfied, then the theory we are building up is not applicable.

The second trap is in the size of the sample: $n = 16, < 30$, so the sampling distribution is normal only if the population is normal. This fact is not explicitly stated; if only approximately true, then the sampling distribution of means is only approximately normal. Unfortunately, you have no way of judging the size of the error caused by assuming the population is normally distributed when it is not.

Example 3

A normal population has mean of 50 and standard deviation of 10. Ninety-five per cent of random samples of size 25 have means in what interval? (Assume the interval has 50 as its midpoint.)

The area under the curve in the desired interval is .95, so the area to the right of $z = 0$ is .475. If $A(z) = .475$, then $z = 1.96$. *(table p.349)*

$$\mu = 50, \sigma = 10, n = 25, \sigma_{\overline{x}} = \frac{10}{\sqrt{25}} = 2$$

$$z = \frac{\overline{X} - \mu}{\sigma_{\overline{x}}} = \frac{\overline{X} - 50}{2}$$

$$\pm 1.96 = \frac{\overline{X} - 50}{2}.$$

.95

.475

| 46.1 | 50 | 53.9 | $\overline{X}$ |
| -1.96 | 0 | 1.96 | z |

$\overline{X} = 50 \pm 3.92 = 53.92$ or 46.08; 95 per cent of the sample means will be between 46.1 and 53.9.

9.5   STATISTICAL INFERENCE

You should now be quite comfortable with the statement, "there is a .95 probability that the mean of a sample of 100 taken at random from a population with $\mu = 400$, $\sigma = 50$ will be between 391.2 and 409.8."

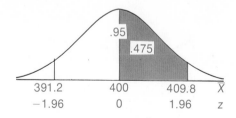

The main interest of the statistician, however, is in studying a sample and making inferences about the population from which the sample is taken. For example, an economist wants to estimate the mean number of miles trucks are driven in a week, the proportion of carpenters with incomes over $10,000 per year, or the differences of mean wages of steelworkers and of men in automobile plants. A psychologist wants to know the mean reaction time of men to a certain stimulus, or the proportion of children whose parents are not college-trained who attend college, or the differences of mean time to solve a maze for rats with and without Vitamin A in their diets. A doctor studies the mean weight of children at birth, or the proportion of leukemia patients who live 5 years after treatment with a certain drug, or the difference in life expectancy of smokers and non-smokers. A political scientist is interested in the mean waiting time for felony trials in New York City, or the proportion of voters who back Candidate A for senator, or the differences in mean age of those favoring a liberal and a conservative opposing each other for Congress.

Three problems are suggested in each case above; one involves a mean, one a proportion, and the third differences of means. But in each case, full information about the population is not likely to be attainable, and the researcher must generalize after examining a sample taken from the population; he must make a statistical inference.

9.6   ESTIMATES OF THE POPULATION
MEAN IF σ IS KNOWN

Suppose that the mean of a random sample of size 36 is 100, and the standard deviation of the (very large) population is 24. An estimate of the population mean can be given in two ways. If a single number is given ("I estimate that $\mu = $ ____"), this is a **point estimate.** If a range of values is given ("I estimate that μ is between ____ and ____"), this is an **interval estimate.**

In the example at the beginning of this section, the best estimate that can be made is that $\mu = 100$. In general, the point estimate of μ is that it equals $\overline{X}$. We have as yet no idea of the accuracy of this estimate, and it is therefore useless to us. Two psychologists, for example, with two different samples may make different point estimates of a population mean, one of which may support a certain theory and the other deny it. We shall find that, as soon as we estimate the accuracy of a point estimate, we are really giving an interval estimate.

Since the sampling distribution of means is normal, we know that 95 per cent of samples will have a mean whose z score is between −1.96 and +1.96, or, stated another way, 95 per cent of sample means will be within 1.96 standard deviation units of the population mean.

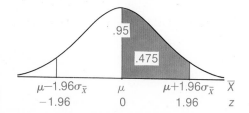

↖ std deviation

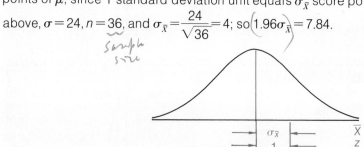

Assume that μ is unknown. Then 95 per cent of sample means will be within $1.96\sigma_{\bar{x}}$ score points of μ, since 1 standard deviation unit equals $\sigma_{\bar{x}}$ score points. In the example stated above, $\sigma = 24$, $n = 36$, and $\sigma_{\bar{x}} = \dfrac{24}{\sqrt{36}} = 4$; so $1.96\sigma_{\bar{x}} = 7.84$.

sample size

Now the researcher who chose the random sample reasons this way: "Out of every 10,000 samples chosen from this population, about 9500 of them will have a mean which is within 7.8 points of the population mean, and about 500 will have a mean which is more than 7.8 points away from the population mean. I don't know for certain which group my particular sample with mean of 100 is in, but I do know that if my sample is in the first group, then the population mean is within 7.8 points of 100 — that is, between 92.2 and 107.8. I also know that if one sample is chosen at random, its probability of being in the first group is .95, and that the sample I am working with was chosen at random. **I am 95 per cent confident that the population mean is between 92.2 and 107.8.**" (Note: If you understand every word of this paragraph, the rest is easy; the real kernel of this chapter is here.)

✱ w/a mean of 100

Same argument

The range 92.2 to 107.8 is called the **confidence interval;** 107.8 is called the **upper confidence limit** and 92.2 is called the **lower confidence limit** for μ. Here are three equivalent statements:

(1)   I am 95 per cent confident that μ falls between 92.2 and 107.8.

(2)   A 95 per cent confidence interval for μ is 92.2 to 107.8.

(3)   The interval estimate for μ, at the 95 per cent confidence level, is 92.2 to 107.8.

The probability that the interval 92.2 to 107.8 contains the unknown population mean is called the **confidence coefficient.** When this coefficient is .95, for example, if many different random samples are taken, and if the confidence interval for each is determined, then 95 per cent of these computed intervals will contain μ.

If you have thoroughly understood how the 95 per cent confidence interval of 92.2 to 107.8 for μ was arrived at, then you should have no trouble understanding the following more formal solution of the same problem:

Example 1

The mean of a random sample of size 36 is 100 and the standard deviation of the (very large) population is 24. Find a 95 per cent confidence interval for μ.

With a .95 confidence coefficient, $z = \pm 1.96$; $\sigma = 24$, $n = 36$; $\sigma_{\bar{x}} = \dfrac{24}{\sqrt{36}} = 4$.

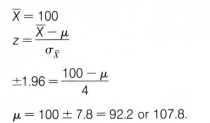

$\bar{X} = 100$

$z = \dfrac{\bar{X} - \mu}{\sigma_{\bar{x}}}$

$\pm 1.96 = \dfrac{100 - \mu}{4}$

$\mu = 100 \pm 7.8 = 92.2$ or 107.8.

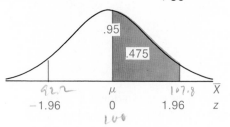

The 95 per cent confidence interval for μ is 92.2 to 107.8.

Always sketch the normal curve representing the sampling distribution of means, and mark appropriate areas, scales, and points.

There is nothing sacred about a .95 confidence coefficient. Try the same problem with other confidence coefficients.

Example 2

The mean of a random sample of size 36 is 100, and the standard deviation of the (very large) population is 24. Make an interval estimate of μ at (a) 90 per cent and (b) 98 per cent confidence levels.

(a)   With a .90 confidence coefficient, $z = \pm 1.65$; $\sigma_{\bar{x}} = 4$, $\bar{X} = 100$ (as before).

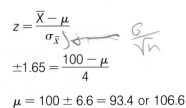

$z = \dfrac{\bar{X} - \mu}{\sigma_{\bar{x}}}$

$\pm 1.65 = \dfrac{100 - \mu}{4}$

$\mu = 100 \pm 6.6 = 93.4$ or 106.6

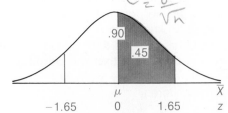

The 90 per cent confidence interval for μ is 93.4 to 106.6.

(b)   With a .98 confidence coefficient, $z = \pm 2.33$.

$\pm 2.33 = \dfrac{100 - \mu}{4}$

$\mu = 100 \pm 9.32 = 90.7$ or 109.3

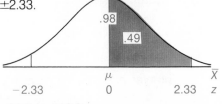

The 98 per cent confidence interval for μ is 90.7 to 109.3.

Here is a summary of results of this problem for different confidence coefficients:

Confidence coefficient	Confidence interval
.90	93.4 to 106.6
.95	92.2 to 107.8
.98	90.7 to 109.3

Note that, as our degree of confidence that the given interval does include the population mean increases, the width of the interval increases.

**General Procedure for Finding a Confidence
Interval for μ When σ Is Known**

Note: This procedure cannot be used unless the population is normal or the sample size n is sufficiently large (at least 30) so that the sampling distribution of means is normally distributed.

1. Find the mean of the random sample, $\overline{X}$.

2. Compute the standard error of the mean (= the standard deviation of the sampling distribution of means) $\sigma_{\overline{x}}$:

$$\sigma_{\overline{x}} = \frac{\sigma}{\sqrt{n}}\sqrt{\frac{N-n}{N-1}} \text{ if } N \text{ is finite } \textbf{and } N < 20\,n \text{ } \textbf{and} \text{ sampling is without replacement.}$$

$$\sigma_{\overline{x}} = \frac{\sigma}{\sqrt{n}} \text{ in all other cases (usually sampling is without}$$

replacement, but still this simplified formula
can be used most of the time since $N > 20n$).

3. Determine the z values for the given confidence level. The most frequently used are:

Confidence level (per cent):	68.3	90	95	98	99	99.7
z:	±1.00	±1.65	±1.96	±2.33	±2.58	±3.00

4. Compute the confidence limits: Solve $\pm z = \dfrac{\overline{X} - \mu}{\sigma_{\overline{x}}}$ for μ.

The interval estimate or confidence interval is given by these upper and lower confidence limits.

Example 3

A random sample of 36 students chosen without replacement from a class of 72 has a mean weight of 140 pounds. It is known that the standard deviation of all students in the class is 5 pounds. Find a 90 per cent confidence interval for the mean weight of all students in the class.

1. $\overline{X} = 140$.

2. $\sigma_{\overline{x}} = \dfrac{5}{\sqrt{36}}\sqrt{\dfrac{72-36}{72-1}} = \dfrac{5}{6}\sqrt{\dfrac{36}{71}} = .59$.

3. With a .90 confidence coefficient, $z = \pm 1.65$.

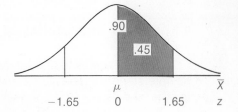

4. $\pm 1.65 = \dfrac{140 - \mu}{.59}$.

$$\mu = 140 \pm 1.65 * .59 = 140 \pm .97$$
$$= 139.0 \text{ or } 141.0$$

I hope you have not just mechanically followed directions but have reasoned as you go along: "At a 90 per cent confidence level, $A(z) = .45$ and $z = \pm 1.65$; there is a 90 per cent probability that the sample mean is within $1.65 * \sigma_{\bar{x}} = 1.65 * .59 = .97$ score units of μ, so I am 90 per cent confident that μ falls between $140 - .97$ and $140 + .97$, that is, between 139.0 and 141.0."

The examples above may seem nonsensical in that σ is known—and if σ, then why not μ? It is true that, upon the occasions you have determined σ, it has been easy (easier, even) to find μ. But there are indeed occasions of great practical interest in which σ can be estimated quite accurately but μ cannot, as the following example demonstrates.

Example 4

Each decade since 1900 the height of army recruits has been determined. Their mean height has gradually increased, but the standard deviation of heights of all recruits each decade has stayed constant at 3.0 in. A random sample of 900 recruits are measured in 1970, and their mean height is found to be 70 in. What is the 95 per cent confidence interval for the mean height of all recruits in 1970?

It is assumed that $\sigma = 3.0$ in 1970 as well.

$\bar{X} = 70$, $n = 900$, $\sigma_{\bar{x}} = \dfrac{3.0}{\sqrt{900}} = .10$; at a 95 per cent confidence level,

$z = \pm 1.96$

$\pm 1.96 = \dfrac{70 - \mu}{.1}$, $\mu = 70 \pm .2 = 69.8$ or 70.2 in.

N.B. ± 1 Std. Deviation
$= 67 < 773$

The 95 per cent confidence interval for μ in 1970 is 69.8 to 70.2 in.

In Example 1 ($\bar{X} = 100$, $n = 36$, $\sigma = 24$; see page 139), we said the point estimate of μ is 100, and the confidence limits are 100 ± 7.8 with 95 per cent confidence. We can now tell how good a guess 100 is as a point estimate, just as an engineer prescribes the diameter of a shaft: .5000 in. $\pm$ .0025 in.—but the engineer hopes for 100 per cent compliance with his prescription and the statistician does not. Without comments on its accuracy a point estimate is pretty useless, but with an idea of its accuracy it becomes an interval estimate.

9.7 BIASED AND UNBIASED ESTIMATORS

Our next aim is to find interval estimates for μ when σ is unknown. Can we use the sample standard deviation s and proceed merrily on our way? Sometimes the answer is "yes"; let us see when and why.

To each sample statistic there corresponds a population parameter: to $\bar{X}$, μ; to s^2, σ^2; to s, σ; we shall soon deal with a sample proportion p and a population proportion P. We shall hope to use $\bar{X}$, s^2, s, p, etc. to estimate μ, σ^2, σ, P, etc. We have already found that the mean $\bar{X}$ of a sample can be used to estimate μ. This does not, of course, mean that the mean of every sample will equal the population mean. But since $\mu_{\bar{x}} = \mu$, we do know that if we take all possible samples, the mean of their means will give us μ. Furthermore, the mean of a large number of sample means will reflect the value of μ pretty well, and a single sample taken at random has a mean which is neither consistently too low nor consistently too high. We say $\overline{X}$ is an unbiased estimator of μ.

> **Definition:** A sample statistic is called an **unbiased estimator** of a parameter if the mean of the sampling distribution of that statistic equals the parameter.

Is s^2 an unbiased estimator of σ^2? The answer is "yes" because it was defined in such a way that it would be. (See the discussion on page 47 and Exercise 19, page 57.)

If σ is unknown **and n is at least 30**, $\sigma_{\overline{x}}$ is approximated by $\dfrac{s}{\sqrt{n}}$. If σ and s are both known, use $\dfrac{\sigma}{\sqrt{n}}$ in preference to $\dfrac{s}{\sqrt{n}}$. Why the restriction that the sample size be at least 30, and what do you do if σ is unknown but n is less than 30? Both these questions will be answered in Chapter 12.

9.8   INTERVAL ESTIMATES OF μ WHEN σ IS UNKNOWN AND $n \geq 30$

Example 1

Find a 90 per cent confidence interval for the height of 2200 students if a random sample of 100 students has a mean height of 67 inches and a standard deviation of 3.0 inches.

$\overline{X} = 67$, $s = 3$, $n = 100$, $N = 2200$; σ is unknown, so

$$\sigma_{\overline{x}} = \frac{s}{\sqrt{n}} = \frac{3}{\sqrt{100}} = .30$$

At a 90 per cent confidence level, $z = \pm 1.65$.

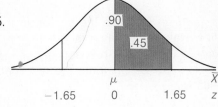

$$\pm 1.65 = \frac{67 - \mu}{.3}$$

$$\mu = 67 \pm .5$$

I am 90 per cent confident that the mean height of all 2200 students is between 66.5 and 67.5 in.

Example 2

The mean weight of a sample of 36 Essex University students is 150 lb. with a standard deviation of 15 lb. It is known, from previous tests, that the standard deviation of the weights of all Essex students is 18 lb. Find the mean weight of all Essex students at a 95 per cent confidence level.

$\overline{X} = 150$, $n = 36$, $s = 15$, $\sigma = 18$.

When s and σ are both known, σ is used to determine $\sigma_{\overline{x}}$.

$$\sigma_{\overline{x}} = \frac{\sigma}{\sqrt{n}} = \frac{18}{6} = 3.0$$

Confidence coefficient = .95, so $z = \pm 1.96$.

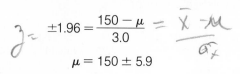

$$\pm 1.96 = \frac{150 - \mu}{3.0} = \frac{\overline{X} - \mu}{\sigma_{\overline{X}}}$$

$$\mu = 150 \pm 5.9$$

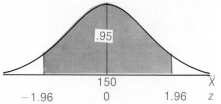

The 95 per cent confidence interval for μ is 144.1 to 155.9 lb.

Let me repeat the warning given at the end of the last section: if σ is unknown, use $s/\sqrt{n}$ as an approximation of $\sigma_{\overline{X}}$ **only** if n is at least 30. If n is less than 30, a different technique will be used. Until you have studied this (in Chapter 12), you will have to increase the sample size if σ is unknown, so that the methods of this section can be used.

9.9   VOCABULARY AND NOTATION

sampling distribution of means   upper confidence limit

$\mu_{\overline{X}}$ lower confidence limit

$\sigma_{\overline{X}}$ confidence coefficient

standard error of the mean   biased estimator

Central Limit Theorem   unbiased estimator

point estimate   statistic

interval estimate   parameter

confidence interval

9.10   EXERCISES

1.   (a) Give the scores in the sampling distribution of means for samples of size 3 taken without replacement from the population whose members A, B, C, D, E have scores 1, 1, 2, 2, 3, respectively.

       (b) Compute the mean and standard error of this sampling distribution of means.

       (c) Compute μ and σ.

       (d) Compare (b) and (c).

2.   (a) Show that the probability is .50 that the mean of a sample of size 40 chosen at random from a population with mean 10 and standard deviation 6 will be between $10 - .67 * 6/\sqrt{40}$ and $10 + .67 * 6/\sqrt{40}$.

       (b) Show that the probability is .50 that the mean of a sample of size n chosen at random from a population with mean μ and standard deviation σ is between $\mu - .67\,\sigma_{\overline{X}}$ and $\mu + .67\,\sigma_{\overline{X}}$. The number $.67\,\sigma_{\overline{X}}$ is called the **probable error of the mean.**

       (c) Find the probable error of the mean in Example 1, Section 9.4 (page 136); interpret.

3.   What proportion of random samples of size 64 taken from a population with mean of 1000 and standard deviation of 160 should have a mean (a) between 975 and 1025? (b) greater than 1040?

4.   The weights of 10,000 ball bearings are normally distributed with mean of 1 oz. and standard deviation of .007 oz. If 100 samples, each of size 49, are chosen at random, how many samples will have a combined weight (a) between 48.9 and 49.1 oz.? (b) more than 49.1 oz.?

5.   What is meant (a) by the phrase **sampling distribution of the median?** (b) by the phrase **standard error of the median?** $= \sigma$ of the sampling distrib. of the med.

6.   Random samples of size 3 are taken without replacement from the population 3, 6, 9, 12.

  (a) Find the mean of each of the four samples of size 3 that can be taken, without replacement, from this population; that is, construct the sampling distribution of means for samples of size 3.

  (b) Show that $\mu_{\bar{x}} = 7.5$, $\sigma_{\bar{x}} = \sqrt{5/4}$.

  (c) Show that $\mu = 7.5$ and $\sigma = \sqrt{45/4}$.

  (d) Does $\mu_{\bar{x}} = \mu$? Does $\sigma_{\bar{x}} = \dfrac{\sigma}{\sqrt{n}} \sqrt{\dfrac{N-n}{N-1}}$?

7.   (a) A population of size 10 consists of the digits 0, 1, . . . , 9. Compute μ and σ. $M = 4.5$; $\sigma = 2.87$

  (b) Sampling with replacement converts this from a population of size 10 to an infinite population, but μ and σ will not be changed. Take the last digit of each of the numbers in column 3 of the random number table, and consider these to be 50 samples of size 1 from this population. Compute the mean and standard deviation of the means of these 50 samples. (Clearly, the mean of a sample of size 1 equals the sample score.) $M_{\bar{x}} = M = 4.5$;

8.   What is the effect on the standard error of the mean if the population is very large and the sample size is changed from (a) 25 to 100? (b) 25 to 2500?

9.   If $\sigma = 100$, $N = 1000$, $n = 100$, and $\bar{X}$ for one random sample taken with replacement is 480, find a 95 per cent confidence interval for μ.

10.   Find a 90 per cent confidence interval for the height of 1200 students if it is found that a random sample of 49 students has a mean height of 67 in., and it is assumed that the standard deviation for all students is 3 in. as it has been in the past.

11.   An economist discovers that the mean number of miles driven per week by a random sample of 64 $1\frac{1}{2}$-ton trucks is 400 miles, with a standard deviation of 100 miles What is his estimate of the mean number of miles driven per week for all $1\frac{1}{2}$-ton trucks, at a 95 per cent confidence level?

12.   A psychologist discovers that the mean reaction time of 36 rats to an 18-volt electric shock is .45 second with a standard deviation of .06 second. Find a 90 per cent confidence interval for the mean reaction time of all rats of the same strain to an 18-volt shock.

13.   A political scientist discovers that the mean waiting time between arrest and trial for 81 men chosen at random from those tried for felonies in New York City is 300 days, with a standard deviation of 30 days. He is 98 per cent confident that the mean interval between arrest and trial for felony cases in New York City has what range?

14.   A college housing officer wants to estimate the mean rental of rooms in the area. He randomly selects 30 of the 300 available rooms and finds a mean monthly rental of $60 per room, with a standard deviation of $10. What is the mean monthly rental per room of all 300 rooms, with 95 per cent confidence?

15.   A random check of 36 factory assemblers shows that the mean number of units completed per worker per hour is 16.2, with a standard deviation of 2.1. What is the 95 per cent confidence interval for the mean number of units completed per hour by all 700 assemblers in the plant?

16.   A national news magazine has 2,500,000 subscribers. A survey of 1,600 random subscribers shows a mean annual income of $12,100 with a standard deviation of $1,000. Find the 98 per cent confidence level for the mean income of all subscribers.

17.   A survey of 49 of the 250 newly appointed district attorneys in the country shows a mean of 13.7 years of legal experience prior to appointment, with a standard deviation of 2.8 years. Find a 90 per cent confidence interval for the mean years of prior legal experience of all the newly appointed district attorneys.

18.   A check of 100 random automobile accidents resulting in death reveals a mean driver's age of 24.5 years, with a standard deviation of 3.5 years. What is the mean age of all drivers involved in fatal accidents, at a 95 per cent confidence level?

19.   There are 2150 supermarkets in the metropolitan area. A random sample of 64 supermarkets indicates a mean of 35 cases of baby food sold per week, with a standard deviation of 6 cases. Find a 95 per cent confidence interval for the mean number of cases of baby food sold per week at all supermarkets in the metropolitan area.

20.   An automobile manufacturer tests a random sample of 81 new Warlock cars and finds a mean of 20.5 miles per gallon in road tests, with a standard deviation of 2.7 mpg. What is the 95 per cent confidence interval for mileage per gallon of all new cars of that model?

21.   A study of 50 households of four people each living in Central City, chosen at random, shows a mean expenditure of $76 a week for food, with a standard deviation of $3. Find the average weekly cost of food for all households of four people in Central City, with 96 per cent confidence.

22.   XYZ tires are known to have a mean life of 20,000 miles, with a standard deviation of 2,000 miles. The manufacturer modifies the tires; he hopes he has increased the mean life but thinks the standard deviation will not be changed. Tests of 60 randomly chosen modified tires give a mean life of 21,500 miles, with a standard deviation of 2100 miles. Find an interval estimate of the mean life of the modified tires, at a 90 per cent confidence level.

23.   A safety consultant tests 36 bumper guards, chosen at random, on Chevrolets by bumping into a concrete wall at 5 miles per hour. The mean cost of repairs is $70, with a standard deviation of $15. With what confidence can he claim the mean cost for all Chevrolet cars under this test would be between $65 and $75?

24.   The mean weekly take-home pay of 10 wage earners in Michigan is $340 with a standard deviation of $60. Find a 95 per cent confidence interval for the mean weekly take-home pay of all American workers.

25.   There are 120 Boy Scout troops in the county. Thirty-six are randomly selected and found to have a mean of 28 boys per troop, with a standard deviation of 3.5 boys. What is the mean number of boys in the Boy Scout troops in the county, at a 95 per cent confidence level?

26.   A survey of 64 White Mountain College students reveals a mean weekly expenditure of $8.10 on books and records, with standard deviation of $2.50. What is the mean expenditure on books and records for all White Mountain College students, with a confidence level of 95 per cent?

27.   Thirty-six men in an army division are picked at random to test an obstacle course. Their mean time for the course is 36 seconds, with 4 seconds standard deviation.

What is the expected mean time for all troops in the division, with 90 per cent confidence?

28.   3600 randomly selected high school students are found to have watched a mean of 10.5 hours of television per week, with a standard deviation of 2 hours. Find the mean television viewing time of all high school students with 98 per cent confidence.

29.   Find a 90 per cent confidence interval for the age of 600 Central High School graduates if a random sample of 100 graduates indicates a mean age of 18.1 years with a standard deviation of .4 years.

30.   Find a 98 per cent confidence interval for radio ownership per family if a random sample of 900 families shows a mean of 2.5 radios per family, with a standard deviation of .33.

31.   There are 120 students in a typing course. A random selection of 49 are tested for speed; the mean speed is 26 words per minute, with a standard deviation of 5 words per minute. Find a 95 per cent confidence interval for the mean speed of the class as a whole.

32.   A random check of 81 of the 640 inmates at the State Prison indicates a mean of 2.1 previous terms per prisoner, with a standard deviation of .4. What is the mean number of previous prison terms for all inmates, with 90 per cent confidence?

33.   A random survey of 25 Ph.D.'s in the social sciences shows a mean of 4.1 years of postgraduate study required for the degree, with a standard deviation of 1.2 years. How many years are required for a graduate degree in social sciences, with 90 per cent confidence?

ANSWERS

1.   (a) Samples: A, B, C; A, B, D; A, B, E; A, C, D; A, C, E; A, D, E; B, C, D; B, C, E; B, D, E; C, D, E. Sample scores: 1, 1, 2; 1, 1, 2; 1, 1, 3; 1, 2, 2; 1, 2, 3; 1, 2, 3; 1, 2, 2; 1, 2, 3; 1, 2, 3; 2, 2, 3. Sample means (scores in the sampling distribution of means): $\frac{4}{3}, \frac{4}{3}, \frac{5}{3}, \frac{5}{3}, 2, 2, \frac{5}{3}, 2, 2, \frac{7}{3}$.

$\quad$ (b) $\mu_{\bar{x}} = 1.8$; $\sigma_{\bar{x}} = \sqrt{7/75} = .306$.

$\quad$ (c) $\mu = 9/5 = 1.8$; $\sigma = \sqrt{14/25} = .748$.

$\quad$ (d) $\mu_{\bar{x}} = \mu$.

$$\sqrt{7/75} = \frac{\sqrt{14/25}}{\sqrt{3}} * \sqrt{\frac{5-3}{5-1}} = \sqrt{\frac{14}{75} * \frac{2}{4}}\left(\textbf{or } .306 = \frac{.748}{\sqrt{3}} * \sqrt{\frac{5-3}{5-1}}\right), \text{ so } \sigma_{\bar{x}} = \frac{\sigma}{\sqrt{n}}\sqrt{\frac{N-n}{N-1}}.$$

3. $\sigma_{\bar{x}} = \dfrac{160}{\sqrt{64}} = 20$

$\quad z = \dfrac{\bar{X} - 1000}{20}$

$\bar{X}$	z
975	−1.25
1025	+1.25
1040	2.00

$\quad$ (a) $p(975 < \bar{X} < 1025) = p(-1.25 < z < 1.25) = 2 * .3944 = .789$; 78.9 per cent of samples should have a mean between 975 and 1025.

$\quad$ (b) $p(\bar{X} > 1040) = p(z > 2) = .5 - .477 = .023$; 2.3 per cent of samples should have a mean greater than 1040.

5.   (a) The distribution of medians of all possible samples of size n taken from a given population.

$\quad$ (b) The distribution described in (a) has its own mean and standard deviation; the latter is the standard error of the median.

148   **confidence intervals for means**

7.   (a) $\mu = 4.5$, $\sigma = 2.9$.

    (b) $\mu_{\bar{x}} = 4.5$, $\sigma_{\bar{x}} \approx 2.8$ ("approximately equals" because you have only 50 sample means).

9. $\pm 1.96 = \dfrac{480 - \mu}{100/\sqrt{100}}$; 460.4 to 499.6.

11. $\pm 1.96 = \dfrac{400 - \mu}{100/\sqrt{64}}$; 375.5 to 424.5 miles per week.

13. $\pm 2.33 = \dfrac{300 - \mu}{30/\sqrt{81}}$; 292.2 to 307.8 days.

15. $\pm 1.96 = \dfrac{16.2 - \mu}{2.1/\sqrt{36}}$; 15.5 to 16.9 units completed per hour by each worker.

17. $\pm 1.64 = \dfrac{13.7 - \mu}{2.8/\sqrt{49}}$; 13.0 to 14.4 years.

19. $\pm 1.96 = \dfrac{35 - \mu}{6/\sqrt{64}}$; 33.5 to 36.5 cases.

21. $\pm 2.05 = \dfrac{76 - \mu}{3/\sqrt{50}}$; $75.10 to $76.90 per week.

23. $z = \dfrac{70 - 65}{15/\sqrt{36}} = 2.00$ or $z = \dfrac{70 - 75}{15/\sqrt{36}} = -200$; $p(-2.00 < z < 2.00) = .954$; 95.4 per cent confidence.

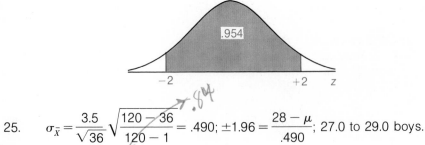

25. $\sigma_{\bar{x}} = \dfrac{3.5}{\sqrt{36}} \sqrt{\dfrac{120 - 36}{120 - 1}} = .490$; $\pm 1.96 = \dfrac{28 - \mu}{.490}$; 27.0 to 29.0 boys.

27. $\pm 1.64 = \dfrac{36 - \mu}{4/\sqrt{36}}$; 34.9 to 37.1 seconds.

29. $\pm 1.64 = \dfrac{18.1 - \mu}{.4/\sqrt{100}}$; 18.03 to 18.17 years.

31. $\pm 1.96 = \dfrac{26 - \mu}{5/\sqrt{49}}$; 24.6 to 27.4 words per minute.

33.   Beware! The population is probably not normally distributed but is skewed to the right, and $n < 30$. Wait until you have studied Chapter 12.

chapter ten   ‸ *skip*

interval estimates of proportions, difference of means, and difference of proportions; sample size

You learned how to find an interval estimate for the population mean after studying the sampling distribution of means. Now you will learn how to make three other kinds of interval estimates—but each one will again be preceded by study of a sampling distribution. There is some new notation: P for proportion in a population, p for proportion in a sample, σ_p for standard error of proportions (= standard deviation of the sampling distribution of proportions), and so on. But once you get used to notation, look for the similarities between **all** *the sampling distributions you study. Since the sampling distributions you will use in this chapter are always normal distributions, you will in every case be using standard scores to find areas under the normal curve. The formula for standard scores will change its appearance but not its essential character as the names of the scores, mean, and standard deviation change:*

Sampling distribution of:	Scores in sampling distribution	Mean	Standard error	Standard scores
Means	$\overline{X}$	μ	$\sigma_{\overline{X}}$	$z = \dfrac{X - \mu}{\sigma_{\overline{X}}}$
Proportions	p	P	σ_p	$z = \dfrac{p - P}{\sigma_p}$
Difference of means	$\overline{X} - \overline{Y}$	$\mu_X - \mu_Y$	$\sigma_{(\overline{X}-\overline{Y})}$	$z = \dfrac{(\overline{X} - \overline{Y}) - (\mu_X - \mu_Y)}{\sigma_{(\overline{X}-\overline{Y})}}$
Difference of proportions	$p_X - p_Y$	$P_X - P_Y$	$\sigma_{(p_X-p_Y)}$	$z = \dfrac{(p_X - p_Y) - (P_X - P_Y)}{\sigma_{(p_X-p_Y)}}$

Once you realize this, confidence intervals for proportions or for difference of means or of proportions will be no more difficult than for means in Chapter 9.

10.1 THE SAMPLING DISTRIBUTION OF PROPORTIONS

Our first aim is to find point and interval estimates of a population proportion from a sample proportion. A typical problem is the following: A poll is conducted to find how many voters will vote for candidate A for President. A random sample is taken of 1000 voters, and it is found that 520 plan to vote for candidate A. Estimate the percentage of the population that will vote for candidate A, at a 95 per cent confidence level.

It should be intuitively apparent that a point estimate is $\dfrac{520}{1000} = 52$ per cent. This is the best single guess at population proportion — but, again, pretty useless without some idea of how good a guess it is. Before finding an interval estimate we must look at the sampling distribution of proportions.

Again we take all possible samples (either with or without replacement) of given size n and find the proportion in each sample, thus forming the distribution of sample proportions — known, commonly but perversely, as the sampling distribution of proportions. Some new notation is involved, but we won't need any new Greek letters.

Notation: P = proportion of population (a parameter)
Q = $1 - P$
p = proportion or percentage of a sample (a statistic) (Warning: Up until now p has been a probability; here it is not a probability but a proportion.
μ_p = mean of the sampling distribution of proportions
$\sigma_p \begin{cases} = \text{standard deviation of the sampling distribution of proportions} \\ = \text{the standard error of proportions} \end{cases}$
N = size of population
n = size of sample

The sampling distribution of proportions has the following characteristics:

1. $\mu_p = P$.

2. $\sigma_p = \sqrt{\dfrac{PQ}{n}} * \sqrt{\dfrac{N-n}{N-1}}$ if the population is finite, and the sample is taken without replacement and $N \leq 20n$.

    $\sigma_p = \sqrt{\dfrac{PQ}{n}}$ in all other cases. As before, this simpler formula can almost always be used, since N is almost always $> 20n$.

3. It is approximately normal if nP and nQ are both 15 or larger. Care is needed here since the proportion p is always between 0 and 1. If $P \approx 1/2$, the sampling distribution is approximately normal if $n \leq 30$. But if p is close to 0 (or 1), the sampling distribution of proportions will tend to have a longer tail to the right (or to the left), and n must be increased to get an approximation to a normal curve.

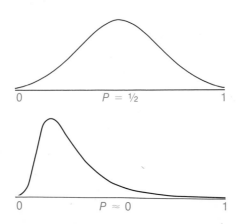

$0 \qquad\qquad P = \frac{1}{2} \qquad\qquad 1$

$0 \qquad\qquad P \approx 0 \qquad\qquad 1$

Example 1

In the town of Peacock there are three registered voters: A, B, and C. A will vote for M for governor, B and C will not vote for M. Consider the sampling distribution of proportions of samples of size 2 taken **with replacement.** Let P be the proportion who will vote for M.

$P = 1/3$, $Q = 2/3$, $N = 3$, $n = 2$.

Sample	p
A, A	1
A, B	1/2
A, C	1/2
B, A	1/2
B, B,	0
B, C	0
C, A	1/2
C, B	0
C, C	0

Sampling distribution of proportions

p	f	fp	fp^2
1	1	1	1
1/2	4	2	1
0	4	$\dfrac{0}{3}$	$\dfrac{0}{2}$
	9		

$$\mu_p = 3/9 = 1/3$$

$$\sigma_p = \sqrt{\frac{2 - 3^2/9}{9}} = \sqrt{\frac{1}{9}} = \frac{1}{3}$$

(a) $\mu_p = P \; (= 1/3)$.

(b) $\sqrt{\dfrac{PQ}{n}} = \sqrt{\dfrac{\frac{1}{3} \cdot \frac{2}{3}}{2}} = \sqrt{\dfrac{1}{9}} = \dfrac{1}{3}$, so $\sigma_p = \sqrt{\dfrac{PQ}{n}}$.

(c) The sampling distribution of proportions is not normal here; nP, which equals 2/3, is too small.

Now let's look at an example without replacement.

Example 2

A population consists of 6 people: D, E, F, G, H, I, of whom two (D, E) wear glasses and four (F, G, H, I) do not wear glasses.

P = proportion of population who wear glasses = 2/6 = 1/3.
Q = proportion of population who do not wear glasses = $1 - P = 1 - 1/3 = 2/3$.

Consider all samples of size 3 taken without replacement from this population. There are $_6C_3 = 20$ of them. For each sample, p = proportion who wear glasses. order not import.

proportion w/ glasses

Sample	p	frequency	fp	$p - \mu_p$	$(p - \mu_p)^2$	$f(p - \mu_p)^2$
FGH FGI FHI GHI	0×4 =	0	$-\dfrac{1}{3}$	$\dfrac{1}{9}$	$\dfrac{4}{9}$
D FG D FH D FI D GH D GI D HI E FG E FH E FI E GH E GI E HI	$\dfrac{1}{3} \times 12$ =	4	0	0	0
DE F DE G DE H DE I	$\dfrac{2}{3} \times 4$ =	$\dfrac{8}{3}$ $\dfrac{\overline{20}}{3}$	$\dfrac{1}{3}$	$\dfrac{1}{9}$	$\dfrac{4}{9}$ $\dfrac{8}{9}$
	20				

The box outlines the sampling distribution of proportions.

1. $\mu_p = \dfrac{20/3}{20} = 1/3 = P.$

2. $\sigma_p = \sqrt{\dfrac{\Sigma f(p - \mu_p)^2}{\Sigma f}} = \sqrt{\dfrac{8/9}{20}} = \sqrt{\dfrac{8}{9 * 20}} = \sqrt{\dfrac{2}{45}}$

   $\sigma_p = \sqrt{\dfrac{PQ}{n}} \sqrt{\dfrac{N - n}{N - 1}} = \sqrt{\dfrac{1/3 * 2/3}{3}} * \sqrt{\dfrac{6 - 3}{6 - 1}} = \sqrt{\dfrac{2}{9 * 3}} * \sqrt{\dfrac{3}{5}} = \sqrt{\dfrac{2}{45}}.$

3. Histogram of the sampling distribution of proportions:
   (*n* is too small for the distribution to be normal)

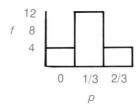

10.2   INTERVAL ESTIMATES OF PROPORTIONS

With the help of a normal curve sketch, you should now be able to solve problems such as this:

Example 1

If $P = .4$ and $N = 1,000$, the probability is .95 that a sample of size 30 will have a proportion p between what two values?

$P = .4, Q = .6, N = 1,000, n = 40.$

$$\sigma_p = \sqrt{\frac{PQ}{n}} = \sqrt{\frac{.4 * .6}{40}} = \sqrt{\frac{.24}{40}} = .077$$

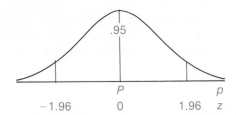

Probability $= .95$, so $A(z) = .475$ and $z = \pm 1.96$.

As always, the transformation from any scores to standard or z scores is made by subtracting the mean of the distribution ($\mu_p = P$) and dividing by its standard deviation (σ_p).

$$z = \frac{p - P}{\sigma_p}$$

$$\pm 1.96 = \frac{p - .4}{.077}$$

$p = .4 \pm 1.96 * .077 = .40 \pm .15 = .25$ or $.55$.

There is a .95 probability that the sample proportion will be between .25 and .55.

A more interesting problem, however, is this one:

Example 2

A poll is conducted to find out how many voters will vote for candidate A for President. A random sample is taken of 1000 voters, and it is found that 520 plan to vote for Candidate A. Find a 95 per cent confidence interval for the proportion that will vote for candidate A.

We start out with the same kind of reasoning as for confidence intervals of means: 95 per cent of the scores of any normal distribution are within 1.96 standard deviation units of the mean of the distribution.

The sampling distribution of proportions is approximately a normal distribution ($np = 1000 * .52 > 15$, so is $nq = 1000 * .48$), so there is a .95 probability that the particular random sample described in this example has a proportion $p = \frac{520}{1000} = .52$ which is within 1.96 standard deviation units or 1.96 σ_p score units of the unknown P.

$$z = \frac{p - P}{\sigma_p}$$

$$\pm 1.96 = \frac{.52 - P}{\sigma_p}$$

$$P = .52 \pm 1.96 \, \sigma_p$$

It is in finding the value of σ_p that we run into trouble. The standard error of the mean did not depend on μ, but the standard error of proportion does depend on P: $\sigma_p = \sqrt{\dfrac{PQ}{n}}$. Fortunately, $\sqrt{pq/n}$ **can be used as an approximation to** σ_p when P is unknown, if n is sufficiently large (≥ 30, say).

In the problem we are solving,

$$P = .52 \pm 1.96\,\sigma_p = .52 \pm 1.96 * \sqrt{\dfrac{.52 * .48}{1000}}$$

$$= .52 \pm 1.96 * \sqrt{.0002497}$$

$$= .52 \pm 1.96 * .0158$$

$$= .52 \pm .03$$

The 95 per cent confidence interval for P is 49 to 55 per cent.

Note that if there are only two candidates for President, this result does not predict the winner, since the 50 per cent needed to win is inside the confidence interval.

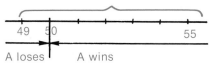

What can be done? Try the same problem if the sample size is 4,000, and again 52 per cent plan to vote for candidate A.

Answer

If $n = 4000$, $\sigma_p = \sqrt{.52 * .48/4000} = .0079$, so $P = .52 \pm 1.96 * .0079 = .52 \pm .015$; the 95 per cent confidence interval for P is 50.5 to 53.5 per cent; you have 95 per cent confidence that candidate A will win the election.

Example 3

One thousand people turn up to see a play. You discover that 42 of the first 90 people to arrive are women. Find a 95 per cent confidence interval for the percentage of women in the entire audience.

OUCH! Find the trap before you read on.

Is this a representative sample? Everything we have done earlier is predicated on random samples. Maybe women always come late to plays. Don't misuse statistics!

Example 4

Work out Example 3, but with a random sample of 90 people of whom 42 are women.

1. $p = \dfrac{42}{90}$, $q = \dfrac{48}{90}$; $n = 90$.

2. $\sigma_p = \sqrt{\dfrac{pq}{n}} \sqrt{\dfrac{N-n}{N-1}} =$

$$\sqrt{\dfrac{42/90 * 48/90}{90}} \sqrt{\dfrac{1000-90}{1000-1}} = .0502.$$

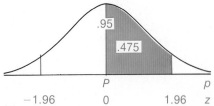

3.   At a 95 per cent confidence level, $z = \pm 1.96$.

4. $z = \dfrac{p-P}{\sigma_p}$

$$\pm 1.96 = \dfrac{42/90 - P}{.050}.$$

$P = 42/90 \pm 1.96 * .050 = .467 \pm .098 = .369$ or $.565$. A 95 per cent confidence interval for the percentage of women in the audience is 37 to 56 per cent.

10.3   THE SAMPLING DISTRIBUTION OF DIFFERENCES OF MEANS

Consider two different populations (one consists of the weights of all men over 17 in the United States, the other of the weights of all women over 17 in the United States, for example). The first population (X) has mean μ_X and standard deviation σ_X, the second (Y) has mean μ_Y and standard deviation σ_Y. (Do not confuse μ_X and $\mu_{\bar{X}}$, in the former, the subscript X distinguishes one population [X scores] from another [Y scores], while in the latter the subscript indicates the mean of the sampling distribution of means.) From the first population take a sample of size n_X and compute its mean $\bar{X}$; from the second population take independently a sample of size n_Y and compute $\bar{Y}$; then determine $\bar{X} - \bar{Y}$. Do this for all possible pairs of samples that can be chosen independently from the two populations. The differences, $\bar{X} - \bar{Y}$, are a new set of scores which form the **sampling distribution of differences of means.**

An example will clarify these ideas.

Example 1

Population X: 1, 3, 5
Population Y: 10, 20, 30, 40

(a)   Taking samples of size 2 from the first population and of size 1 from the second, form the sampling distribution of differences of means.

(b)   Compute μ_X, σ_X^2, $\sigma_{\bar{X}}^2$, μ_Y, σ_Y^2, $\sigma_{\bar{Y}}^2$ (note that variances are asked for, not standard deviations), the mean of the sampling distribution of differences of means $\mu_{(\bar{X}-\bar{Y})}$ and the standard error of differences of means $\sigma_{(\bar{X}-\bar{Y})}$. (Fractions will be used instead of decimals so that relations between these will be more visible.)

(a)

From Population X	From Population Y	$\bar{X}$	$\bar{Y}$	$\bar{X} - \bar{Y}$
	Sample			
1, 3	10	2	10	− 8
1, 3	20	2	20	−18
1, 3	30	2	30	−28
1, 3	40	2	40	−38
1, 5	10	3	10	− 7
1, 5	20	3	20	−17
1, 5	30	3	30	−27
1, 5	40	3	40	−37
3, 5	10	4	10	− 6
3, 5	20	4	20	−16
3, 5	30	4	30	−26
3, 5	40	4	40	−36

$n_X = 2$, $n_Y = 1$. The sampling distribution for differences of means for samples of sizes 2 and 1 from populations X and Y, respectively, is enclosed in a box at the right above.

(b) $\mu_X = 3$, $\sigma_X^2 = 8/3$, $n_X = 2$, $\sigma_{\bar{X}}^2 = \dfrac{8/3}{2} \cdot \dfrac{3-2}{3-1} = 2/3$. (Sampling is without

replacement, so the correction factor $\dfrac{N-n}{N-1}$ is needed here; note that the correction factor is squared since the variance, not the standard deviation, is being found.)

$$\mu_Y = 25, \ \sigma_Y^2 = 125, \ n_Y = 1, \ \sigma_{\bar{Y}}^2 = \frac{125}{1} \cdot \frac{4-1}{4-1} = 125$$

$$\mu_{\bar{X}} - \mu_{\bar{Y}} = -22, \ \sigma_{(\bar{X}-\bar{Y})} = \sqrt{377/3}$$

You should keep familiar with means and standard deviations by checking all these values. Remember that, for a sampling distribution, the denominator of the standard error is the number of scores in the distribution, as for a population—not 1 less than this number, as for a sample.)

Now check the following relations:

1. $\mu_{(\bar{X}-\bar{Y})} = \mu_X - \mu_Y$. ($-22 = 3 - 25$.)

2. $\sigma_{(\bar{X}-\bar{Y})} = \sqrt{\sigma_{\bar{X}}^2 + \sigma_{\bar{Y}}^2}$, since $\sqrt{377/3} = \sqrt{2/3 + 125}$.

3. n_X and n_Y are too small to hope for a normal distribution.

The characteristics of the sampling distribution of differences of means are:

1. The mean of the sampling distribution of differences of means equals the difference of the population means:

$$\mu_{(\bar{X}-\bar{Y})} = \mu_X - \mu_Y$$

2. The standard deviation of the sampling distribution of differences of means, also called the standard error of differences of means, is denoted by $\sigma_{(\bar{X}-\bar{Y})}$.

$$\sigma_{(\bar{X}-\bar{Y})} = \sqrt{\sigma_{\bar{X}}^2 + \sigma_{\bar{Y}}^2}$$

where $\sigma_{\bar{X}}$ is the standard error of the mean of the first population and $\sigma_{\bar{Y}}$ is the standard error of the mean of the second population.

Note the **plus** sign between $\sigma_{\bar{X}}^2$ and $\sigma_{\bar{Y}}^2$, even though we are dealing with **differences** of means.

Note: Since you need $\sigma_{\bar{X}}^2 = \dfrac{\sigma_X^2}{n}$, don't waste time taking a square root and finding $\sigma_{\bar{X}}$; you will just have to square again.

3. The distribution is close to normal if $n_X \geq 30$ and $n_Y \geq 30$ or if both populations are normal.

Have you the courage to work out an example with both n_X and n_Y equal to 30 or more?!

10.4 USE OF THE SAMPLING DISTRIBUTION OF DIFFERENCES OF MEANS

In using this sampling distribution, it must be emphasized very strongly that the two samples must be picked independently. If a random sample of 36 students is tested for reaction time before and after drinking an ounce of alcohol, you have two sets of scores, but you do not have two independent samples. Techniques for making inferences from two sets of related or paired scores will be studied in Chapters 12, 15, 16, and 17.

Another limitation on the use of this sampling distribution of means is that n_X and n_Y must both be at least 30 if s_X^2 and s_Y^2 are used in finding $\sigma_{\bar{X}}^2$ and $\sigma_{\bar{Y}}^2$. What if this requirement is not met? A different method will be used, using a distribution which is not normal. Wait until Chapter 12.

Example 1

The mean IQ of 1200 Defoe College students is 122 with a standard deviation of 6, while 2000 students from Daniel College have a mean IQ of 118 with a standard deviation of 5. What is the probability that the mean IQ of a random sample of 36 Defoe students will be at least 6 points higher than the mean IQ of a random sample of 49 Daniel students? Try working this out before you read on.

$$\text{Defoe: } \mu_X = 122, \sigma_X = 6, n_X = 36, \sigma_{\bar{X}}^2 = \frac{6^2}{36} = 1$$

$$\text{Daniel: } \mu_Y = 118, \sigma_Y = 5, n_Y = 49, \sigma_{\bar{Y}}^2 = \frac{5^2}{49} = \frac{25}{49}$$

$$\mu_X - \mu_Y = 122 - 118 = 4$$

$$\sigma_{(\bar{X}-\bar{Y})} = \sqrt{1 + \frac{25}{49}} = 1.23$$

$$\bar{X} - \bar{Y} = 6$$

$$z = \frac{(\bar{X}-\bar{Y}) - (\mu_X - \mu_Y)}{\sigma_{(\bar{X}-\bar{Y})}} = \frac{6-4}{1.23} = 1.63$$

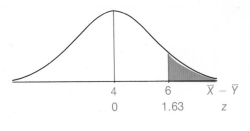

$$p(\bar{X} - \bar{Y} > 6) = p(z > 1.63) = .5 - p(0 < z < 1.63) = .5 - .448 = .052.$$

The more interesting problem, as before, puts the emphasis the other way around: If the means of two samples, each chosen independently and at random from a different population, are known, and if the standard deviations, either of the populations or of the samples, are known, what statement can be made about the differences of the means of the populations from which the samples are taken?

Example 2

A psychologist tests in a maze 36 rats whose diet contains no Vitamin A, and finds their mean time to "solve" the maze is 4.0 minutes with a standard deviation of .50 minute; then he tests in the same maze 50 rats whose diet is rich in Vitamin A, and finds a mean time of 3.2 minutes with a standard deviation of .40 minute. At a 95 per cent confidence level what can he say about the difference in mean times for solving the maze of rats with Vitamin A – free and Vitamin A – rich diets?

Population X (rats with no Vitamin A):

$$\overline{X} = 4.0,\ n_X = 36,\ \sigma_X \text{ is unknown},\ s_X = .50,\ \sigma_{\overline{X}}{}^2 = \frac{s_X{}^2}{n_X} = \frac{.50^2}{36} = .0069.$$

Population Y (rats with Vitamin A):

$$\overline{Y} = 3.2,\ n_Y = 50,\ s_Y = .40,\ \sigma_{\overline{Y}}{}^2 = \frac{s_Y{}^2}{n_Y} = \frac{.40^2}{50} = .0032.$$

$$\overline{X} - \overline{Y} = 4.0 - 3.2 = .8$$

$$\sigma_{(\overline{X}-\overline{Y})} = \sqrt{\sigma_{\overline{X}}{}^2 + \sigma_{\overline{Y}}{}^2} = \sqrt{.0069 + .0032} = .101$$

At a 95 per cent confidence level, $z = \pm 1.96$.

$$z = \frac{(\overline{X} - \overline{Y}) - (\mu_X - \mu_Y)}{\sigma_{(\overline{X}-\overline{Y})}}$$

$$\pm 1.96 = \frac{.8 - (\mu_X - \mu_Y)}{.101}$$

$$\mu_X - \mu_Y = .8 \pm .198 = .60 \text{ or } 1.00.$$

The 95 per cent confidence interval for the difference in mean times of rats with and without Vitamin A in their diets is .60 to 1.00 minute (36 to 60 seconds).

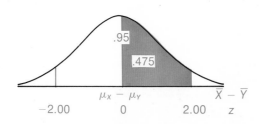

Example 3

If a random sample of 50 non-smokers has a mean life of 76 years with a standard deviation of 8 years, and a random sample of 65 smokers live 68 years with a standard deviation of 9 years, find a 95 per cent confidence interval for the difference of mean lifetime of non-smokers and smokers.

Population X (non-smokers):

$$n_X = 50, \bar{X} = 76, s_X = 8, \sigma_{\bar{X}}^2 = \frac{8^2}{49} = 1.31.$$

Population Y (smokers):

$$n_Y = 65, \bar{Y} = 68, s_Y = 9, \sigma_{\bar{Y}}^2 = \frac{9^2}{64} = 1.27.$$

$$\bar{X} - \bar{Y} = 76 - 68 = 8$$
$$\sigma_{(\bar{X}-\bar{Y})} = \sqrt{1.31 + 1.27} = 1.61$$

At a 95 per cent confidence level, $z = \pm 1.96$.

$$z = \frac{(\bar{X} - \bar{Y}) - (\mu_X - \mu_Y)}{\sigma_{(\bar{X}-\bar{Y})}}$$

$$\pm 1.96 = \frac{8 - (\mu_X - \mu_Y)}{1.61}$$

$$\pm 3.2 = 8 - (\mu_X - \mu_Y)$$

$$\mu_X - \mu_Y = 8 \pm 3.2$$

$$\mu_X - \mu_Y = 4.8 \text{ or } 11.2.$$

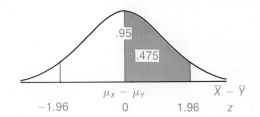

The upper confidence limit is 11.2; the lower confidence limit is 4.8; the 95 per cent confidence interval is 4.8 to 11.2. On the basis of these two samples, I am 95 per cent confident that the mean lifetime of non-smokers is 4.8 to 11.2 years longer than the mean lifetime of smokers.

10.5 SAMPLING DISTRIBUTION OF DIFFERENCE OF PROPORTIONS

Consider two different populations again (voters in New York City and those in Chicago, for example). The first population has a proportion P_X who fit in a certain category (New York City voters who will vote for candidate A for President) and a proportion $Q_X = 1 - P_X$ who do not fit in that category (will not vote for candidate A for President); the same proportions in the second population are P_Y and Q_Y. Take a sample of size n_X from population X (all voters in New York City) and find the proportion p_X who fit a given category (will vote for candidate A for President). Then take **independently** a sample of size n_Y from population Y (all voters in Chicago) and find the proportion p_Y who fit the given category. Find the difference $p_X - p_Y$, and you have **one** score in the sampling distribution of differences of proportions. To find all the scores, go through the same process with all possible combinations of samples of sizes n_X and n_Y, respectively, chosen independently from the two populations. None of this should surprise you, nor should the following results:

1. The mean of the sampling distribution of proportions, $\mu_{(p_X-p_Y)}$, equals the difference in the proportions P_X and P_Y in the X and Y populations:

$$\mu_{(p_X-p_Y)}=P_X-P_Y.$$

2. $\sigma_{(p_X-p_Y)}$, the standard deviation of this new distribution (also called the standard error of differences of proportions), equals $\sqrt{\sigma_{p_X}{}^2 + \sigma_{p_Y}{}^2}$, where σ_{p_X} is the standard error of the proportion of the X population: $\sigma_{p_X} = \sqrt{\dfrac{P_XQ_X}{n_X}}$†; similarly, $\sigma_{p_Y} = \sqrt{\dfrac{P_YQ_Y}{n_Y}}$‡

    Note that

$$\left(\sqrt{\frac{P_XQ_X}{n_X}}\right)^2 + \left(\sqrt{\frac{P_YQ_Y}{n_Y}}\right)^2 = \frac{P_XQ_X}{n_X} + \frac{P_YQ_Y}{n_Y}$$

so don't bother to find square roots since you are about to square. Therefore,

$$\sigma_{(p_X-p_Y)} = \sqrt{\frac{P_XQ_X}{n_X} + \frac{P_YQ_Y}{n_Y}}$$

Again note the + sign under the square root, even though this is a distribution of differences.

3. This new distribution is approximately normal if both n_X and n_Y are ≥ 30.

Example 1

There are 3 voters in the town of Peacock: A, B, and C. A will vote for M for governor; B and C will not vote for M. The town of Boots (quite a metropolis) also has 3 voters: D, E, and F. D and E will vote for M, and F against him. Consider the sampling distribution of differences of proportions if each sample consists of 2 voters from Peacock and 2 from Boots.

$$P_X = 1/3, \; Q_X = 2/3, \; N_X = 3, \; n_X = 2, \; \sigma_{p_X} = \sqrt{\frac{P_XQ_X}{n_X}} \sqrt{\frac{N_X - n_X}{N_X - 1}}$$

$$= \sqrt{\frac{1/3 * 2/3}{2}} \sqrt{\frac{3 - 2}{3 - 1}} = \sqrt{\frac{1}{18}}$$

$$P_Y = 2/3, \; Q_Y = 1/3, \; N_Y = 3, \; n_Y = 2, \; \sigma_{p_Y} = \sqrt{\frac{P_YQ_Y}{n_Y}} \sqrt{\frac{N_Y - n_Y}{N_Y - 1}}$$

$$= \sqrt{\frac{2/3 * 1/3}{2}} \sqrt{\frac{3 - 2}{3 - 1}} = \sqrt{\frac{1}{18}}$$

---

†Or $\sqrt{\dfrac{P_XQ_X}{n_X}} \sqrt{\dfrac{N_X - n_X}{N_X - 1}}$.

‡Or $\sqrt{\dfrac{P_YQ_Y}{n_Y}} \sqrt{\dfrac{N_Y - n_Y}{N_Y - 1}}$.

Sample		Proportion for M		
FROM PEACOCK	FROM BOOTS	FROM PEACOCK (p_1)	FROM BOOTS (p_2)	$p_1 - p_2$
AB	DE	1/2	1	−1/2
AB	DF	1/2	1/2	0
AB	EF	1/2	1/2	0
AC	DE	1/2	1	−1/2
AC	DF	1/2	1/2	0
AC	EF	1/2	1/2	0
BC	DE	0	1	−1
BC	DF	0	1/2	−1/2
BC	EF	0	1/2	−1/2

The sampling distribution of differences of proportions is enclosed in a box at the right above; it is shown in a more palatable fashion in the box below.

$p_X - p_Y$	f	$f(p_X - p_Y)$	$f(p_X - p_Y)^2$
−1	1	−1	1
−1/2	4	−2	1
0	4	0	0
	9	−3	2

1. $\mu_{(p_X - p_Y)} = \dfrac{-3}{9} = \dfrac{-1}{3}$; $P_X - P_Y = \dfrac{1}{3} - \dfrac{2}{3} = -\dfrac{1}{3}$; $\mu_{(p_X - p_Y)} = P_X - P_Y$.

2. $\sigma_{(p_X - p_Y)} = \sqrt{\dfrac{2 - (-3)^2/9}{9}} = \sqrt{2/9 - 1/9} = 1/3$;

   $\sqrt{\sigma_{p_X}^2 + \sigma_{p_Y}^2} = \sqrt{1/18 + 1/18} = 1/3$; $\sigma_{(p_X - p_Y)} = \sqrt{\sigma_{p_X}^2 + \sigma_{p_Y}^2}$.

3. As before, n_X and n_Y are too small for the sampling distribution to be a normal distribution.

Example 2

Each of two groups consists of 100 patients who have tuberculosis. A new drug is given to the first group but not to the second (the control group). It is found that in the first group 75 people recover, but only 60 in the second group. Find 95 per cent confidence limits for the difference in the proportion of all patients with tuberculosis who recover. Try this problem before you read on.

$$p_X = .75, \; q_X = .25, \; n_X = 100, \; \sigma_{p_X}^2 = \frac{p_X q_X}{n_X} = \frac{.75 * .25}{100} = .001875$$

$$p_Y = .60, \; q_Y = .40, \; n_Y = 100, \; \sigma_{p_Y}^2 = \frac{p_Y q_Y}{n_Y - 1} = \frac{.60 * .40}{100} = .002400$$

$$p_X - p_Y = .75 - .60 = .15$$

$$\sigma_{(p_X - p_Y)} = \sqrt{\sigma_{p_X}^2 + \sigma_{p_Y}^2} = \sqrt{.001875 + .00240} = .065$$

At a 95 per cent confidence level, $z = \pm 1.96$.

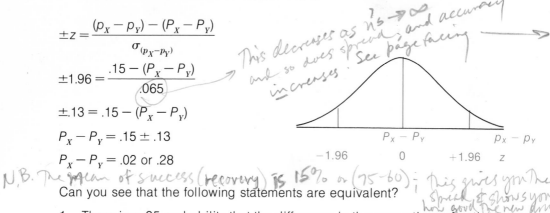

This decreases as n's → ∞ and accuracy and so does spread; and see page facing → increases.

$$\pm z = \frac{(p_X - p_Y) - (P_X - P_Y)}{\sigma_{(p_X - p_Y)}}$$

$$\pm 1.96 = \frac{.15 - (P_X - P_Y)}{.065}$$

$$\pm .13 = .15 - (P_X - P_Y)$$

$$P_X - P_Y = .15 \pm .13$$

$$P_X - P_Y = .02 \text{ or } .28$$

N.B. The mean of success (recovery) is 15% or (75-60); This gives you the spread & shows you how good the new drug is.

Can you see that the following statements are equivalent?

1. There is a .95 probability that the difference in the proportions of patients treated with the new drug who recover and of those without it is between .02 and .28.

2. There is a .95 probability that between 2 and 28 per cent more patients recover when treated with the new drug.

3. The 95 per cent confidence interval for the difference in the proportions of patients treated with the new drug who recover and of those without it is 2 to 28 per cent.

10.6 SUMMARY OF FORMULAS FOR STATISTICAL INFERENCE

Inference about	Formula	Standard error†
Mean		
If σ is known	$z = \dfrac{\overline{X} - \mu}{\sigma_{\overline{X}}}$	$\sigma_{\overline{X}} = \dfrac{\sigma}{\sqrt{n}}$
If σ is unknown	$z = \dfrac{\overline{X} - \mu}{\sigma_{\overline{X}}}$	$\sigma_{\overline{X}} = \dfrac{s}{\sqrt{n}}$
Proportion		
If P is known	$z = \dfrac{p - P}{\sigma_p}$	$\sigma_p = \sqrt{\dfrac{PQ}{n}}$
If P is unknown	$z = \dfrac{p - P}{\sigma_p}$	$\sigma_p = \sqrt{\dfrac{pq}{n}}$
Differences of means	$z = \dfrac{(\overline{X} - \overline{Y}) - (\mu_X - \mu_Y)}{\sigma_{(\overline{X} - \overline{Y})}}$	$\sigma_{(\overline{X} - \overline{Y})} = \sqrt{\sigma_{\overline{X}}^2 + \sigma_{\overline{Y}}^2}$
Differences of proportions	$z = \dfrac{(p_X - p_Y) - (P_X - P_Y)}{\sigma_{(p_X - p_Y)}}$	$\sigma_{(p_X - p_Y)} = \sqrt{\sigma_{p_X}^2 + \sigma_{p_Y}^2}$

†The factor $\sqrt{\dfrac{N - n}{N - 1}}$ has been omitted from the first four standard error formulas; you should know when it is needed.

10.7 SIZE OF A SAMPLE

In the problems you have solved until now, the size of the sample has been stated. In real life, however, a research worker realizes that in general† a larger sample may increase his accuracy—if the whole population is tested there is no need for statistical inference—but that testing a larger sample will take more time and cost more money. How large a sample is absolutely necessary for a given degree of accuracy is a very practical and necessary question for a statistician. How many people should be polled if a political scientist is to estimate the proportion who will vote for candidate X within 1 per cent and with 98 per cent confidence? How many rats should a psychologist test if he needs to know their mean reaction time within .6 second at 95 per cent confidence?

Note that in each of these questions there is a statement of the expected accuracy of the answer ("within 1 per cent," "within .6 second"). Before sample size can be determined, the desired accuracy must be determined. But even when this is done, the desired accuracy cannot be guaranteed: "With 98 per cent confidence" or "with 95 per cent confidence" means that, for every 100 samples that might be chosen, the confidence intervals computed will in 2 or 5 cases **not** include the population parameter; in these cases, the desired accuracy will not be attained. The only way of guaranteeing the stated accuracy is to measure the whole population.

We shall take up in turn determination of sample size for estimating population mean, proportion, differences of means, and differences of proportions. Do not get lost in the woods. There is a common pattern for the four cases, and each involves the following steps:

1. Determine the required accuracy—or, to put it another way, decide what error E in the results is permissible. Do you need to know the population mean within 1 or 3 score units, or within .1 or .05 standard deviation unit?

2. Set up an equation which involves both sample size n and the permissible error E. This equation will include population parameters that must be estimated.

3. Solve the equation for n.

10.8 SAMPLE SIZE FOR ESTIMATING POPULATION MEANS

If I measure the length of this line _____ to the nearest inch, I might record its length as 2 in. $\pm$ .5 in. This means the true length of the line is between 1.5 in. and 2.5 in., or that the true length is 2 in. with a possible error of .5 in. (in either direction—that is, 2 in. may be too large or too small, but is within .5 in. of the actual length of the line).

Similarly, we have been finding confidence limits in the form $\mu = 70 \pm .8$; the population mean is 70 with a possible error of .8 in either direction. In the general case,

$\mu = \overline{X} - z\sigma_{\overline{x}} = \overline{X} - z\dfrac{\sigma}{\sqrt{n}}$ where z has two possible values—one is positive, the other

negative. If we give z only a positive value, then this equation can be written as μ

$= \overline{X} \pm |z|\dfrac{\sigma}{\sqrt{n}}$ ‡ The error in the value of μ, then, is $|z|\dfrac{\sigma}{\sqrt{n}}$.

---

†Surprisingly, there are a few very special cases when increasing the sample size will <u>not</u> <u>increase the accuracy</u> of the statistical inference. We shall not consider these cases, however.

‡See page 46 for a reminder about absolute values.

A psychologist, economist, or other scientist setting up a study decides the maximum error E he is willing to accept, and then solves the equation $|z| * \sigma_{\bar{x}} = E$ to find the necessary sample size. But since $\sigma_{\bar{x}} = \sigma/\sqrt{n}$, this becomes

$$|z| \frac{\sigma}{\sqrt{n}} = E$$

$$z^2 \frac{\sigma^2}{n} = E^2$$

$$n = \frac{z^2 \sigma^2}{E^2}$$

Since n must be an integer (you can't have 44.2 scores), take the next larger whole number ($n = 45$ in this example).

Example 1

You are a manufacturer of light bulbs, and know that the standard deviation of the lifetimes of your light bulbs is 100 hours. How large a sample must you test to be 95 per cent confident that the error in the estimated mean life of your light bulbs is less than 10 hours? (or ±2.5% of mean life)
$\sigma = 100$ hours; at a 95 per cent confidence level, $z = \pm 1.96$; $E = 10$ hours.

$$n = \frac{z^2 \sigma^2}{E^2} = \frac{(1.96)^2 100^2}{10^2} = 384.2; \; n = 385$$

Does this seem to be a very large number of light bulbs to test? It is because the error (10 hours) you are willing to allow is small. If the mean life of your bulbs is approximately 2,000 hours you are asking that it be determined within ±.5 per cent. If you are willing to accept an error twice as large (20 hours), then the size of the sample will be only one-fourth as large: you need test only 97 bulbs.

Example 2

You are still manufacturing light bulbs, but don't know the standard deviation of the lifetimes of your bulbs. How large a sample must you test to be 95 per cent confident that the error in the estimated mean life of all your light bulbs is less than $.1\sigma$?

$$|z| \frac{\sigma}{\sqrt{n}} = .1\sigma = E$$

$$n = \frac{z^2 \sigma^2}{E^2} = \frac{(1.96)^2 \sigma^2}{(.1\sigma)^2} = \frac{3.842\sigma^2}{.01\sigma^2} = 384.2; \; n = 385$$

If σ is unknown, the appropriate sample size may be determined if the error is expressed in standard deviation units. The difficulty is that you don't know the error in score units until after you have chosen your sample, and tested it, and determined s. Then $\sigma_x = \frac{\sigma}{\sqrt{n}} \approx \frac{s}{\sqrt{n}}$, so now σ (and therefore E) may be estimated: $\sigma \approx s$, and now $E = .1\sigma$ may be determined.

Example 3

Continue to manufacture light bulbs, again with σ unknown. How large a sample must you test to be 95 per cent confident that the error in the estimated mean life of all your light bulbs is less than 10 hours?

$$n = \frac{(1.96)^2 \sigma^2}{10^2}$$

Here the strategem that helped in the second example cannot be used. There are 2 unknowns, σ and n, and no knowledge of either. What is to be done? If you can estimate σ on the basis of previous experience, then you can use this estimated σ to determine n. If not, the answer is to make a small pilot study—test a sample of, say, 30 bulbs and determine s for that sample. Then use s as an approximation of σ. If 30 bulbs are tested and have a standard deviation of 90 hours, then

$$n \approx \frac{(1.96)^2 90^2}{10^2} = 311.2; \; n = 312$$

Note that the mean $\overline{X}$ of the pilot sample is not used.

You have already tested 30 and need to test 282 more. What if you don't have the time or money to test this many? You must either give up the experiment or you must realize that, if your sample is smaller, you cannot achieve the accuracy you had hoped to have.

10.9 SAMPLE SIZE FOR PROPORTIONS

Using the same reasoning as for means, the equation $\pm z = \dfrac{p - P}{\sigma_p}$ is solved for P: $P = p \pm z\sigma_p$. The error term is $|z|\sigma_p$, so you would like to solve the equation $|z|\sqrt{\dfrac{PQ}{n}} = E$ for n:

$$n = \frac{z^2 PQ}{E^2}$$

Again you have a population parameter, P, which is not known until after you have completed the experiment. There are three things you can do:

1. If you can estimate the range of P on the basis of previous experience, use the maximum value of PQ for P in this range. This means use for P that value in the range which is closest to 1/2. (If you think P is between .2 and .3, let $P = .3$, $Q = .7$.)

2. As before, you can make a pilot study, determine the proportion p in this sample, and use pq as an approximation to PQ in the equation, thus determining how much the sample should be enlarged.

3. The maximum value of PQ is 1/4. Algebra or calculus is needed to prove this, but a graph should be quite persuasive: recall that $PQ = P(1 - P)$.

P	PQ
0	.00
.1	.09
.2	.16
.3	.21
.4	.24
.5	.25 = 1/4
.6	.24
.7	.21
.8	.16
.9	.09
1.0	.00

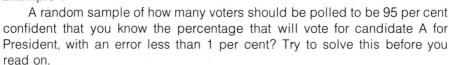

If $PQ \le 1/4$, then $n = \dfrac{z^2 PQ}{E^2} \le \dfrac{z^2}{4E^2}$; use

$$n = \frac{z^2}{4E^2}$$

as your estimate of n. Unless your sample proportion p turns out to be quite close to .5, the value of n you compute will be too large and you will do unnecessary work, as the next example illustrates. So use this method only if the first two cannot be used.

Example 1

A random sample of how many voters should be polled to be 95 per cent confident that you know the percentage that will vote for candidate A for President, with an error less than 1 per cent? Try to solve this before you read on.

At a 95 per cent confidence level, $z = \pm 1.96$; $E = .01$.

$$n = \frac{(1.96)^2}{4(.01)^2} = 9,600$$

$PQ = .5(.5)$ maximized

If, on the other hand, a pilot study of 200 voters shows that only 10 per cent will vote for A, then 3,460 is a large enough sample to be 95 per cent confident that the error is less than 1 per cent:

$$PQ \approx pq = .1 * .9 = .09$$

$$n = \frac{(1.96)^2 * .09}{(.01)^2} = 3,460$$

10.10★ SAMPLE SIZE FOR DIFFERENCES OF MEANS

Assume that both samples have the same size: $n_X = n_Y$ (=n); $z = \dfrac{(\bar{X} - \bar{Y}) - (\mu_X - \mu_Y)}{\sigma_{(\bar{X} - \bar{Y})}}$, or $\mu_X - \mu_Y = (\bar{X} - \bar{Y}) \pm z\sigma_{(\bar{X} - \bar{Y})}$, so here the error term is $|z| * \sigma_{(\bar{X} - \bar{Y})}$. (Compare with $|z|\sigma_{\bar{X}}$ for means and $|z|\sigma_p$ for proportions.)

$$|z|\sigma_{(\bar{X}-\bar{Y})} = z\sqrt{\sigma_{\bar{x}}^2 + \sigma_{\bar{y}}^2} = z\sqrt{\frac{\sigma_x^2}{n} + \frac{\sigma_y^2}{n}} = \frac{z\sqrt{\sigma_x^2 + \sigma_y^2}}{\sqrt{n}} = E, \text{ or}$$

$$n = \frac{z^2(\sigma_x^2 + \sigma_y^2)}{E^2}$$

Example 1

An economist wishes to learn the difference in salaries of registered nurses in public and in private hospitals at the present time. From past studies he knows that the standard deviations are $900 and $1200 respectively for nurses in public and in private hospitals. How large should his samples be if he wants to determine the difference within $300 with 98 per cent confidence? At a 98 per cent confidence level, $z = \pm 2.33$.

$$n = \frac{(2.33)^2(900^2 + 1200^2)}{300^2}$$

$$= \frac{(2.33)^2(3^2 * 300^2 + 4^2 * 300^2)}{300^2}$$

$$= (2.33)^2(9 + 16) = 135.7; \, n = 136$$

If σ_x and σ_y are unknown, and reasonable estimates cannot be made, then it is again necessary to carry out pilot studies with samples (of size 30, say?) for this purpose. Remember that, to use the methods we have studied up until now, your samples must be of size at least 30 if σ_x and σ_y are unknown; otherwise the sampling distribution of means is not a normal distribution. All the conclusions we have been able to draw, however, are based on the assumption that it is a normal distribution.

10.11★ SAMPLE SIZE FOR DIFFERENCES OF PROPORTIONS

Again, **assume that both samples have the same size:** $n_X = n_Y \, (= n)$;

$$z = \frac{(p_X - p_Y) - (P_X - P_Y)}{\sigma_{(p_X - p_Y)}}, \text{ or } P_X - P_Y = (p_X - p_Y) \pm z\sigma_{(p_X - p_Y)}.$$

$$|z|\sigma_{(p_X - p_Y)} < E.$$

The error term is $|z|\sigma_{(p_X - p_Y)}$.

$$z\sqrt{\frac{P_X Q_X}{n} + \frac{P_Y Q_Y}{n}} = E, \text{ or}$$

$$n = \frac{z^2(P_X Q_X + P_Y Q_Y)}{E^2}$$

There are, as with sample size for proportions, three things you can do:

1. If you can approximate P_X and P_Y, well and good; if you estimate a range for each, take the value nearest 1/2. (If you think P_X is between .1 and .3 and P_Y is between .6 and .8, then let $P_X = .3$, $P_Y = .6$, so $Q_X = .7$, $Q_Y = .4$.)
2. Make a pilot study, and approximate P_X and P_Y by the p_X, p_Y found in this preliminary study.
3. Assume $P_X Q_X = 1/4$ and $P_Y Q_Y = 1/4$; the formula then becomes

$$n = \frac{z^2}{2E^2}$$

Example 1

An editor suspects that a higher percentage of *New York Crow* readers read the sports page than do readers of the *New York Eagle*. How large should samples be if he wants to be 95 per cent confident that he knows the difference within 10 per cent?

At a 95 per cent confidence level, $z = \pm 1.96$; $E = .10$.

$$n = \frac{(1.96)^2}{2(.1)^2} = 192.1; \; n = 193$$

Each of his 2 samples (readers of the *Crow* and readers of the *Eagle*) must be at least 193.

Example 2

At a football game you find that 200 of a random sample of 600 fans are women; at a baseball game you find that 20 of a random sample of 100 fans are women. (a) Are your samples large enough to be 95 per cent confident that you know the difference in the proportion of women at the football and base-ball games with an error less than 5 per cent? (b) What is the minimum size of each sample for a 5 per cent error with 95 per cent confidence?

(a) $n_X = 600$, $p_X = 1/3$; $n_Y = 100$, $p_Y = 1/5$.
At a 95 per cent confidence level, $z = \pm 1.96$.
Is $|z| \sigma_{(p_X - p_Y)} < .05$?

$$|z| \sigma_{(p_X - p_Y)} = 1.96 \sqrt{\frac{1/3 * 2/3}{600} + \frac{1/5 * 4/5}{100}} = 1.96 \sqrt{\frac{2}{5400} + \frac{4}{2500}} = .09 > .05$$

No. The samples are too small.

(b) $$n = \frac{1.96^2 (1/3 * 2/3 + 1/5 * 4/5)}{(.05)^2} = 587.$$

Each sample should include 587 or more people if the error is to be at most 5 per cent with 95 per cent confidence.

10.12 VOCABULARY AND NOTATIONS

P	q
Q	μ_p
p	σ_p

N
standard error of proportion

$\sigma_{(\bar{X}-\bar{Y})}$
standard error of
    differences of means

$\mu_{(p_X-p_Y)}$
$\sigma_{(p_X-p_Y)}$
standard error of
    differences of proportions
error term

10.13   EXERCISES

1.   A population consists of 5 people: A, B, and C who drive VW's, and D and E who drive Cadillacs. Let *P* equal the proportion of the population which drives VW's.
        (a) Find the sampling distribution of proportions for samples of size 2 taken without replacement from this population.
        (b) Find μ_p and *P*, and show that they are equal.
        (c) Compute the standard error of proportions σ_p and show that it equals

$$\sqrt{\frac{PQ}{n}} \ \sqrt{\frac{N-n}{N-1}}.$$

2.   If 80 per cent of all college freshmen take a course in English composition, what is the probability that, in a random sample of 100 college freshmen, 75 to 85 per cent study English composition?

3.   A psychologist is studying the proportion of college students neither of whose parents attended college, and would like to estimate this proportion with 95 per cent confidence. He discovers that, in a random sample of 4,000 college students, the parents of 400 did not attend college. Help him with his estimate.

4.   A random sample of 200 Madison wage earners shows that 80 have an income over $9,000. Find a 90 per cent confidence interval for the percentage of all Madison wage earners who have incomes over $9,000.

5.   (a) Postcards were sent to 10,000,000 voters asking their choices for President. One-fifth of them answered; 56 per cent of those who returned the postcard reported they would vote for Mr. L. Find a 68 per cent confidence interval for the proportion of all voters who would vote for Mr. L.
        (b) In the election which followed, only 34 per cent voted for Mr. L. What is the probability that, in a random sample of 2,000,000, 56 per cent or more will vote for Mr. L.?
        (c) Can you explain the contradiction between (a) and (b)?

6.   A doctor tests a new drug on 100 patients with leukemia, chosen at random. After six years, 60 patients are alive. What proportion of all leukemia patients will be alive after six years of treatment with this drug, at a 95 per cent confidence level?

7.   The editors of *The Eagle Rises,* student newspaper of Oxo College, question a random sample of 200 subscribers about the editorials in the paper. 155 subscribers express dissatisfaction. At a 90 per cent confidence level, what proportion of subscribers are dissatisfied with the editorials?

8.   In a random sample of 100 city physicians, 77 have an income over $30,000. Find a 95 per cent confidence interval for the proportion of all city physicians who have incomes over $30,000.

9.   A drug treatment center tests a new withdrawal repressant on 100 heroin users, chosen at random. Of these, 36 patients display marked relief from withdrawal symptoms. Find a 95 per cent confidence interval for the proportion of all addicts who would find relief.

10.   A random sample of 150 General Motors workers shows that 105 own a General Motors vehicle. What proportion of all General Motors employees own a General Motors vehicle, at a 90 per cent confidence level?

11.   An economist finds that a random sample of 100 steelworkers have a mean weekly wage of $300 with a standard deviation of $50, while 80 automobile workers chosen at random have a mean wage of $250 with a standard deviation of $45. Find a 95 per cent confidence interval for the difference in wages of steelworkers and automobile workers.

12.   Two women, whose heights are 62 in. and 66 in., live in Dormitory X. Three men, whose heights are 70 in., 71 in., and 72 in., respectively, live in Dormitory Y. Samples of size 1 are taken from each dormitory independently and the difference of heights is computed.

(a) Find the sampling distribution of differences of means for the height of students if samples of size 1 are taken from each dormitory.

(b) Compute $\mu_{(\bar{X}-\bar{Y})}$, μ_X and μ_Y, and show that $\mu_{(\bar{X}-\bar{Y})} = \mu_X - \mu_Y$.

(c) Compute $\sigma_{(\bar{X}-\bar{Y})}$, $\sigma_{\bar{X}}^2$, and $\sigma_{\bar{Y}}^2$, and show that $\sigma_{(\bar{X}-\bar{Y})} = \sqrt{\sigma_{\bar{X}}^2 + \sigma_{\bar{Y}}^2}$.

13.   A political scientist finds the mean age of 50 voters who favor Jerry P. Goodman, a liberal, for Congressman is 36 with a standard deviation of 4 years, while 80 voters who favor Alexander J. Whitehead, a conservative, have a mean age of 46 years with a standard deviation of 5 years. Both are random samples. Find the differences in ages of those who favor the liberal and the conservative, at a 98 per cent confidence level.

14.   Professor Think taught the same psychology course two years in a row, the first year at 8 A.M. and the second year at 1 P.M. A random sample of 50 morning students had a mean quiz grade of 82 with a standard deviation of 8, while the mean was 87 with a standard deviation of 6 in a random sample of 60 from the afternoon class. Find a 95 per cent confidence interval for the difference in grades between the two classes.

15.   At Johnson University, 200 students chosen randomly were signed up for a mean of 15 credits per semester with a standard deviation of .4. At Jameson University, 150 students chosen randomly were taking an average of 16 credits per semester with a standard deviation of .3. What is the difference in mean number of credits taken per semester at the two universities, at a 90 per cent confidence level?

16.   The mean age of 50 listeners to radio station WOLD is 62 with a standard deviation of 12, while the mean age of 60 persons listening to station WZAP is 19 with a standard deviation of 3. Both samples are random. Find a 95 per cent confidence interval for the difference in ages of listeners to the two radio stations.

17.   The mean length of 50 Beatle songs chosen randomly is 3 minutes with a standard deviation of 30 seconds. A random sample of 55 Rolling Stones' songs has a length of 4 minutes 6 seconds with a standard deviation of 45 seconds. At a 98 per cent confidence level, what is the difference in song length between all songs sung by the Beatles and by the Rolling Stones?

18.   One hundred basketball players have a mean height of 6.3 feet, with a standard deviation of .4, while 100 jockeys have a mean height of 4.8 feet with a standard deviation of .3. Find, with 98 per cent confidence, the difference in heights of all basketball players and all jockeys.

19.   Population X consists of four numbers: 1, 2, 4, 8; population Y consists of the numbers 5, 10, 20. Let P_X and P_Y be the proportion of even numbers in these populations.

(a) Find the sampling distribution of difference of proportions if samples of size 2 are taken from the X population and of size 1 from the Y population, independently and with replacement.

(b) Compute $\mu_{(p_X - p_Y)}$ and $P_X - P_Y$, and show they are equal.

(c) Compute $\sigma_{(p_X - p_Y)}$ and show that it equals $\sqrt{\dfrac{P_X Q_X}{n_X} + \dfrac{P_Y Q_Y}{n_Y}}$.

20.   A random sample of 1,000 Democrats shows that 400 favor capital punishment, whereas 250 of a random sample of 500 Republicans favor capital punishment. Find a 90 per cent confidence interval for the difference in the proportion of Democrats and Republicans who favor capital punishment.

21.   A random sample of 800 adults and 400 teenagers at 9 P.M. on November 16 shows that 200 adults and 50 teenagers are watching T.V. (a) Find the difference in the proportion of all adults and all teenagers who were watching T.V. at that time, at a 95 per cent confidence level. (b) With what degree of confidence can it be said that the difference in the proportions is .25 ± .05?

22.   In a random sample of 800 city residents, 480 favor day care centers, but only 200 of a random sample of 500 suburban residents favor day care centers. Find a 98 per cent confidence interval for the difference in the proportion of urban and suburban residents who favor day care centers.

23.   Study of a random sample of 1200 male chemistry majors shows that 480 went to work for large corporations immediately after graduation, but the same is true of only 60 of 300 female chemistry majors, also a random sample. Find the difference in the proportions of men and women chemistry majors who get such jobs, with 90 per cent confidence.

24.   Two countries plan to form one political unit, and appropriate resolutions are passed unanimously by the general assemblies in each country. In random polls of 1000 people in each country, those in one country approve the proposal 9 to 1, but in the other country the choice is 3 to 1 for approval. Find the difference, with 95 per cent confidence, in the proportion of those who favor the merger.

25.   In a random sample of 100 amateur British Egyptologists, 5 believe the pyramids were built by levitation. This belief is shared by 4 of a random sample of 56 amateur American Egyptologists. At a 98 per cent confidence level, what is the difference in the proportion of British and American amateurs who believe this is how the pyramids were built?

26.   At a national fraternity convention, random samples were taken of 330 delegates from western states and 300 from eastern states. Among the westerners who were questioned, 231 wanted an increase in the number of chapters, but 180 easterners favored the opposite, a consolidation of chapters. Find a 95 per cent confidence interval for the difference in the proportion of fraternity members who prefer consolidation of chapters.

27.   An irate citizen passed around a petition to 212 homeowners and 44 store owners. Only 10 per cent of the store owners signed, but 90 per cent of the homeowners whom he approached did so. His petition demanded closing of their street to trucks. At a 95 per cent confidence level, find the difference in the proportion of homeowners and merchants who favor his proposal.

28.   The makers of HELTHPUP, a diet supplement for dogs, wish to determine the mean gain in weight per month of 1-year-old Eskimo dogs whose diet includes HELTHPUP, with 95 per cent confidence that the error is less than 2.5 oz. It is known that

the standard deviation of weight gain for such dogs is 6 oz. How many dogs should be tested?

29.   The chairman of the French department at Norwich College wants to compare students in his department with those from other colleges who are studying French. A national examination has a mean of 500 and a standard deviation of 100. To how many of his students should he give the test if he wishes to know their mean grade within 1σ with 95 per cent confidence?   *a tenth of a deviation of the population* $n = (1.96)^2 \dfrac{\sigma^2}{(.1\sigma)^2} = 384$, *too many for his class*

30.   At present only 60 per cent of patients with leukemia survive six years. A doctor has developed a new drug, and wants to know with 95 per cent confidence the 6-year survival rate with an error of less than 20 per cent. On how many patients should he test the new drug?

31.   A political scientist investigating the difference in the proportion of registered Republicans among well-to-do and poor voters should take samples of what size if he wishes to determine the difference within 10 per cent with 90 per cent confidence?

$n = \dfrac{z^2}{4E^2}$

32.   A reading test is to be given to random samples of boys and girls in New Jersey public schools. To how many boys and girls should the test be given if their mean grades are to be compared within 4 score points, with 90 per cent confidence?   *(for difference of means)*

33.   The registrar of Petersen University wishes to compare the grade-point averages of married and unmarried students this year. He would like to know the difference within .1, with 95 per cent confidence. He knows that last year the GPA of 100 randomly chosen married students was 2.85, with a standard deviation of .4, while 100 randomly chosen unmarried students had a mean GPA of 2.73 with a standard deviation of .3. He should take random samples of what size?   *This is past data to base guess on* $n = z^2 \dfrac{(\sigma_x^2 + \sigma_y^2)}{E^2} = 96$ *for each*

34.   The number of robberies in West Side and East Side retail stores is to be compared. It is known that last year 12 per cent of West Side and 15 per cent of East Side stores were robbed. Data from how many retail stores should be collected this year
      (a) to determine, with an error less than 10 per cent and with 90 per cent confidence, the difference in the proportion of stores robbed on the West and East Sides?
      (b) to determine the difference in the number of stores robbed on the West and East Sides?

35.   You suspect a die is not fair, and wish to find out. You decide to toss the die and count the number of times a 6 comes up. How many times should you toss it?

ANSWERS

1.

	p	f	fp	fp^2
AB AC BC	1	3	3	3
AD AE BD BE CD CE	1/2	6	3	3/2
DE	0	$\dfrac{1}{10}$	$\dfrac{0}{6}$	$\dfrac{0}{9/2}$

(a) $\mu_p = 6/10 = 3/5 = P.$

(b) $\sigma_p = \sqrt{\dfrac{9/2 - 6^2/10}{10}} = \sqrt{\dfrac{9}{100}}; \sqrt{\dfrac{PQ}{n}} \sqrt{\dfrac{N-n}{N-1}} = \sqrt{\dfrac{3/5 * 2/5}{2}} * \dfrac{5-2}{5-1}$

$= \sqrt{\dfrac{9}{100}}.$

3. $\pm 1.96 = \dfrac{.10 - P}{\sqrt{\dfrac{.10 * .90}{4000}}}$; 9.1 to 10.9 per cent.

5. (a) These figures are (roughly) those obtained in the *Literary Digest* poll for Landon vs. Roosevelt in 1936. If the sample were random:

$\pm 1.0 = \dfrac{.56 - P}{\sqrt{\dfrac{.56 * .44}{2,000,000}}}$; 55.95 to 56.05 per cent.

(b) $z = \dfrac{.56 - .34}{\sqrt{\dfrac{.34 * .66}{2,000,000}}} = 657.$ The probability is incredibly small.

$3 = \dfrac{p - P}{\sqrt{\dfrac{P*(1-P)}{n}}}$

(c) The sample was not a random one, of course. Ballots were sent to those who owned an automobile or had a telephone during a great depression.

7. $\pm 1.64 = \dfrac{.775 - P}{\sqrt{\dfrac{.775 * .225}{200}}}$; 73 to 82 per cent of subscribers are dissatisfied with

the editorials. (The editors must be careful to see that the sample is random; voluntary responses are probably biased toward dissatisfaction.)

9. $\pm 1.96 = \dfrac{.36 - P}{\sqrt{\dfrac{.36 * .64}{100}}}$; the 95 per cent confidence interval for the proportion of

all *heroin* addicts who will find relief is .27 to .45. Be careful not to generalize about all addicts if only heroin addicts have been tested.

11. $\pm 1.96 = \dfrac{(300 - 250) - (\mu_x - \mu_y)}{\sqrt{\dfrac{50^2}{100} + \dfrac{45^2}{80}}}$; \$36 to \$65.

13. $\pm 2.33 = \dfrac{(36 - 46) - (\mu_x - \mu_y)}{\sqrt{\dfrac{4^2}{50} + \dfrac{5^2}{80}}}$; voters for the liberal are 8.1 to 11.9 years

younger than voters for the conservative, at a 98 per cent confidence level.

15. $\pm 1.64 = \dfrac{(15 - 16) - (\mu_x - \mu_y)}{\sqrt{\dfrac{.4^2}{200} + \dfrac{.3^2}{100}}}$; students at Johnson University are taking .93

to 1.07 fewer credits per semester than students at Jameson U., at a 90 per cent confidence level.

17. $\pm 2.33 = \dfrac{(3 - 4.1) - (\mu_x - \mu_y)}{\sqrt{\dfrac{.5^2}{50} + \dfrac{.75^2}{55}}}$; I have 98 per cent confidence that Beatle

songs take .81 to 1.39 minutes (49 to 83 seconds) less than those sung by the Rolling Stones.

19.

Samples			Samples			Samples		
X	Y	$p_X - p_Y$	X	Y	$p_X - p_Y$	X	Y	$p_X - p_Y$
1,1	5	0	1,1	10	−1.0	1,1	20	−1.
1,2	5	.5	1,2	10	− .5	1,2	20	− .5
1,4	5	.5	1,4	10	− .5	1,4	20	− .5
1,8	5	.5	1,8	10	− .5	1,8	20	− .5
2,1	5	.5	2,1	10	− .5	1,8	20	− .5
2,2	5	1.0	2,2	10	0	2,2	20	0
2,4	5	1.0	2,4	10	0	2,4	20	0
2,8	5	1.0	2,8	10	0	2,8	20	0
4,1	5	.5	4,1	10	− .5	4,1	20	− .5
4,2	5	1.0	4,2	10	0	4,2	20	0
4,4	5	1.0	4,4	10	0	4,4	20	0
4,8	5	1.0	4,8	10	0	4,8	20	0
8,1	5	.5	8,1	10	− .5	8,1	20	− .5
8,2	5	1.0	8,2	10	0	8,2	20	0
8,3	5	1.0	8,4	10	0	8,4	20	0
8,4	5	1.0	8,8	10	0	8,8	20	0

(a)

$p_X - p_Y$	f	$f(p_X - p_Y)$	$f(p_X - p_Y)^2$
−1.0	2	−2	2
−.5	12	−6	3
0.0	19	0	0
.5	6	3	3/2
1.0	9	9	9
	48	4	31/2

(b) $\mu_{(p_X - p_Y)} = 4/48 = 1/12; \; P_X - P_Y = 3/4 - 2/3 = 1/12.$

(c) $\sigma_{(p_X - p_Y)} = \sqrt{\dfrac{31/2 - 4^2/48}{48}} = \sqrt{\dfrac{93/6 - 2/6}{48}} = \sqrt{\dfrac{91}{288}}$

$\sqrt{\dfrac{P_X Q_X}{n_X} + \dfrac{P_Y Q_Y}{n_Y}} = \sqrt{(3/4 * 1/4)/2 + (2/3 * 1/3)/1} = \sqrt{91/288}.$

21. (a) $\pm 1.96 = \dfrac{1/8 - (P_X - P_Y)}{\sqrt{\dfrac{1/4 * 3/4}{800} + \dfrac{1/8 * 7/8}{400}}};$ 8.1 to 16.9 per cent.

(b) $|z| * \sigma_p = |z| * .023 = .05$ finding confidence of population proportion
$|z| = 2.17$
97 per cent confidence

23. $p_X = .40, n_X = 1200, p_Y = .20, n_Y = 300, z = \pm 1.645.$

$\sigma_{(p_X - p_Y)} = \sqrt{\dfrac{.40 * .60}{1200} + \dfrac{.20 * .80}{300}} = .027$

$\pm 1.645 = \dfrac{.20 - (P_X - P_Y)}{.027}$

$P_X - P_Y = .20 \pm .044 = .156 \text{ or } .244$

The 90 per cent confidence interval for the difference in the proportion of male and female chemistry majors who get jobs with large corporations is 15.6 to 24.4 per cent.

25. $p_X = .05$, $p_Y = .071$, $n = 100$, $n_Y = 56$, $z = \pm 2.33$.

$$\sigma_{(p_X - p_Y)} = \sqrt{\frac{.05 * .95}{100} + \frac{.071 * .929}{56}} = \sqrt{.00165} = .041$$

$$\pm 2.33 = \frac{(.05 - .071) - (P_X - P_Y)}{.041}$$

$$P_X - P_Y = -.02 \pm .09 = -.11 \text{ or } +.07$$

With 98 per cent confidence, the difference in the percentage of British and American amateur archeologists who believe in levitation for the pyramids is between -11 and $+7$ per cent. The samples are too small to determine which country wins the prize for foolishness.

27.   Did you waste time on this one? Nonsense! The samples are not random at all.

29. $E = .1$, so $n = \dfrac{1.96^2 \sigma^2}{(.1\sigma)^2} = 384$. If he doesn't have this many students in his department, he must increase the error or decrease his confidence. (At a 75 per cent confidence level and if $E = .2\sigma \approx 20$ score points, then $n = 33$, for example.)

31. $n = \dfrac{z^2}{4E^2} = \dfrac{1.64^2}{4 * .10^2} = 68$ in each sample.

33.   Assume the standard deviations have not changed, so $\sigma_X = .4$, $\sigma_Y = .3$, $n = \dfrac{z^2(\sigma_X^2 + \sigma_Y^2)}{E^2} = \dfrac{1.96^2(.4^2 + .3^2)}{.1^2} = 96$ in each sample.

35.   Neither error nor the confidence you wish to have is stated, so it is hopeless to work out the answer. (In chapter 11 you will learn how to test the hypothesis "the die is fair" after tossing the die a certain number of times — but it will be necessary to state first the probability of error in rejecting this hypothesis when it is true.)

chapter eleven

hypothesis testing

In this chapter you will learn how to state and test a hypothesis about the population mean or proportion, or about the difference between the means or proportions of two populations. You will learn how to formulate an alternative hypothesis which is accepted if your original one is rejected, how to decide which one to accept, and what is the probability of error. This is not difficult if you realize early that you are simply using your old friends the sampling distributions for a different purpose.

$$z = \frac{\bar{x} - \mu}{\sigma/\sqrt{n}} \longrightarrow \overset{@95\%}{1.96 = \frac{100 - \mu}{5/\sqrt{50}}}$$

11.1 INTRODUCTION

In Chapters 9 and 10 you learned how to make an inference about the population mean or proportions from knowledge of the mean or properties of a sample. For example, if a random sample of size 50 has mean 100 and standard deviation 5, we can then claim with 95 per cent confidence that the population mean is between 98.6 and 101.4.

Now the emphasis will be turned around. We shall make a guess or assumption, or **hypothesis,** about the population mean or proportion (or about the difference between the means or proportions of two different populations), and then test that hypothesis by looking at a sample (or samples) drawn at random from the population(s). For example, we might form the hypothesis that a coin is fair (that $P = 1/2$). If you flip the coin 100 times and it comes up heads 50 times you will certainly accept the hypothesis that $P = 1/2$. If heads turn up 49 times you will probably say, "I expect some fluctuation between samples. If I flip the coin another 100 times, maybe it will come up heads 51 times. So I still accept the hypothesis that $P = 1/2$." But if heads turn up only once in 100 flips, you would no doubt be pretty suspicious of the coin and reject the hypothesis that $P = 1/2$. Note, however, that if $P = 1/2$ (if the coin is fair), the probability is not 0 that heads will turn up exactly once in 100 throws. It is $_{100}C_1(1/2)^1(1/2)^{99}$, which is an extremely small number but **not** 0. There is a very small probability that you are making an error if you reject the hypothesis that $P = 1/2$ when heads turn up exactly once. Intuitively, you probably accept the hypothesis that $P = 1/2$ if 49 or 50 or 51 heads turn up in 100 throws; you reject it if 1 or 0 or 99 heads turn up. What do you do if heads turn up 37 times? 59 times? 43 times? It is clear that a method of making the decision whether or not to accept the hypothesis $P = 1/2$ must be decided upon.

In general, hypothesis testing in statistics involves the following steps:

1. Choosing the hypothesis that is to be tested.

2. Choosing an alternative hypothesis which is accepted if the original hypothesis is rejected.

3. Choosing a rule for making a decision about which hypothesis to accept and which to reject.

(4.) Choosing a random sample from the appropriate population and computing appropriate statistics; that is, mean, variance, and so on.

(5.) Making the decision.

You are already familiar with the fourth step. The fifth step will be simple if you thoroughly understand the first four.

11.2 CHOOSING THE HYPOTHESIS TO BE TESTED (H₀)

The hypothesis you wish to test is called the **null hypothesis,** since acceptance of it commonly implies "no effect" or "no difference." It is denoted by the symbol H_0. (Read "H naught"; "naught" is a word about as old-fashioned as a horse and carriage. Both have their uses on certain occasions, and here is the place for "naught," which means "zero.")

In the coin-flipping experiment in the preceding section, the hypothesis that the coin is fair may be expressed symbolically:

$$H_0 : P = 1/2$$

H_0 is **always** a statement about some aspect (mean, proportion, etc.) of population(s). It is not about a sample, nor are sample statistics used in formulating the null hypothesis. H_0 might be $\mu = 100$ or $\mu_1 - \mu_2 = 4$ or $P = .6$ or $P_1 - P_2 = 0$, but **not** $\bar{X} = 4, p = .36, \bar{X}_1 - \bar{X}_2 = 0$, etc. "A random sample of size 50 has $\bar{X} = 100, s = 4.5$. At a .05 level of significance,† can μ be 101.5?" The numbers 100 and 4.5 will **never** occur in H_0. Here, H_0 is "$\mu = 101.5$." H_0 is usually easy to determine.

There is one convention that, for the moment, you will have to accept; the reason for it will be shown later (Section 11.7). H_0 is an equality ($\mu = 14$) rather than an inequality ($\mu \geq 14$ or $\mu < 14$). For the present there are four possible choices of H_0:

$$\mu = \underline{\quad}$$
$$P = \underline{\quad} \quad probability$$
$$\mu_X - \mu_Y = \underline{\quad}$$
$$P_X - P_Y = \underline{\quad}$$

where the blanks are filled in by constants. In later chapters you will learn how to test other types of hypotheses: that a population has a certain type of distribution, or that two populations have the same variance, for example. But for now, we'll stick to hypotheses about the mean(s) or proportion(s).

Example 1

You manufacture hammocks and buy ropes for them from the Nylon Coil Co. A sample of 36 ropes chosen at random from a Nylon Coil delivery has a mean breaking strength of 127 lb. with a standard deviation of 5 lb. The desired mean breaking strength is 130 lb. Should you accept the shipment, at a .05 level of significance?

What is H_0?

---

†The meaning of the phrase "level of significance" will be explained later (Section 11.4).

If you answered $\bar{X} = 127$ you forgot that H_0 is an assumption about populations; if $\mu = 127$, you forgot that <u>sample figures do not determine the form of the null hypothesis</u>; if $\mu \geq 130$, you forgot our convention that inequalities are not allowed in H_0. The correct answer is $H_0: \mu = 130$.

Try to determine H_0 in the following examples:

Example 2

At present only 60 per cent of patients with leukemia survive six years. A doctor develops a new drug. Of 40 patients, chosen at random, on whom the new drug is tested, 26 are alive after six years. Is the new drug better than the former treatment (.10 level of significance)?   *1.65*

$p = 26/40$ refers to a sample; $P = 26/40$ depends on sample values, $P \leq .60$ is an inequality. The correct answer is $H_0: P = .60$.

Example 3

A botanist has developed a new strain of rice which he thinks will increase the yield by 30 bushels per acre over the kind presently sown in Madras, India. Seeds of the new and old strains are randomly selected, and 60 acres of roughly the same fertility, moisture, and exposure are divided into one-acre plots. Thirty of these are randomly chosen; when planted with the new strain they produce 180 bushels per plot with a standard deviation of 4 bushels. The remaining 30 acres, planted with the old strain, produce an average of 152 bushels per plot with a standard deviation of 5 bushels. Does the new strain produce at least 30 bushels more per acre, at a .05 level of significance?   *New*   *Old — old*

Let the new strain be X, the old Y. $\mu_X - \mu_Y = 27$ uses sample values, $\bar{X} - \bar{Y} = 27$ refers to samples, $\mu_X \geq \mu_Y + 30$ is an inequality, $\mu_X = \mu_Y$ is not the hypothesis that is to be tested. The correct answer is $H_0: \mu_X - \mu_Y = 30$.

Mean New — Mean old
(Hypothesis)

Example 4

A machine fills milk bottles; the mean amount of milk in each bottle is supposed to be 32 oz. with a standard deviation of .06 oz. In a routine check to see that the machine is operating properly, 36 filled bottles are chosen at random and found to contain a mean of 32.1 oz. At a .05 level of significance, is the machine operating properly?   *−1.96 ≤ z ≤ 1.96*

Look only at the information about the population and at the question being asked when deciding on H_0. $H_0: \mu = 32$.

Example 5

It is suspected that the mean salary of professors in public colleges is $1,000 more than in private colleges. A random sample of 100 public college professors shows a mean salary of $15,000 with a standard deviation of $1,000, while 100 private college professors selected at random have a mean salary of $14,200 with a standard deviation of $900. At a .01 level of significance, is the mean salary of public college professors at least $1,000 more than that of private college professors?

$H_0: \mu_X - \mu_Y = 1000$. (Assume X refers to all public college professors, Y to all private college professors.)

11.3   CHOOSING THE ALTERNATIVE HYPOTHESIS (H₁)

The notation H_1 (read "H one") is used for the hypothesis which will be accepted if H_0 is rejected. H_1 must also be formulated before a sample is tested, so it, like H_0, does not depend on sample values. If the fairness of a coin is questioned, H_0 is "$P = 1/2$". H_0 probably is "$P \neq 1/2$", but other alternatives are "$P < 1/2$", "$P > 1/2$", or "$P = .1$". An example of the last will be given very shortly (page 181). With that exception the hypotheses that we shall consider are severely limited, since in practice the alternative hypothesis cannot be specified exactly. Here are the possible choices:

If H_0 is	then H_1 is
$\mu = A$	$\mu \neq A$ or $\mu < A$ or $\mu > A$
$P = B$	$P \neq B$ or $P < B$ or $P > B$
$\mu_X - \mu_Y = C$	$\mu_X - \mu_Y \neq C$ or $\mu_X - \mu_Y < C$ or $\mu_X - \mu_Y > C$
$P_X - P_Y = D$	$P_X - P_Y \neq D$ or $P_X - P_Y < D$ or $P_X - P_Y > D$

A, B, C, and D are constants.

If you are alert and intelligent you are asking, "Why do I have to choose an alternative hypothesis before I test the one I've got? Why can't I accept or reject the null hypothesis, and think of an alternative only if it is rejected?" The reason is that the form of the alternative hypothesis will affect the decision to accept or reject the null hypothesis. This matter will be discussed in detail in Section 11.4, but see if you can begin to find some answers yourself as you look again at the examples in the preceding section (Section 11.2); now try to determine H_1 in each example.

Answers

Example 1. $H_0: \mu = 130$; $H_1: \mu < 130$. If the sample mean is over 130, you will be happy to accept the shipment. But if the sample mean is much lower than the expected population mean—that is, lower not just because of sampling fluctuations, but because it is drawn from a whole shipment whose mean is less than 130—then you will reject the shipment.

Example 2. $H_0: P = .60$; $H_1: P > .60$. The doctor is trying to reach a decision on whether to make further tests on the new drug. If the proportion of patients who live at least six years is not increased under the new treatment or is increased only by an amount due to sampling fluctuation, he will look for another drug. But if the proportion who are aided is significantly larger—if he is able to conclude that the population proportion is greater than .60—then he will continue his tests.

Example 3. $H_0: \mu_X - \mu_Y = 30$; $H_1: \mu_X - \mu_Y < 30$. Here μ_X is the mean yield per acre of the new strain. Unless it is at least 30 bushels per acre higher than the yield of the strain used now, the botanist will not encourage the expense of producing large quantities of the new strain or the trouble of inducing farmers to use it. If, however, production is increased by 30 bushels per acre or more, he will be willing to encourage the expense and trouble.

Example 4. $H_0: \mu = 32$; $H_1: \mu \neq 32$. Look only at the information about the population and at the question being asked when deciding on H_0 and H_1.

You are deciding whether the amount of milk differs from 32 oz., and will get the machine adjusted if it is too high or too low.

Example 5. $H_0: \mu_X - \mu_Y = 1000; H_1: \mu_X - \mu_Y > 1000.$

11.4   CHOOSING A RULE FOR MAKING A DECISION

A method for making a decision must be agreed upon. Either H_0 is accepted (and H_1 rejected) or H_0 is rejected (and H_1 is accepted). Every sample chosen at random will not have the same mean or proportion as the population from which it is taken. It seems reasonable to accept the null hypothesis if a sample value is relatively close to the value predicted for it if the null hypothesis is true. But if the outcome of tests on a sample is highly improbable on the basis of sheer chance, we say the difference is significant and reject the null hypothesis. The method of making a decision should be decided on **before** the test on a sample is carried out. But how is a "significant" difference defined?

A null hypothesis is either true or false, and it is either accepted or rejected. No error is made if it is true and accepted, or if it is false and rejected. An error is made, however, if it is true but rejected, or if it is false and accepted.

Definitions: A **Type I error** is made when H_0 is true but is rejected.

A **Type II error** is made when H_0 is false but is accepted.

Time out to learn two more Greek letters:

α (alpha) is the Greek lower case a. *"level of significance" — z-score*

β (beta) is the Greek lower case b.

Notation: α is the probability of a Type I error; α is called the **level of significance.** *if .05 then $-1.96 < z < 1.96$*   *95%*

β is the probability of a Type II error.   *1.96 1.96*

The following table summarizes these relationships.

		If H_0 is Accepted	If H_0 is Rejected
Where H_0 is	True	No error	Type I error (α)
	False	Type II error (β)	No error

Suppose you plan to test whether a coin is fair ($H_0: P = 1/2; H_1: P \neq 1/2$) by tossing it 100 times. You must decide on what basis H_0 should be accepted or rejected. You will soon learn how to make a wise decision. At the moment, let us consider four different arbitrary choices and the effect of each on α and β.

Example 1

$H_0: P = 1/2$ is always accepted, no matter how many times heads come up in 100 tosses. (a) What is α? (b) What is β if P actually equals (i) .35, (ii) .40, (iii) .45?

(a) $\alpha = 0$. (Since you never reject H_0, the probability is 0 that it is rejected when true.)

(b) $\beta = 1.00$ (no matter what P equals). Since you always accept H_0 whether it is true or false, you are certain to accept it when false.

Example 2

(handwritten: $z_{scan} = \frac{x-\mu}{\sigma}$; (i) $z_1 = \frac{39.5-35}{4.77} = .94$; $P = .5\Theta .94 (35) = .33$; $z_2 = \frac{60.5-35}{4.77} = 5.35$ can't use)

$H_0: P = 1/2$ is accepted if heads come up between 40 and 60 times (inclusive) in 100 tosses, and rejected if there are fewer than 40 or more than 60 heads. (a) What is α? (b) What is β if P actually equals (i) .35, (ii) .40, (iii) .45?

(a) The probability of getting less than 40 or more than 60 heads when $P = 1/2$ is .04 (see Exercise 15, page 120). $\alpha = .04$. *(handwritten: $\sqrt{npq}$)* *(handwritten: β)*

(b) (i) If $P = .35$, then $\mu = .35 * 100 = 35$, $\sigma = \sqrt{100 * .35 * .65} = 4.77$, and $\beta = p(39.5 < X < 60.5) = p(.94 < z < 5.35) = .5 - .33 = .17$.

(ii) If $P = .40$, then $\mu = .40 * 100 = 40$, $\sigma = \sqrt{100 * .40 * .60} = 4.90$, and $\beta = p(39.5 < X < 60.5) = p(-.10 < z < 4.18) = .50 + .04 = .54$.

(iii) If $P = .45$, $\mu = .45 * 100 = 45$, $\sigma = \sqrt{100 * .45 * .55} = 4.97$, and $\beta = p(39.5 < X < 60.5) = p(-1.11 < z < 3.12) = .87$.

Example 3

(handwritten: (iii) $z_1 = \frac{39.5-45}{4.97} = 1.1$; $z_2 = \frac{60.5-45}{4.97} = 3.12$) from table $p = .364 + .499 = .87$)

$H_0: P = 1/2$ is accepted if heads come up between 43 and 57 times (inclusive) in 100 tosses. (a) What is α? (b) What is β if P actually equals (i) .35, (ii) .40, (iii) .45?

(a) The probability of getting less than 43 or more than 57 heads in 100 tosses if $P = 1/2$ is $1 - .87 = .13$ (see Exercise 15, page 120). $\alpha = .13$.

(b) (i) $\beta = p(42.5 < X < 57.5) = p(1.57 < z < 4.72) = .50 - .44 = .06$.

(ii) $\beta = p(42.5 < X < 57.5) = p(.51 < z < 3.57) = .50 - .20 = .30$.

(iii) $\beta = p(42.5 < X < 57.5) = p(-.50 < z < 2.52) = .192 + .494 = .69$.

Example 4

$H_0: P = 1/2$ is always rejected, no matter how many times heads come up in 100 tosses. (a) What is α? (b) What is β?

(a) $\alpha = 1.00$. (If you always reject H_0, then you are certain to reject it when true.) *(handwritten: i.e. 100%)*

(handwritten: p.120/1) †To refresh your memory: $p = 1/2$, $n = 100$, and the binomial assumptions are satisfied. $p(40 \leq X \leq 60) = ?$ Approximate with the normal distribution, with $\mu = np = 100 * \frac{1}{2} = 50$, $\sigma = \sqrt{npq}$

$= \sqrt{100 * \frac{1}{2} * \frac{1}{2}} = 5$. Find the area for scores from 39.5 to 60.5 since the normal distribution is continuous. $p(39.5 < X < 60.5) = p(-2.10 < z < 2.10) = .964$. Therefore, p(number of heads is less than 40 or more than 60) $= 1 - .964 = .036$. *(handwritten: .04)*

(b) $\beta = 0$. (Since you never accept H_0, you never accept it when it is false.)

These examples should have renewed your acquaintance with the binomial distribution and its approximation by the normal distribution, but also—and far more important—should illustrate some important points. Here is a summary of the results:

incr β ——→ incr # Heads in 100 tosses

Example	Accept H_0 and reject H_1	(a) α	(b) $H_1: P = .35$	(c) $H_1: P = .4$	(d) $H_1: P = .45$
1	Always	0	1.0	1.0	1.0
2	if 40–60 heads	.02	.17	.54	.87
3	if 43–57 heads	.13	.06	.30	.69
4	Never	1.00	0	0	0

(error) over column (a); *β (error)* over columns (c) (d)

% reject when Ho true → Sm #3 Below → % accept when Ho false → β incr.

Note the following:

1. Once H_0 and H_1 are chosen, then a decision choice has to be made: how is one to decide whether to accept or reject the null hypothesis? This, you will discover, is a straightforward matter for the type of problems you will be asked to solve. In this example, arbitrary decision choices were made to show the effect of different choices on α and β.

2. α can be easily determined. β can be determined **only** if H_1 is specified exactly ($P = .4, P = .45$, etc.) and varies as this specific choice varies; one value of β cannot be determined when H_1 is simply the general statement "$P \neq 1/2$." (The exceptions are cases in which H_0 is always accepted or never accepted, no matter what the results of the experiment—then there is no need to carry out the experiment at all.)

3. As α increases, β decreases. Compare column (a) in turn with columns (b), (c), and (d). As the probability of rejecting H_0 when it is true increases, the probability of accepting H_0 when it is false decreases. It is not true in general, however, that $\beta = 1 - \alpha$. α is determined by assuming the population has a binomial distribution with $P = .5$, while β is computed under different assumptions about the population distribution (that $P = .35, .40, .45$, respectively). The relation between α and β is much more complicated; this will be discussed in Section 11.7.

As α incr, an rejecting when true, As β incr.

In this example, various decision choices were made arbitrarily: it was decided to accept H_0 if the number of heads is between two limits, and to reject H_0 if the number of heads is outside those limits. Then α was determined for each decision choice. In practice, however, the emphasis is put the other way around: α is chosen arbitrarily, and then limits for acceptance of H_0 are determined; if a sample statistic is outside those limits, H_0 is rejected (and H_1 is accepted).

The form of H_1 will determine the kind of limits to be set up. Let us first consider how to make a decision about accepting or rejecting H_0 when H_1 includes the symbol "$\neq$": when H_1 is $\mu \neq$ ____, $P \neq$ ____, $\mu_X - \mu_Y \neq$ ____, or $p_X - p_Y \neq$ ____, where the blank is filled in with a constant. Then a decision is made in the following fashion:

1. α is arbitrarily chosen, equal to a small number (usually .01, .05 or .10). On page 189 this choice will be discussed. α is called the **level of significance.**

(margin notes: incr. reject when True 'α' ; decr. accept when false β ; worse 'α' ; better β)

2. z values are determined so the area in each of the tails of the normal distribution is $\alpha/2$. The most common values of z are, then,

α	.10	.05	.01
z	± 1.64	± 1.96	± 2.58

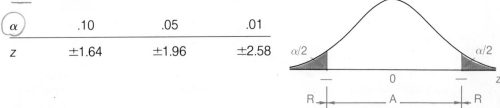

The form of H_1 we are considering now leads to a "two-tail test"; in Section 11.5 we shall find that when H_1 is of the form $\mu < \_\_\_\_, P > \_\_\_\_, \mu_X - \mu_Y > \_\_\_\_$, etc., then a "one-tail test" will be used. Remember that H_1 is decided on before the experiment is carried out, and therefore a two-tail or one-tail test is also decided upon before the experiment is carried out.

3. The experiment is carried out and the z value of the appropriate sample statistic $(\overline{X}, p, \overline{X} - \overline{Y}, \text{ or } p_X - p_Y)$ is determined. If this computed z value falls within the limits determined in Step 2 above, H_0 is accepted; if the computed z value is outside those limits, H_0 is rejected (and H_1 is accepted). Since they separate the acceptance and rejection regions, the limits determined in Step 2 will be referred to as **the critical values of z.**

Study the following examples with care, to discover the logic behind this method of making a decision.

Example 1

Assume that the mean IQ of all high school seniors is 110 with a standard deviation of 10. A random sample of 49 seniors at Swampscott High School has a mean IQ of 112. Does the mean IQ of all seniors at Swampscott High School differ† from that of high school students in general at a .05 level of significance?

$H_0: \mu = 110$. (The population consists of all seniors at Swampscott High School. The hypothesis to be tested states that their mean IQ is the same as the mean IQ of all high school seniors.)

$H_1: \mu \neq 110$

$\alpha = .05 \rightarrow -1.96 < z < 1.96$

Sample: $n = 49, \overline{X} = 112$

.025   .025 by def.

110

-1.96 0 1.96 z

Reject H_0 — Accept H_0 — Reject H_0

Look at the sampling distribution of means, with mean of 110 and standard error $= \dfrac{\sigma}{\sqrt{n}} = \dfrac{10}{\sqrt{49}} = \dfrac{10}{7}$. The probability of rejecting H_0 when it is true is

---

†Questions such as "Is the mean IQ of Swampscott students **higher**?" will be discussed in Section 11.5.

to be .05. The area of each shaded "tail" is .025, and the corresponding z scores at the boundaries are ± 1.96 (see Table 4, Appendix C). This means that H_0 is to be rejected if the z score of a sample mean is greater than 1.96 or smaller than -1.96: the probability of getting such a sample is so small (less than 5 per cent) that it is unlikely to occur because of sampling fluctuation and likely to occur because H_0 is false. If, on the other hand, a sample mean has a z score which falls between -1.96 and $+1.96$, then we reason as follows: "Every sample will not have a mean exactly equal to the population mean. But if H_0 is true, there is a .95 probability that the mean of a sample will have a z score between -1.96 and $+1.96$. If this occurs for the particular random sample chosen, then H_0 cannot be rejected; rather, it is accepted." The z score for the random sample of 40 Swampscott High School students is $\dfrac{112 - 100}{10/\sqrt{49}} = 1.40$. This score falls inside the "acceptance region" from -1.96 to $+1.96$, so H_0 is accepted; the mean IQ of all seniors at Swampscott High School is the same as the mean IQ of all high school students.

Example 2

A political scientist knows that a certain district voted 54 per cent Democratic in the last Congressional election. He also knows, however, that drastic changes in the types of employment in this district have occurred since the last election. He believes that these changes may have changed the percentage of voters who will vote for the Democratic candidate for Congress, but doesn't know which way it will be changed. He polls 1000 voters in the district and finds that 480 of these plan to vote for the Democratic candidate. Is his belief supported, at a .01 level of significance?

$$H_0: P = .54; \quad H_1: P \neq .54$$

Here $\alpha = .01$; that is, we find rejection regions such that the probability of rejecting H_0 when it is true is .01. This means that if the sampling distribution of proportions does have a mean of .54 (as is the case when H_0 is true), then we must find a region or regions such that the probability of a random sample's proportion falling there is .01. Since H_1 has the form "$P \neq .54$" we find 2 regions, each of area $.01/2 = .005$.

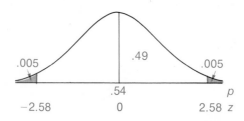

Referring to Table 4, Appendix C, we find that the z value corresponding to an area of $.5 - .005 = .495$ is 2.58. The null hypothesis will be rejected then (and H_1 accepted) if the z score of the sample proportion is greater than 2.58 or less than -2.58; it will be accepted if this z score is between -2.58 and $+2.58$.

Of 1000 voters polled, 480 will vote for the Democratic candidate; $p = 480/1000 = .48$; $n = 1,000$.

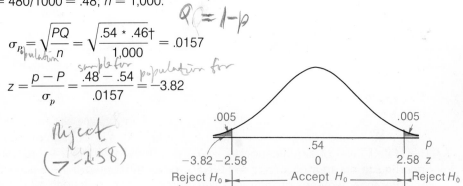

$$Q = 1 - p$$

$$\sigma_{P_{\text{population}}} = \sqrt{\frac{PQ}{n}} = \sqrt{\frac{.54 * .46\dagger}{1,000}} = .0157$$

$$z = \frac{p - P}{\sigma_p} = \frac{.48 - .54}{.0157}_{\text{ sample for } / \text{ population for}} = -3.82$$

Reject (> -258)

.005   .005

.54
$-3.82 \;\; -2.58$ 0   2.58 z

Reject H_0 ⊢——— Accept H_0 ———⊣ Reject H_0

This z score for the proportion in the random sample falls in the rejection region (it is less than -2.58), so H_0 is rejected; it is concluded that there is a significant change in the proportion of voters who will vote Democratic, at a .01 level of significance.

Note, however, that if 100 different random samples were drawn, probably 1 of them would have a z value for the proportion voting Democratic that would be less than -2.58 or more than $+2.58$ **if the null hypothesis were true;** the probability is .01 that H_0 is rejected when it is true. But, faced with uncertainty, the political scientist says "I am aware of this probability of error, but I think that, if my sample proportion falls in a tail, it is more likely to do so because H_0 is not true; the difference from the population proportion is so great that I think it arises not from sampling fluctuation but because the sample comes from a different population."

Example 3

An economist in 1965 compared the wages of electricians and of carpenters in New Jersey; he discovered that the mean monthly wage of electricians is $100 more than that of carpenters. He repeats the study in 1975, since he wonders whether or not there is still the same difference in their pay. He finds that the mean wage of 100 electricians, chosen at random, is $1,000, with a standard deviation of $200, while a random sample of 81 carpenters has a mean wage of $950 with a standard deviation of $100. At a .05 level of significance, should he conclude that the difference in wages of electricians and carpenters is unchanged?

$H_0: \mu_X - \mu_Y = 100$. (He is asking whether the mean monthly wage of all electricians in New Jersey is still $100 more than that of all carpenters in the state.)

$H_1: \mu_X - \mu_Y \neq 100$. (He doesn't know whether the difference in wages of electricians and carpenters has decreased or increased; he simply wants to find out whether it has changed.)

---

†Note that in hypothesis testing P is given in the null hypothesis, so there is no difficulty in determining σ_p; testing a hypothesis about a proportion differs in this way from finding a confidence interval for a proportion.

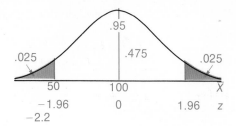

As in Example 1, the form of H_1 ("$\ne$") tells us to make a two-tail test: H_0 is rejected if the z score for difference of sample means is too high **or** too low. The stated level of significance ($\alpha = .05$) tells us to find the z score corresponding to an area of .475; $z = 1.96$. H_0 will be accepted if the z score for the difference of means of samples taken from the two populations is between -1.96 and $+1.96$, and rejected if this z score is outside those limits.

The sampling distribution of differences of means is needed here; its mean is $\mu_X - \mu_Y = 100$ (this information comes from H_0).

$$n_X = 100,\ s_X = 200,\ \sigma_{\bar{x}} = \frac{200}{\sqrt{100}} = 20$$

$$n_Y = 81,\ s_Y = 100,\ \sigma_{\bar{y}} = \frac{100}{\sqrt{81}} = 11.1$$

$$\sigma_{(\bar{X}-\bar{Y})} = \sqrt{20^2 + 11.1^2} = \sqrt{523.21} = 22.9$$

$$\bar{X} = 1000,\ \bar{Y} = 950$$

$$z = \frac{(\bar{X}-\bar{Y}) - (\mu_X - \mu_Y)}{\sigma_{(\bar{X}-\bar{Y})}} = \frac{(1000-950)-100}{22.9} = \frac{-50}{22.9} = -2.2$$

H_0 is rejected, since $-2.2 < -1.96$; we accept the alternative hypothesis that the mean difference in monthly wages of electricians and carpenters in New Jersey is no longer $100. On the basis of this test we do not conclude that the mean difference is now less than $100, nor that it is more; we only conclude that it is one or the other.

11.5   ONE-TAIL TESTS

In Section 11.4, you learned how to set up a null hypothesis H_0 and an alternative hypothesis H_1, and how to reach a decision whether to accept or reject the null hypothesis at a given level of significance when H_1 has one of the following forms: $\mu \ne$ ____, $P \ne$ ____, $\mu_1 - \mu_2 \ne$ ____, $P_1 - P_2 \ne$ ____. This decision was reached by following a specific convention: Let the area in each tail of the sampling distribution be $\alpha/2$; this determines two values of z which separate an acceptance region from two rejection regions. Then the z value of a statistic is computed for a random sample, and H_0 is accepted if this z value falls in the acceptance region previously determined; otherwise, H_0 is rejected.

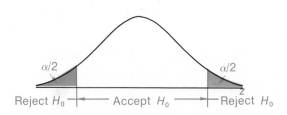

Now let's look at "one-tail tests" in which H_1 has the form $\mu < \underline{\quad}, P > \underline{\quad}, \mu_1 - \mu_2 > \underline{\quad}$, etc.

Suppose H_0 has the form $\mu = 100$. If H_1 has the form $\mu > 100$, this means that it is reasonable to believe the observed sample is from a population with mean of 100 rather than from a population whose mean is greater than 100 if the mean of the sample is itself less than 100. The rejection region is therefore concentrated in one tail of the sampling distribution of means, rather than separated into two parts or tails as when H_1 has the form $\mu \neq 100$. The area of this one tail is now α, rather than the $\alpha/2$ used previously, and the critical value of z changes accordingly.

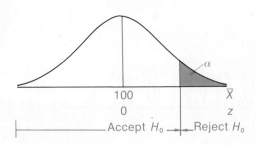

Two-tail test

$H_0: \mu = 100; H_1: \mu \neq 100; \alpha = .05$

One-tail test

$H_0: \mu = 100; H_1: \mu > 100; \alpha = .05$

The most frequently used values of α and the corresponding critical values of z are:

		z		
		Two-tail	One-tail, $<$	One-tail, $>$
$\alpha =$	.10	±1.645	−1.28	1.28
	.05	±1.96	−1.64	1.64
	.01	±2.58	−2.33	2.33

Remember that α is still being emphasized and chosen arbitrarily, as with two-tail tests; try to keep an open mind on which value of α to use, and on whether this is even a good way to decide to accept or reject H_0, until you have read the next section.

Example 1

It is known that 1-year-old dogs have a mean gain in weight of 3.0 pounds per month, with a standard deviation of .40 pound. A special diet supplement, HELTHPUP, is given to a random sample of 50 1-year-old dogs for a month; their mean gain in weight per month is 3.15 pounds, with a standard deviation of .3 lbs.

of .30 pound. Does weight gain in 1-year-old dogs increase if HELTHPUP is included in their diet? (.01 level of significance.)

$H_0: \mu = 3.0$; $H_1: \mu > 3.0$; $\alpha = .01$; <u>one-tail test.</u>

Reject H_0 if $z > 2.33$.

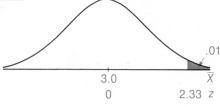

$\sigma_{Sample} = Std. Error = \frac{6}{\sqrt{n_{sample}}}$

$\frac{u_s - u}{}$

$$z = \frac{3.15 - 3.0}{.4/\sqrt{50}} = 2.7.$$ Reject H_0. At a .01 level of significance, weight gain is increased if pups have HELTHPUP.

Example 2   *1 Random Sample. Don't know Std. Deviation of Sample:*

An economist working for the Bureau of Labor Statistics knows that last month 7.1 per cent of those in the labor force in the United States were unemployed. This month he discovers that 350 are unemployed in a random sample of 5,000. At a .05 level of significance, has unemployment decreased this month?   *$\bar{X} = \frac{350}{5000} = .07$*

$H_0: P = .071$; $H_1: P < .071$; $\alpha = .05$; <u>one-tail test.</u>

Reject H_0 if $z < -1.645$.

$$z = \frac{.070 - .071}{\sqrt{.071 * .929/5000}} = -.28$$

proportion

Accept H_0; the unemployment rate has not changed significantly this month.

Example 3   *2 random Samples Only w/ Std. Deviations Known:*

The registrar of Peterson University is comparing the grade-point averages of married and unmarried students. He finds that 100 married students, chosen at random, have a mean GPA of 2.85 with a standard deviation of .4, while a random sample of 100 unmarried students has a mean GPA of 2.73 with a standard deviation of .3. At a .10 level of significance, do married students have a higher GPA?

$H_0: \mu_X - \mu_Y = 0$; $H_1: \mu_X - \mu_Y >= 0$; $\alpha = .10$; <u>one-tail test.</u>

H_0 is rejected if $z > 1.28$.

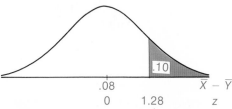

$\mu_X - \mu_Y$

$$z = \frac{(2.85 - 2.73) - 0}{\sqrt{.4^2/100 + .3^2/100}} = 2.4$$

$\sigma_x^2/n + \sigma_s^2/n_s$

H_0 is rejected; married students do have a higher GPA.

Example 4 *Z Random Samples – Don't know Std. deviation(s) :*

A political scientist is investigating the difference in the proportion of registered Republicans among well-to-do and poor voters. He takes random samples of 1000 voters whose families have incomes over $14,000 and 1,000 whose families have incomes under $6,000. He finds that in the two groups 25.3 per cent and 22.0 per cent, respectively, are registered Republicans. At a .05 level of significance, is the proportion of registered Republicans higher among well-to-do than among poor voters?

$H_0: P_X - P_Y = 0; H_1: P_X > P_Y; \alpha = .05;$ one-tail test.

H_0 is rejected if $z > 1.645$.

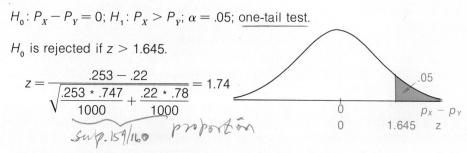

$$z = \frac{.253 - .22}{\sqrt{\dfrac{.253 * .747}{1000} + \dfrac{.22 * .78}{1000}}} = 1.74$$

Sup. 159/160   proportion

H_0 is rejected; more well-to-do voters are registered Republicans.

11.6 TYPE I AND TYPE II ERRORS

Example 1

A defendant being tried by a jury may be either not guilty or guilty; the jury may find him either not guilty or guilty. The four possible outcomes of the trial may be illustrated as follows:

		Defendant is	
		Not guilty	Guilty
Jury finds him	Not guilty	A	B (Type II: β)
	Guilty	C (Type I: α)	D

Justice is done if outcome A or D occurs; errors are made if B or C occurs. The null hypothesis is that the defendant is not guilty. Outcome C (rejection of the null hypothesis when it is true) means a Type I error has been made; α is the probability of a Type I error. Outcome B (acceptance of the null hypothesis when it is false) means a Type II error has been made; β is the probability of a Type II error.

A defense attorney will try to get a jury unlikely to make a Type I error; he certainly wants α to be small, and may not be unhappy if β is large. A prosecuting attorney, on the other hand, wants a jury chosen so that the probability of finding the defendant not guilty when he has committed the crime is small; β should be minimized in the prosecutor's view.

A doctor who accepts the hypothesis that a new drug has the same effect on a disease as the drug presently used, when in fact the new drug is better (Type II error), may be missing a new cure for his patients and may cause great loss to a pharmaceutical company. If, on the other hand, he concludes that the new drug looks promising when it really is no improvement (Type I error), he may use much energy, time, and money on further tests which prove fruitless.

A manufacturer of television sets who rejects a shipment of TV tubes, because after testing a sample he concludes their mean life is too short, is making a Type I error if this is not the case; he may have production problems for lack of an essential part and he may have poor relations in the future with his supplier. But if he accepts the shipment when the mean life is indeed less than he expects (Type II error), he may have trouble with his guarantee to his customers.

These examples indicate that emphasis on a Type I or a Type II error depends on the hypothesis that is being tested and the reason for testing it. In our convention for testing H_0 using the sampling distribution of a statistic, however, it is only a Type I error (whose probability is measured by the level of significance, α) that is important; β (the probability of a Type II error) is ignored. Some examples will illustrate further the relationship between α and β in testing H_0 using the sampling distribution of a statistic. As you study these examples, concentrate on the following results:

1. As α increases (or decreases), β decreases (or increases). The relationship is not a simple one, however. It is not true that doubling α will cut β in half, for example.

2. For fixed α, β is easily determined if the alternative H_1 has the form $\mu = 80$, $P = .40$, etc., but the value of β depends upon the actual constant selected. For the kind of alternative hypotheses we have been testing ($\mu < 80$, $P \neq .40$, etc.), one value of β cannot be determined.

3. For fixed α, the value of β can be decreased by increasing the size of the sample.

Example 2

A manufacturer claims his light bulbs have a mean life of 1000 hours with a standard deviation of 40 hours. A big department store tests a random sample of 64 bulbs from a large shipment.

(a)   The shipment should be accepted if the mean life of the bulbs in the sample is over what value, if the level of significance is (i) .05, (ii) .01?

(b)   What is the probability of accepting the shipment when in fact the mean life of the bulbs in the shipment is 987 hours if (i) $\alpha = .05$, (ii) $\alpha = .01$? Assume the standard deviation is still 40 hours.

90% confd.   987 confid.

Answer

(a) $H_0: \mu = 1000$; $H_1: \mu < 1000$; one-tail test.   *use this when $n > 30$; it is an approximation*

(i) $\alpha = .05$, so $z = -1.645$; $\sigma_{\bar{x}} = \dfrac{40}{\sqrt{64}} = 5.$

.05

90% Confidence

	992	1000	$\bar{X}$
	−1.645	0	z

$$z = \frac{\overline{X} - \mu}{\sigma_{\overline{X}}};\ -1.645 = \frac{\overline{X} - 1000}{5};\ \overline{X} = 992$$

907. The shipment is accepted if $\overline{X} > 992$.

(ii) $\alpha = .01$, so $z = -2.33$.

$$-2.33 = \frac{\overline{X} = 1000}{5};\ \overline{X} = 988.$$

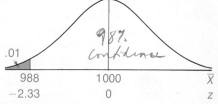

.01

987. confidence

988   1000 $\overline{X}$
−2.33   0 z

987. The shipment is accepted if $\overline{X} > 988$.

Note that, up until now, we have determined a critical value or values of z which separate an acceptance region from a rejection region; then the z score for a sample statistic was computed to find in which region it falls. Here the critical value is expressed not in z units or standard deviation units but rather in original score units.

(b) $H_0: \mu = 1000$; $H_1: \mu = 987$. The sampling distributions of means corresponding to these two hypotheses are the normal distributions which overlap as shown below.

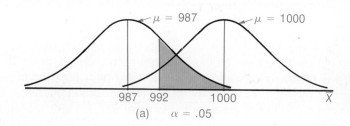

$\mu = 987$ $\mu = 1000$

987 992   1000 $\overline{X}$
(a) $\alpha = .05$

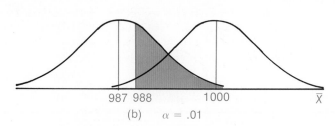

987 988   1000 $\overline{X}$
(b) $\alpha = .01$

The probability of accepting the shipment when the mean life of the light bulbs is actually 987 hours is represented by the shaded area, in figure (a) when $\alpha = .05$ and in figure (b) when $\alpha = .01$. The shaded area, then, represents the probability of a Type II error, β, if the null hypothesis states that $\mu = 1000$, and H_1 that $\mu = 987$.

(i) If $\alpha = .05$, β is the area of the left-hand normal curve in figure (a) to the right of $\overline{X} = 992$. But since $\sigma_{\overline{X}} = 5$ − see (a) above − and since for this curve the mean is 987, an $\overline{X}$ score of 992 corresponds to a z score of $\dfrac{992 - 987}{5} = 1$; $\beta = .5 - .341 = .16$.

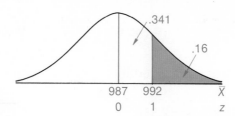

(ii) If $\alpha = .01$, β is the area of the left-hand curve in figure (b), to the right of 988. An $\overline{X}$ score of 988 has a z score of $\dfrac{988 - 987}{5} = .2$, so $\beta = .5 - .079 = .42$.

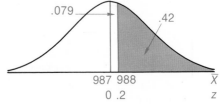

z score	
α	β
.05	.16
.01	.42

less chance of error as $\beta \to 1$

Notice that β increases as α decreases.

Example 3

Again the manufacturer claims his light bulbs have a mean life of 1000 hours with a standard deviation of 40 hours. Again the department store tests a random sample of 64 bulbs at a .05 level of significance. Compute β if in fact the mean life of the light bulbs is (a) 977, (b) 982, (c) 987, (d) 992, (e) 997, (f) 1002 hours.

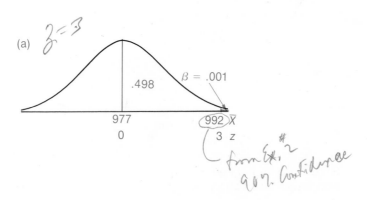

(a) $z = 3$

.498 $\beta = .001$

977   992 $\overline{X}$
0   3 z

from Ex. 2
90% Confidence

Answer

(a)   At a .05 level of significance, the department store accepts the shipment if the mean life of the bulbs in the sample is 992 hours. [See example 1(a).] But if the true population mean is 977, 992 corresponds to a z score

z of $\dfrac{992 - 977}{5} = 3.00$; $\beta = .5 - .499 = .001$.

$\sigma/\sqrt{n} = \dfrac{40}{\sqrt{64}}$

The remaining parts of this example are solved in a similar fashion, as the following diagrams should make clear:

$$3 = \frac{992-982}{5} = 2$$

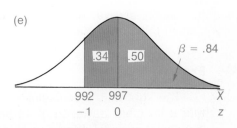

(b)

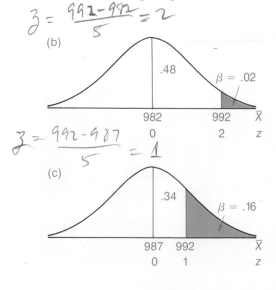

$$3 = \frac{992-987}{5} = 1$$

(c)

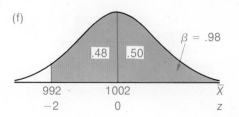

(e)

(d)

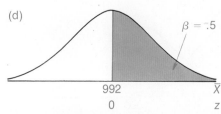

(f)

These results can be summarized and serve as the basis for graphing an **operating characteristic curve** (OC curve), which shows the probability of a Type II error under various forms of H_1 if H_0 is $\mu = 1000$ and $\alpha = .05$.

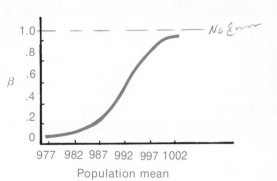

$H_1: \mu =$	β
977	.001
982	.02
987	.16
992	.50
997	.84
1002	.98

It is seen from this curve that the probability of a Type II error is close to 1 if the true population mean is 997 or above, but rapidly diminishes if the true population mean is less than 997; β is almost 0 if the true population mean is 977 or less, but the null hypothesis states that this mean is 1,000.

The value of β, then, depends on the form of the particular alternative hypothesis that you will accept if H_0 is rejected. In practice, however, this is seldom known. As a form for H_1, "$\mu > 100$" is far more likely than "$\mu = 108$." For purely practical reasons,

α is emphasized rather than β. But bear in mind that α is emphasized rather than β often for reasons for safety, too. The new drug being tested is rejected unless it almost certainly is an improvement rather than accepted because it might be an improvement. Remember constantly that choice of α for accepting or rejecting the null hypothesis is a convention; always consider in a particular experiment whether it is really a small α or a small β that is needed.

Example 4

The light bulb manufacturer is still claiming a mean life of 1000 hours with a standard deviation of 40 hours. If $\alpha = .05$, find β for the alternative hypothesis $\mu = 987$ if the size of the sample is (a) 64, (b) 100.

(a)   See Example 3(b) in this section: $\beta = .16$ if $n = 64$.

(b)   If $n = 100$, $\sigma_{\bar{x}} = \dfrac{40}{\sqrt{100}} = 4$.

$\alpha = .05$, so $z = -1.64$.

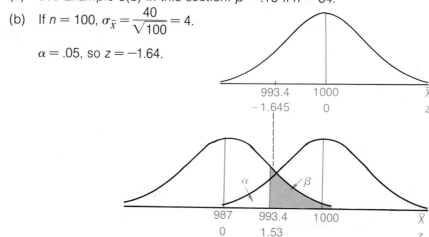

Since $-1.64 = \dfrac{\bar{X} - 1000}{4}$, $\bar{X} = 993.4$, $H_0 : \mu = 1000$ is rejected if the mean of a sample of size 100 is less than 993.4. But if H_1 states that $\mu = 987$, $\beta = p(\bar{X} > 993.4$, given that $\mu = 987) = p(z > \dfrac{993.4 - 987}{40/\sqrt{100}} = 1.61) = .5 - .445 = .06$.

For a fixed α (=.05), β gets smaller as n gets larger: if $n = 64$, $\beta = .16$; if $n = 100$, $\beta = .06$.

How is β determined if H_1 calls for a two-tail test?

Example 5

The manufacturer has given up the making of light bulbs and has become president of a company that makes electrical resistors. A supplier who purchases parts from him is wary of resistors that are either too high or too low. The manufacturer claims that part P36483E has a mean resistance of 1000 ohms with a standard deviation of 40 ohms. The supplier chooses a random sample of 64 from his order, and is testing the manufacturer's claim at a .05 level of significance.

(a)   The manufacturer's claim should be accepted if the sample mean is between what values?

(b)   What is the probability of a Type II error if the mean resistance of all the P36483E resistors is actually 980 ohms? 985? 990? 995? 999? 1005? 1010? 1015? 1020?

    (c) Draw the operating characteristic curve.

    (a) $H_0: \mu = 1000$, $H_1: \mu \neq 1000$, $\sigma = 40$, $n = 64$, $\alpha = .05$, so $z = \pm 1.96$. [.025 handwritten]

$$\pm 1.96 = \frac{\overline{X} - 1000}{40/\sqrt{64}}; \overline{X} = 990 \text{ or } 1010.$$

H_0 is accepted if $\overline{X}$ is between 990 and 1010.

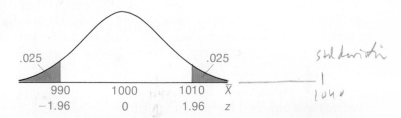

.025   .025

| 990 | 1000 | 1010 | $\overline{X}$ |
| -1.96 | 0 | 1.96 | z |

[handwritten right of curve: std deviation / 1 over 1040]

    (b) If H_1 is $\mu = 980$ [or 985 or . . . 1020], what is the probability that a random sample of size 64 will have a mean between 990 and 1010?

[handwritten left margin: $z_1 = \frac{980-990}{5} = 2$; $z_2 = \frac{980-1010}{5} = 6$]

    (i) $\mu = 980$, $\sigma_{\overline{x}} = 40/\sqrt{64} = 5$, $p(990 < \overline{X} < 1010) = p(2 < z < 6) = .5 - .4772 = .02$.

    (ii) $\mu = 985$, $\sigma_{\overline{x}} = 5$, $p(990 < \overline{X} < 1010) = p(1 < z < 5) = .5 - .341 = .16$.

    (iii) $\mu = 990$, $\sigma_{\overline{x}} = 5$, $p(990 < \overline{X} < 1010) = p(0 < z < 4) = .50$. [handwritten: $-0 = .5$]

    (iv) $\mu = 995$, $\sigma_{\overline{x}} = 5$, $p(990 < \overline{X} < 1010) = p(-1 < z < 3) = .341 + .479 = .82$.

    (v) $\mu = 999$, $p(990 < \overline{X} < 1010) = p(-1.8 < z < 2.5) = .464 + .494 = .96$.

    (vi) $\mu = 1005$, $p(990 < \overline{X} < 1010) = p(-3 < z < 1) = .341 + .478 = .82$.

Can you see the symmetry of these results? Without further work, can you determine β when $\mu = 1010$, 1015, 1020?

    (c) Here is the operating characteristic curve (OC curve) for this example:

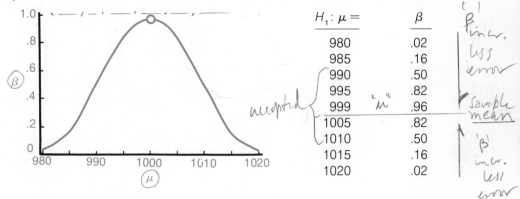

$H_1: \mu =$	β
980	.02
985	.16
990	.50
995	.82
999	.96
1005	.82
1010	.50
1015	.16
1020	.02

[handwritten annotations: finer, less error / sample mean / β incr., less error]

11.7 WARNINGS

    1. Under our rules of law, a defendant who is found not guilty by a jury cannot be tried for the same crime again. The same kind of thing is not true with hypothesis testing. If the manufacturer claims the mean life of his light bulb is 1000 hours with $\sigma = 40$ hours, a test of a random sample of 64 light bulbs showing their mean life is 998 hours gives

a result consistent with the hypothesis at a .05 level of significance. The result is, however, also consistent with the hypothesis that the mean life of all light bulbs is 998 or 996 hours. Such an outcome, then, does not prove that the null hypothesis is true; it can only be concluded that the null hypothesis may be true.

If, on the other hand, the random sample of 64 bulbs has a mean life of 960 hours, then it can be shown that such a sample is extremely unlikely to be chosen if the null hypothesis is true; we conclude that H_0 is probably false. Acceptance of H_0 implies that it **may be** true; rejection of H_0 implies that it is very likely to be false.

2.   In testing hypotheses, the assumption is always made that the sample being tested is a random sample. If a sample is deliberately chosen with only high scores, for example—that is, the sample mean occurs in the extreme right tail of the distribution of sampling means—then no conclusions can be drawn about the probability of such a mean occurring at random, and no fair test of the null hypothesis has been made.

3.   It is assumed that the sampling distribution of the statistic found is known. This distribution may, however, be known only if certain assumptions are satisfied. Up until now, we have assumed the sampling distribution is normal (the population is normal and σ is known, or else $n \geq 30$). In the next chapter we shall find another kind of distribution if σ is not known and n is small.

4.   Always remember that H_0, H_1, and α are determined before the test is carried out, and that the form of H_1 determines whether a one-tail or two-tail test is carried out.

11.8   VOCABULARY AND NOTATION

null hypothesis H_0 — Type II error β
alternative hypothesis H_1 — one-tail test
level of significance — two-tail test
Type I error α — operating characteristic (OC) curve

11.9   EXERCISES

1.   A machine fills milk bottles; the mean amount of milk in each bottle is supposed to be 32 oz. with a standard deviation of .06 oz. In a routine check to see that the machine is operating properly, 36 filled bottles are chosen at random and found to contain a mean of 32.1 oz. At a .05 level of significance, is the machine operating properly? (See Example 4, page 178.) $= 1-.05 = .95 \quad -1.96 < z < 1.96$

2.   It is known that 1-year-old dogs have a mean gain in weight of 1.0 pound per month with a standard deviation of .40 pound. A special diet supplement, HELTHPUP, is given to a random sample of 50 1-year-old dogs for a month; their mean gain in weight is 3.15 pounds, with a standard deviation of .30 pound. Does HELTHPUP affect weight gain in 1-year-old dogs? (.01 level of significance.)

3.   An economist working for the Bureau of Labor Statistics knows that last month 7.1 per cent of those in the labor force in the United States were unemployed. This month he discovers that 350 are unemployed in a random sample of 5000. At a .05 level of significance, has there been a change in the unemployment rate? $-1.96 < z < 1.96$

4.   The registrar of Peterson University is comparing the grade-point averages of married and unmarried students. He finds that 100 married students, chosen at random, have a mean GPA of 2.85 with a standard deviation of .4, while a random sample of 100 unmarried students has a mean GPA of 2.73, with a standard deviation of .3. At a .10 level of significance, is there a difference between the grade-point averages of married and unmarried students?

5.   A political scientist is investigating the difference in the proportion of registered Republicans among well-to-do and poor voters. He takes random samples of 1000 voters whose families have incomes over $14,000 and of 1000 whose families have incomes under $6,000. He finds that in these 2 groups 25.3 per cent and 22.0 per cent, respectively, are registered Republicans. At a .05 level of significance, do well-to-do and poor voters register Republican in different proportions? *(handwritten note)*

6.   The chairman of the French department at Norwich College finds that a random sample of 100 Norwich students who have studied French for one year have a mean score of 470 on a language achievement test; the national average on this test after one year of French in college is 500 with a standard deviation of 100. Do students at Norwich differ significantly from the national average? (.01 level of significance.)

(handwritten margin note)

7.   The mean number of years of school completed by adults over 21 in the United States is 10.4 with a standard deviation of 2.0 years. The Board of Education in Madison, New Jersey, surveys a random sample of 200 adult residents and finds that their mean number of years of school completed is 11.3 with a standard deviation of 1.8 years. At a .05 level of significance, do Madison adults differ from the national average in years of schooling completed? *Don't need Std. Deviation of sample!*

8.   Is the proportion of students applying for financial aid the same among men and women students? A university Dean of Admissions randomly chooses from his files the folders of 60 male and 50 female applications for admission. He finds that, in this sample, 36 men and 35 women had applied for aid. What is his conclusion, at a .01 level of significance?

9.   Twelve out of 100 randomly selected West Side retail stores were robbed in 1975, as were 24 out of 150 randomly selected East Side stores. At a .10 level of significance, do robbers prefer the East Side? *One tail test - 2 Random Samples* $\left(\begin{array}{c} \text{see p.14} \\ \text{p. 189}\end{array}\right)$

10.   A test of a random sample of 100 True Blue cigarettes by the Madder Laboratories shows an average of 17.2 mg. of tar, with a standard deviation of 2 mg. (a) Do these results support the hypothesis that the amount of tar in True Blue cigarettes does not differ from 18 mg.? (b) Do these results support the claim that the amount of tar is at most 17 mg.? (.05 level of significance.)

11.   In 1970, 25 per cent of the fire alarms in Edgerton were false alarms. A reporter for the Edgerton News finds that this year 42 of a random sample of 150 fire alarms have been false. Has the proportion of alarms which are false increased since 1970? ($\alpha = .01$.) *proportion See p.183 similar problem*

12.   A flu epidemic caused many employees of the Firethorn Tire Co. to be absent in January. In a random sample of 400 employees it was found that the mean number of days absent for 150 employees who had had flu shots was 2.4 with a standard deviation of .3, while the mean number of days absent for the rest of the sample (who had had no flu shots) was 6.5 days with a standard deviation of .6. Did the flu shots decrease the mean number of days absent by at least 2 days? ($\alpha = .05$.)

Wk-2wcy

Sue #3
_Ex #3_
_p.188_

13.   It is suspected that the mean salary of professors in public colleges is $1,000 more than in private colleges. A random sample of 100 public college professors shows a mean salary of $15,000 with a standard deviation of $1,000, while 100 private college professors selected at random have a mean salary of $13,800 with a standard deviation of $900. At a .01 level of significance, is the salary of public college professors at least $1,000 more than that of private college professors? _2 Random Samples w/ known σ's_

14.   Supermarkets throughout the United States sell an average of 460 quarts of milk per day. Over a sample of 36 days chosen at random, a supermarket sells an average of 470 quarts per day, with a standard deviation of 20 quarts. Does the store sell more milk than the national average, at a .05 level of significance?

15.   A group of 40 sophomores and a group of 40 seniors were given a geography test with the following results: _(See Ex #3 p.188)_ _2 Random Samples w/ Std. Dev. Known_

from Calc. Sophomores				Seniors			
Score	mid pt. f	f	fx̄	Score	midpt f	f	fx̄ _from Calc._
90-99	94.5	8	756	90-99	94.5	7	661.5 x̄=77.5
80-89	84.5	12	1014	80-89	84.5	8	676
70-79	74.5	13	968.5	70-79	74.5	15	1117.5 S²=108.7
60-69	64.5	7	451.5	60-69	64.5	10	645.

x̄=79.75
S²ₛₚₕ = 102.5

Assume both groups were chosen at random. Did seniors do less well than sophomores on the test, at a .01 level of significance?

16.   A pharmaceutical company claims that a drug which it manufactures relieves cold symptoms for a period of 10 hours in 90 per cent of those who take it. In a random sample of 400 people with colds who take the drug, 350 find relief for 10 hours. At a .05 level of significance, is the manufacturer's claim correct?

17.   You are suspicious that a certain die is weighted. You roll the die 1200 times; a 3 comes up 230 times. At a .05 level of significance, are your suspicions justified? _(See Ex #2 p.184/5) 1 Random Sample, Don't know Std. Dev. 2 tail typ_

18.   A reading test is given to random samples of 200 boys and 100 girls in the third grade in Michigan public schools. The boys have a mean score of 60 with a standard deviation of 10; the girls have a mean of 65 with a standard deviation of 8. Do boys read less proficiently than girls in the third grade in Michigan schools, at a .01 level of significance?

19.   Random samples of 400 axles manufactured by machine A and 300 axles manufactured by machine B showed 10 and 3, respectively, to be defective. At a .05 level of significance, (a) do the two processes differ in the proportion of defective axles? (b) Does machine A produce a higher proportion of defective axles than B? _(a) 2 tail typ + (b) 1 tail typ = 1/2 The confidence 90%. vs .95%._

20.   In 1974 the mean expenditure for food per week for a family of four in Central City was $68 with a standard deviation of $6.00. A random sample of 50 families of four people in the first quarter of 1975 shows a mean expenditure of $72 a week with a standard deviation of $5.00. Has the mean expenditure on food for a family of four in Central City risen, at a .05 level of significance?

21.   In 1971 there were 54,700 motor vehicle deaths in the U.S.; in 1972 there were 56,600. In Oregon there were 695 motor vehicle deaths in 1971 and 734 in 1972. At a .01 level of significance, does the rate of change in the death incidence in Oregon differ from that of the nation?

22.   In 1960, a random survey of 1000 American families with at least one child under 18 revealed that 100 families had 4 or more children. A random survey of 1000 similar families in 1970 revealed that 98 had four or more children. At a .01 level of significance, did the proportion of U.S. families with 4 or more children fall during the decade 1960 to 1970?

23.   A passenger railway has been carrying a mean of 12,000 passengers per day with a standard deviation of 2000. The railway is reorganized and within the first year a random sample of 30 days shows a mean of 12,500 passengers a day, with standard deviation of 1000. Has the number of passengers carried per day increased, at a .02 level of significance?

24.   Over a long period of time, the percentage of A's in Economics 121 at Bulwich University has been 20 per cent. During one Fall term there were 24 in a class of 100 students. Has the proportion of A's increased significantly? (.05 level of significance.)

25.   A new cafeteria is planned for Claudius College. The manager thinks 1800 students can be served dinner between 5:30 and 7:00 if the average time a student spends eating dinner is 30 minutes or less. He times a random sample of 64 students and finds their mean time for dinner is 32 minutes with a standard deviation of 6 minutes. Does the manager conclude that the cafeteria will serve 1800 students in the given hours? (.02 level of significance.)

26.   A piece of land is divided into 60 plots of equal size, fertility, sunlight, and so on. A new fertilizer, GROWCO, is tested by spreading it on 30 plots chosen at random, while 5-10-5 fertilizer is used on the other 30 plots as a control. Each plot is planted with the same number of tomato seedlings. The mean number of pounds of tomatoes from plots with GROWCO is 130 pounds, with a standard of deviation of 10 pounds; the mean for the 5-10-5 plots is 125 pounds with a standard deviation of 8 pounds.
   (a) Is there a difference in the weight of tomatoes produced on the GROWCO and 5-10-5 plots, at a .05 level of significance?
   (b) Do the GROWCO plots produce more tomatoes, at a .05 level of significance?

27.   Nationwide, the mean grade on a college entrance examination was 75 with a standard deviation of 8. A random sample of 144 students in New Mexico had a mean of 76 with a standard deviation of 8.
   (a) At a .05 level of significance, did the students in New Mexico do better than the national average?
   (b) What is the probability of accepting $H_0: \mu = 75$ when the mean of New Mexico students taking the test is actually (i) 78, (ii) 77, (iii) 76, (iv) 75, (v) 74?
   (c) Draw the operating characteristic curve.

28.   A scuba diver finds a trunk full of gold and silver coins. He samples 10 coins, with replacement, and has adopted the following decision rule: accept the hypothesis that half the coins are gold if 3 to 7 of the coins he samples are gold, and reject it otherwise.
   (a) Find the probability of rejecting the hypothesis when it is true.
   (b) Find the probability of accepting the hypothesis that half the coins are gold when the true proportion of gold coins is (i) .1, (ii) .2, (iii) .3, (iv) .4, (v) .6.
   (c) Draw the operating characteristic curve by making a graph with P on the horizontal axis and β on the vertical axis.

ANSWERS

mean

95% prob.

1. $H_0: \mu = 32.0$; $H_1: \mu \neq 32.0$. $\alpha = .05$; H_0 is accepted if z is between -1.96 and $+1.96$.

$$z = \frac{32.09 - 32.0}{.06/\sqrt{36}} = 3.00.$$ H_0 is rejected; the machine needs adjustment.

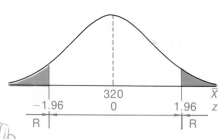

Sample Prob = .07
350/5000

probability

$\sqrt{\frac{P(1-P)}{n}}$

3. $H_0: P = .071$; $H_1: P \neq .071$. $\alpha = .05$; H_0 is rejected if $z > 1.96$ or if $z < -1.96$;

$$z = \frac{.070 - .071}{\sqrt{.071 * .929/5000}} = -.28.$$ Accept H_0; compare with Example 2, Section 11.5.

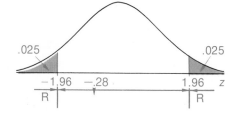

5. $H_0: P_X - P_Y = 0$; $H_1: P_X - P_Y \neq 0$. $\alpha = .05$; H_0 is rejected if $z > 1.96$ or if $z < -1.96$.

$$z = \frac{(.253 - .22) - 0}{\sqrt{\dfrac{.253 * .747}{1,000} + \dfrac{.22 * .78}{1,000}}} = 1.74.$$

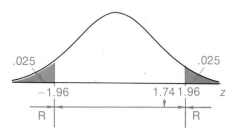

There is no difference in the proportion of registered Republicans in the two groups.

7. $H_0: \mu = 10.4$; $H_1: \mu \neq 10.4$. $\alpha = .05$; the critical values of z are ± 1.96.

$$z = \frac{11.3 - 10.4}{2/\sqrt{200}} = 6.36$$

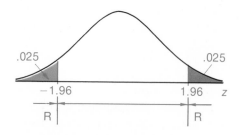

Reject H_0.

9. W = West Side, E = East Side. $H_0: P_W = P_E$; $H_1: P_W < P_E$. $\alpha = .10$ so $z = \pm 1.64$.

$p_W = \dfrac{12}{100} = .12$, $p_E = \dfrac{24}{150} = .16$.

$$\sigma_{(p_W - p_E)} = \sqrt{\frac{.12 * .88}{100} + \frac{.16 * .84}{150}} = .044$$

$z = \dfrac{(.12 - .16) - 0}{.044} = -.91$, which falls in the acceptance region; robbers treat West and East Sides equally, at a .10 level of significance.

11. $H_0: P = .25$, $H_1: P > .25$, $p = .28$, $\sigma_p = \sqrt{\dfrac{.25 * .75}{150}} = .0354$, $z = \dfrac{.28 - .25}{.0354} = .85$, < 2.33. H_0 is accepted; the proportion of false alarms has not increased significantly.

13. X = public colleges, Y = private colleges. $H_0: \mu_X - \mu_Y = 1000$; $H_1: \mu_X - \mu_Y > 1000$. $\alpha = .01$, $z = 2.33$.

$$\sigma_{(\bar{X} - \bar{Y})} = \sqrt{\frac{1000^2}{100} + \frac{900^2}{100}} = 134$$

$$z = \frac{(15,000 - 13,800) - 1000}{134} = 1.49$$

H_0 is accepted; professors at public colleges are paid at least $1,000 more than those at private colleges.

15. The mean of the sophomores = $\bar{X} = 79.8$, $s_X^2 = 102.5$.
The mean of the seniors = $\bar{Y} = 77.5$, $s_Y^2 = 108.8$.
$H_0: \mu_X = \mu_Y$ (or $\mu_X - \mu_Y = 0$); $H_1: \mu_X > \mu_Y$; $\alpha = .01$, one-tail test.
Reject H_0 if $z > 2.33$.

$n_X = 40$; $\sigma_{\bar{X}}^2 = \dfrac{102.5}{40} = 2.56$

$n_Y = 40$, $\sigma_{\bar{Y}}^2 = \dfrac{108.8}{40} = 2.72$

$\sigma_{(\bar{X} - \bar{Y})} = \sqrt{2.56 + 2.72} = 2.30$

$z = \dfrac{(79.8 - 77.5) - 0}{2.30} = 1.00$

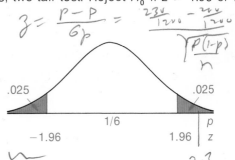

Accept H_0; there is no difference between seniors' and sophomores' grades at a .01 level of significance.

17. $H_0: P = 1/6$; $H_1: P \neq 1/6$. $\alpha = .05$, two-tail test. Reject H_0 if $z > 1.96$ or if $z < -1.96$.

$p = 230/1200$; $\sigma_p = \sqrt{\dfrac{1/6 * 5/6}{1200}} = .011$

$z = \dfrac{230/1200 - 1/6}{.011} = 2.3$

Reject H_0; your suspicions are well-founded.

19.   (a) $H_0: P_X = P_Y;$ $H_1: P_X \neq P_Y.$ $\alpha = .05;$ two-tail test. Reject H_0 if $z > 1.96$ or if $z < -1.96.$

$$\sigma_{p_X}{}^2 = \frac{10/400 * 390/400}{400} = .000061; \quad \sigma_{p_Y}{}^2 = \frac{3/300 * 297/300}{300} = .000033$$

$$\sigma_{(p_X - p_Y)} = \sqrt{.000061 + .000033} = .00969; \quad p_X = .025, \quad p_Y = .010$$

$$z = \frac{(.025 - .010) - 0}{.0097} = 1.55$$

[handwritten: $z = \frac{\bar{X} - \mu_0}{\sigma}$]

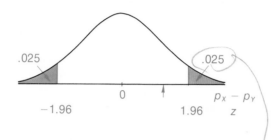

There is no difference between the two machines at a .01 level of significance.

(b) $H_1: P_X > P_Y;$ one-tail test. Reject H_0 if $z > 1.64;$ the computed value of z is 1.55 as in (a).

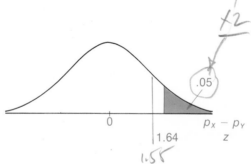

[handwritten: 1.55]

Machine A does not produce a higher proportion of defective axles.

21.   These are census rather than sample figures; statistical inference is not needed. There was a 5.6 per cent increase in motor vehicle deaths in Oregon, which differs from the 3.5 per cent increase in the United States.

23. $H_0: \mu = 12,000;$ $H_1: \mu > 12,000.$ $\alpha = .02,$ so H_0 is accepted if $z < 2.05.$

$$z = \frac{12,500 - 12,000}{2000/\sqrt{30}} = 1.37$$

H_0 is accepted; there has been no significant increase in the number of passengers.

25. $H_0: \mu = 30;$ $H_1: \mu > 30.$ $\alpha = .02;$ the critical value of z is 2.05.

$$z = \frac{32 - 30}{6/\sqrt{64}} = 2.67$$

H_0 is rejected. He concludes that 1800 students cannot be served between 5:30 and 7 p.m.

27. (a) H_0: $\mu = 75$, H_1: $\mu > 75$. $\sigma_{\bar{x}} = 8/\sqrt{144} = 2/3$; $\alpha = .05$, so $z = 1.64$.

$$z = \frac{76 - 75}{2/3} = 3/2 = 1.5, < 1.64$$

H_0 is accepted; the students in New Mexico did not differ significantly.

  (b) If $\alpha = .05$, H_0: $\mu = 75$ is accepted if $z < 1.64$, which implies $X < 76.1$
   (i) $\beta = p(X < 76.1/\mu = 78) = p(z < -2.85) = .002$.
   (ii) $\beta = p(X < 76.1/\mu = 77) = p(z < -1.35) = .09$.
   (iii) $\beta = p(X < 76.1/\mu = 76) = p(z < \quad .15) = .56$.
   (iv) $\beta = p(X < 76.1/\mu = 75) = p(z < \quad 1.64) = .95$.
   (v) $\beta = p(X < 76.1/\mu = 74) = p(z < \quad 3.15) = 1.00$.

  (c)

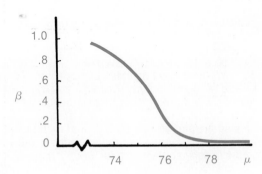

chapter twelve

Guinness
employee
for √ ≤30 dyms
it freedom

student's *t* distributions

Only it you suspect both are Normal

You know how to make inferences about the population mean if it can be assumed that the sampling distribution of means is a normal distribution. What can be done, however, if σ is unknown and n is small so that this assumption of normality can not be made? This question will be answered in this chapter.

You will meet a family of t distributions (a different one for each value of n), but fear not, because you will soon discover that, as n gets larger, the corresponding t distribution gets very close to the normal distribution, and, even when n is small, a t distribution is used in much the same familiar way as a normal distribution.

You will learn when to use t scores and when to use z scores, and ***how*** *to use t scores, both for a population mean and for the differences in means of two populations. Finally, you will discover how to treat paired scores in samples which are not independent.*

12.1 THE NEED FOR *t* DISTRIBUTIONS

By now you should be quite comfortable with the sampling distribution of means, and you know that it is a normal distribution if (a) the population is normally distributed or (b) n is sufficiently large ($n \geq 30$). It is essential to recognize that if the sample means are normally distributed, then their standard or z scores $\dfrac{X - \mu}{\sigma_X}$ are normally distributed, and the table which gives areas under the standard normal curve (Table 4, Appendix C) may be consulted. But if a distribution is not normal, then the corresponding distribution of z scores is not normal either. For example, the probability distribution which consists of the scores 0, 3, 12, each with probability 1/3, has mean 5, and standard deviation $\sqrt{26}$, and is not normally distributed. This distribution has z scores $-5/\sqrt{26}$, $-2/\sqrt{26}$, $7/\sqrt{26}$, which are also not normally distributed.

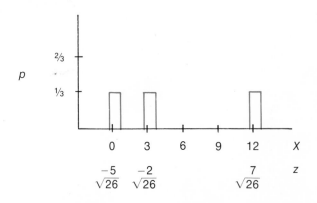

In discussing a confidence interval for a mean or for a null hypothesis about a mean, four possible cases arise. You are familiar with three of them.

Case 1.   **If _n_ is large and σ is known:**   There is no problem. The sample means $\overline{X}$ are normally distributed and so are their standard scores $\dfrac{\overline{X} - \mu}{\sigma/\sqrt{n}}$. In practice "_n_ is large" means $n \geq 30$. If the population distribution is very much skewed (incomes of all U.S. heads of household, say) it would be wise if you, as a psychologist or economist setting up an experiment, were to increase the size of your sample — if you can afford the time and money.

Case 2.   **If _n_ is small (< 30) and σ is known:**   Here one relies on the statement that the sampling distribution of means is normal if the population has a normal distribution (even if _n_ is small). Thus there is again no difficulty **if** the population is normally distributed. (If it is not, however, then you have no choice but to increase the size of your sample, regardless of time or expense. If you don't, your inferences are meaningless.)

Example 1

Past studies have shown that the heights of all American 18-year-olds are normally distributed, that σ is always 2.3 inches, but that mean height is gradually increasing. In 1970 a random sample of 16 American 18-year-olds has a mean height of 68.0 inches. Find a 95 per cent confidence interval for the height of all American 18-year-olds in 1970.

At a 95 per cent confidence level, $z = \pm 1.96$.

$$z = \pm 1.96 = \frac{68 - \mu}{2.3/\sqrt{16}} = \frac{\overline{X} - \mu}{\sigma\sqrt{n}}$$

$$\mu = 68 \pm 1.96 * 2.3/\sqrt{16} = 68 \pm 1.1.$$

The 95 per cent confidence interval for μ is 66.9 to 69.1 inches.

This solution cannot be criticized for its use of standard scores and areas under the normal curve. The problem itself can be questioned, however, on two grounds: First, how is the random sample selected? (Is it truly **random?**) Instead of "all Americans" aged 18, the population may really consist of "all seniors at Ishkosh High School" or "all apprentice plumbers in Essex County, Massachusetts." The second reason for questioning the problem is the size of the confidence interval when _n_ is small. If the same example is carried out ($\overline{X} = 68$ in., $\sigma = 2.3$ in.) but with $n = 100{,}000$, the new 95 per cent confidence interval is 67.95 to 68.05 in. (compared to 66.9 to 69.1 in. when $n = 16$). If you wish to make a comparison with the height of American 18-year-olds in 1940, a small sample may hide any real difference.

Case 3.   **If _n_ is large and σ is unknown:**   In this case, $\sigma_{\overline{x}}$ cannot be computed by using $\sigma/\sqrt{n}$. Earlier (page 143) we discovered that s^2 is an unbiased estimator of σ^2 and that the scores $\dfrac{\overline{X} - \mu}{s/\sqrt{n}}$ are approximately normally distributed if $n \geq 30$. You will better understand the reason for this approximation shortly.

Case 4.   **If _n_ is small and σ is unknown:** Here _s_ is known or can be computed.

Take all possible samples of size *n,* compute the mean of each sample, and then for each compute a "*t* score":

$$t = \frac{\bar{X} - \mu}{s/\sqrt{n}}$$

If the population is normally distributed (and only in this case!), the distribution of all these *t* scores for different samples of size *n* is called a *t* distribution; it is not a normal distribution. (An Irishman named William Gosset worked out the *t* distribution. The brewery for which he worked would not allow him to publish his results, so he did so anonymously with the pseudonym "Student"; hence the term "Student's *t* distribution," or, simply, "*t* distribution.") Actually it would be much better to use a name other than "*t* distribution." You can determine a *t* score for any sample, and all possible *t* scores for different samples certainly have a distribution, but not necessarily a *t* distribution. You will see the distinction more clearly if you think of *z* scores (such as $-5\sqrt{26}$, $-2/\sqrt{26}$, as on page 204; they have a distribution which might be called a *z* distribution, but it is not a normal distribution. The phrase "*t* distribution" for *t* scores is used so commonly by statisticians, however that its use will be continued here. But always remember that *t* scores must be taken from a normally distributed population if they are to form a *t* distribution.

Example 2

　Two samples, each of size 100, are taken from a population with $\mu = 100$, $\sigma = 10$. The samples have the same mean: $\bar{X}_1 = \bar{X}_2 = 110$, but different standard deviations: $s_1 = 9$, $s_2 = 8$. Compute the *z* and *t* scores for each sample.

$$z = \frac{\bar{X} - \mu}{\sigma/\sqrt{n}}: \qquad z_1 = \frac{110 - 100}{10/\sqrt{100}} = 10, \qquad z_2 = \frac{110 - 100}{10/\sqrt{100}} = 10$$

$$t = \frac{\bar{X} - \mu}{s/\sqrt{n}}: \qquad t_1 = \frac{110 - 100}{9/\sqrt{100}} = 11.1, \qquad t_2 = \frac{110 - 100}{8/\sqrt{100}} = 12.5$$

Note that $z_1 = z_2$, but $t_1 \neq t_2$.

　This example illustrates that if two samples from the same distribution have the same mean, they have the same *z* scores (even if they have different standard deviations). But two samples with the same mean and different standard deviations will have different *t* scores. The *z* scores of means of all possible samples are normally distributed if the population is normal or if *n* is large. The corresponding *t* scores are not identical with the *z* scores; the *t* scores are **not** normally distributed. We must look at *t* distributions to see what kind of curves they have, just as earlier we looked at normal distributions and normal curves.

12.1　CHARACTERISTICS OF *t* DISTRIBUTIONS

　Student's *t* distributions have the following characteristics:

1.　There is not just one *t* distribution but a different one for each value of *n*. There is one standard normal curve, and a one-page table can give areas under this curve for most of the values of *z* that we are interested in. There is a whole family of "standard" *t* curves, however. If a table were given for areas under the *t* curve for **each** *n* (comparable to the table for areas under the standard normal curve), it would take quite a volume. The *t* tables will be very much abbreviated. In section 12.5 you will learn what they look like and how to use them.

2. Every *t* curve is symmetric about 0. This means that if the right-hand half of a *t* curve looks like this

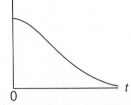

then the left-hand half must be its reflection; the whole curve will look like this:

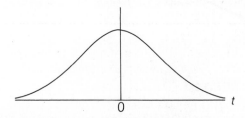

Because of this, the **mean of every t distribution is 0.**

3. The highest point on the curve occurs when $t = 0$.

4. As *n* gets larger, the *t* curves get closer and closer to the normal curve. It's easy to see this by looking at some superimposed graphs:

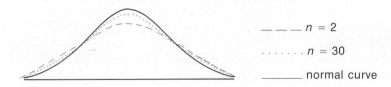

$-\,-\,-\,-$ $n = 2$

$\cdots\cdots$ $n = 30$

_____   normal curve

If *n* is 30 or larger, a *t* distribution and the standard normal distribution are sufficiently close so that areas under the latter may be used to approximate areas under the former. It is this closeness of *t* and *z* curves for large *n* which justifies the use of $s/\sqrt{n}$ to approximate $\sigma_{\bar{x}}$ when *n* is large and σ is unknown (Case 3, page 205).

5. Every *t* distribution is a probability distribution—that is, the area under the whole curve is 1, and the probability that a *t* score is between *a* and *b* equals the area under the curve between the lines $t = a$ and $t = b$.

12.3   WHEN IS A t DISTRIBUTION USED?

A *t* distribution is used when:

(a) the population is normally distributed **and**

(b) σ is unknown (but *s* is known or can be computed) **and**

(c) $n \leq 30$ (above N=30+, use Normal Distrib.)

If the population is not normally distributed, then you will be making an error. How large is the error? It depends on how far from being a normal distribution the population is. If it is almost normal, then your error is small. If it is very far from being normal, then the *t* distribution is pretty useless. In practice, the problem is that you often don't know whether or not the population follows a normal curve.

What do you do then? One answer is to use a different test—one which does not require the assumption that the population is normal. Such tests are discussed in Chapter 17. Another answer is to increase the size of the sample. The less confident you are that the population is normal, the larger your sample size should be. Be very careful not to use very small samples unless you are quite certain that the population is normally distributed. This is the main trouble with *t* tests: they are most questionable when they are most useful (that is, with small samples), but many a researcher has forgotten to question the population distribution. Beware!

If σ is known, then use standard or *z* scores and the normal distribution.

If $n > 30$, then the *t* and *z* curves are so close together that it doesn't much matter which you use if the population is normal.

12.4 DEGREES OF FREEDOM

How is a *t* distribution used? Before going on, we must sidetrack for a moment and consider degrees of freedom.

Imagine three children playing a very simple game: three cards are marked, respectively, 0, 10, and 20. The cards are shuffled, and each child in turn chooses one card. Jimmy is impulsive and grabs a card first; he gets 10 points. Mary chooses next, and gets 0. Tommy takes the last card, he must get 20 points. If, however, Jimmy drew 0 and Mary 20, then Tommy would get 10. The point is that, once two children have chosen their cards, the third child's card is fixed. Two of the children choose freely: there are two **degrees of freedom.**

If 4 scores have a mean of 50, how many of the scores can be freely chosen? Try it and see: fill in the blanks with whatever scores you like, so long as their mean is 50:

——— ——— ——— ———

I'm going to choose 40 for the first score, then 30 for the second, and 70 for the third. I choose these numbers arbitrarily; because 40 happened to be first, I feel no inhibitions about choosing 30 or 40 or 1,000,000 or −32 for the second. The same is true for the third. But once three have been chosen, then the fourth is fixed by the requirement that the mean be 50. If my first three choices are 40, 30, 70, then my fourth choice can **only** be 60:

$$\underline{40} \quad \underline{30} \quad \underline{70} \quad \underline{60} \qquad \overline{X} = 50$$

If I try anything else for the fourth score, then the mean no longer equals 50:

$$\underline{40} \quad \underline{30} \quad \underline{70} \quad \underline{20} \qquad \overline{X} = 40$$

I have freedom in choosing three of the four scores, and then the fourth is determined. There are three **degrees of freedom.**

If n scores have a mean $\overline{X}$, $n - 1$ can be freely chosen and then the last is determined; there are **$n - 1$ degrees of freedom.**

For finding a confidence interval or testing a hypothesis about the mean using a _t_ distribution for one sample of size _n,_ the number of degrees of freedom is $n - 1$. The letter _D_ is used to denote the number of degrees of freedom:

$$D = (n - 1) \text{ \# Degrees of freedom} \Big\} \text{ for } N \leq 30$$

12.5 TABLE OF _t_ DISTRIBUTIONS   = # sampling – 1

For the standard normal distribution, one table (Table 4, Appendix C) supplies enough information to find the area under any normal curve. Each _t_ distribution, however, depending on the number of degrees of freedom, requires a separate table if the same amount of information is to be given. As a result, information about _t_ distributions is given in very concentrated form.

Table 5, Appendix C, shows values of _t_ for selected areas under the _t_ curve. Different values of _D_ appear in the first column. The table is adapted for efficient use for either one- or two-tail tests. Note very carefully, however, that areas under _t_ curves are given for one or both tails, whereas areas under the normal curve are given from 0 to _z_.) ✳

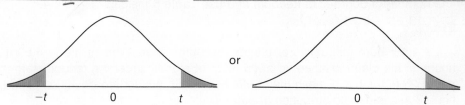

Values of _t_ (See Table 5, Appendix C)

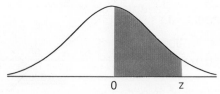

Areas Under the Normal Curve
(See Table 4, Appendix C)

Finding a confidence interval with a _t_ distribution will, of course, always involve a two-tail test. It is necessary to subtract the confidence coefficient from 1 to find the area in the two tails. The confidence interval is wide for small values of _n,_ and therefore a _t_ distribution is used more commonly for testing a hypothesis.

Example 1

If $D = 8$, 5 per cent of _t_ scores are above what value?

p.350 Look in Table 5, Appendix C, along the row labeled "One tail" to the value .05; the intersection of the .05 column and the row with 8 in the _D_ column gives the value of _t: t = 1.860._

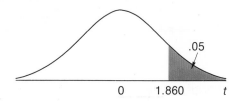

Example 2

If $D = 5$, what is the probability that a t score is above 2.015 or below −2.015?

A two-tail test is implied. Look along the "$D = 5$" row to find the entry 2.015. The probability is .10.

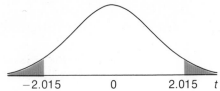

−2.015   0   2.015 t

12.6 CONFIDENCE INTERVALS FOR, AND TESTS OF HYPOTHESES ABOUT, THE MEAN

Relax; confidence intervals and tests of hypotheses about the mean are carried out with the t distribution just as for the normal distribution, except that you must consider the number of degrees of freedom and use Table 5 instead of Table 4.

Example 1

A light bulb manufacturer claims his bulbs have a mean life of 1,000 hours. Is his claim justified, at a .05 level of significance, if a random sample of 25 bulbs has a mean life of 994 hours with a standard deviation of 30 hours? Assume the distribution of burning life for all bulbs is normal.

$H_0: \mu = 1{,}000; H_1: \mu < 1{,}000$; one-tail test; $\alpha = .05$.

$D = 24; H_0$ is rejected if $t < -1.71$.

$$t = \frac{\bar{X} - \mu}{s/\sqrt{n}} = \frac{994 - 1{,}000}{30/\sqrt{25}} = -1.00.$$

std. Dev.

His claim is justified; H_0 is accepted.

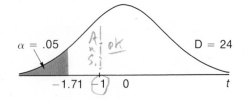

$\alpha = .05$ ok $D = 24$

−1.71   −1   0 t

Example 2

two-tail @ .02

Find a 98 per cent confidence interval for μ if a random sample of size 9 from a normally distributed population has a mean of 14 and a standard deviation of 2.

$$\bar{X} = 14, s = 2, n = 9, D = 8$$

With a .98 confidence coefficient when $D = 8, t = \pm2.90$ (note that the area in the two tails is $1 - .98 = .02$).

$$t = \frac{\bar{X} - \mu}{s/\sqrt{n}}$$

$$\pm2.90 = \frac{14 - \mu}{2/\sqrt{9}}$$

$$\mu = 14 \pm 1.93$$

A 98 per cent confidence interval for μ is 12.1 to 15.9.

Example 3

The specifications for a certain drug call for 30 per cent aspirin in each pill. Sixteen pills are chosen at random and analyzed; their mean content of aspirin is 30.4 per cent with a standard deviation of .8 per cent. Does the drug satisfy the specifications at a .01 level of significance? Assume the errors are normally distributed.

$H_0: \mu = .30$; $H_1: \mu \neq .30$. Two-tail test; $\alpha = .01$.

$\overline{X} = .304$, $s = .008$, $n = 16$, $D = 15$

$t = \dfrac{.304 - .30}{.008/\sqrt{16}} = 2.00.$

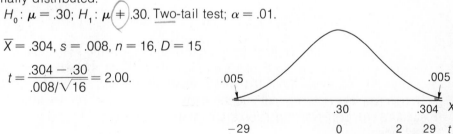

$D = 15$, $\alpha = .01$; H_0 is rejected if t is > 2.95 or < -2.95.

H_0 is accepted; the specifications are satisfied at a .01 level of significance.

12.7 DIFFERENCES OF MEANS FOR INDEPENDENT SAMPLES

A t distribution can be used for testing hypotheses about differences of means for independent samples **only if both populations are normal.** If you suspect this is not the case, then sample sizes should be increased (at least 30 for each) and the sampling distribution of $\overline{X} - \overline{Y}$ approximates a normal distribution. It is also assumed that the samples are both random and are chosen independently.

There are two cases to consider. In the first, the variances of the two populations are equal; in the second, they are not equal. How do you know whether they have the same variance? If you actually knew the variance of each, you would use the z test as on page 157 instead of a t test. And, because of sampling fluctuations, the samples may have different variances even if the population variances are the same. In Chapter 14 we'll learn how to test the hypothesis that two populations have the same variance. For the moment, assume that you can distinguish the two cases.

Case 1. The Populations Have the Same Variance

Confidence intervals are determined and hypotheses are tested using t tests just as with z tests except that

(a) $\sigma_{\overline{X}-\overline{Y}} = \sqrt{\dfrac{(n_X - 1)s_X^2 + (n_Y - 1)s_Y^2}{n_X + n_Y - 2}\left(\dfrac{1}{n_X} + \dfrac{1}{n_Y}\right)}.$

This formula is arrived at by "pooling" the two sample variances to arrive at an estimate of the common population variance σ^2; its derivation is shown in Section 12.8, page 214.

(b) $D = (n_X - 1) + (n_Y - 1) = n_X + n_Y - 2.$

(c) Table 5 is used instead of Table 4.

Example 1

A random sample of 17 third graders who read poorly has a mean IQ of 98 with a standard deviation of 10; a random sample of 10 third graders who read well has a mean IQ of 101 with a standard deviation of 9. At a .05 level of significance, is there a difference in mean IQ of poor and good readers? Assume the IQ's of both groups are normally distributed and have the same variance.

$$H_0: \mu_X - \mu_Y = 0; \; H_1: \mu_X - \mu_Y \neq 0; \; \text{two-tail test}; \; \alpha = .05.$$

$$\bar{X} = 98, \; s_X = 10, \; n_X = 17; \; \bar{Y} = 101, \; s_Y = 9, \; n_Y = 10$$

$$\sigma_{\bar{X}-\bar{Y}} = \sqrt{\frac{16 * 10^2 + 9 * 9^2}{17 + 10 - 2}\left(\frac{1}{17} + \frac{1}{10}\right)} = \sqrt{\frac{2329}{25} * \frac{27}{170}} = 3.85$$

$$t = \frac{(\bar{X} - \bar{Y}) - (\mu_X - \mu_Y)}{\sigma_{\bar{X}-\bar{Y}}} = \frac{(98 - 101) - 0}{3.85} = -.78$$

$D = 25, \; \alpha = .05$, two-tail test, so $t = \pm 2.060$.

A t score of $-.78$ falls in the acceptance region.

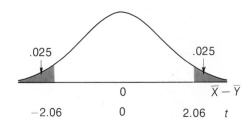

.025   .025

0     $\bar{X} - \bar{Y}$

-2.06 0   2.06 t

There is no difference in the mean IQ of poor and good third-grade readers at a .05 level of significance.

Example 2

Cotton threads are made by two different processes which are to be compared. A random sample of 20 threads manufactured by the first process has a mean breaking strength of 12 ounces with a standard deviation of 1.5 ounces, while a sample of 5 threads manufactured by the second process has a mean breaking strength of 10 ounces with a standard deviation of 2.0 ounces. Assume both populations are normal and have the same variance.

(a) Test the hypothesis that there is no difference in the mean breaking strength of cotton threads made by the two processes at a .05 level of significance.

(b) Find, with 95 per cent confidence, the difference in mean breaking strengths of threads made by the two processes.

Answer

(a) $H_0: \mu_X - \mu_Y = 0; H_1: \mu_Y \neq 0$.

$\overline{X} = 12, s_X = 1.5, n_X = 20; \overline{Y} = 10, s_Y = 2.0, n_Y = 5$

$D = 20 + 5 - 2 = 23, \alpha = .05$, so $t = \pm 2.07$

$$t = \frac{(\overline{X} - \overline{Y}) - (\mu_X - \mu_Y)}{\sqrt{\frac{(n_X - 1)s_X^2 + (n_Y - 1)s_Y^2}{n_X + n_Y - 2}\left(\frac{1}{n_X} + \frac{1}{n_Y}\right)}}$$

$$= \frac{(12 - 10) - 0}{\sqrt{\frac{19 * 1.5^2 + 4 * 2.0^2}{23}\left(\frac{1}{20} + \frac{1}{5}\right)}} = \frac{2.0}{.80} = 2.5$$

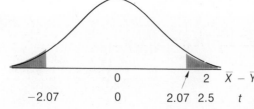

	0	2	$\overline{X} - \overline{Y}$
-2.07	0	2.07 2.5	t

H_0 is rejected: there is a difference.

(b) $\pm 2.07 = \dfrac{(12 - 10) - (\mu_X - \mu_Y)}{.80}$

$\mu_X - \mu_Y = 2 \pm 1.66 = .34$ or 3.66

The 95 per cent confidence interval for the difference in mean breaking strengths is .3 to 3.7 ounces.

Case 2.   The Populations Do Not Have the Same Variance

In this case it is impossible to "pool" the sample variances to estimate a common population variance, so s_X and s_Y are used to estimate σ_X and σ_Y, respectively, and

$$\sigma_{(\overline{X} - \overline{Y})} = \sqrt{\frac{s_X^2}{n_X} + \frac{s_Y^2}{n_Y}}.$$

This seems much simpler than the first case; the trouble comes in determining the degrees of freedom. Suppose the sample variances are fairly close but n_X is much smaller than n_Y. Then s_X^2/n_X is much larger than s_Y^2/n_Y and therefore plays the dominant part in determining σ_{X-Y}, but it would be misleading to use $n_X + n_Y - 2$ degrees of freedom. The formula used to compute the degrees of freedom is

$$D = \frac{\left(\dfrac{s_X^2}{n_X} + \dfrac{s_Y^2}{n_Y}\right)^2}{\left(\dfrac{s_X^2}{n_X}\right)^2 \dfrac{1}{n_X + 1} + \left(\dfrac{s_Y^2}{n_Y}\right)^2 \dfrac{1}{n_Y + 1}} - 2$$

Round off D to the nearest integer.

Example 3

Test again the hypothesis that the cotton threads of Example 2 do not differ in mean breaking strength. Assume that both populations are normal, but this time do **not** assume they have the same variance.

$$H_0: \mu_X - \mu_Y = 0; \; H_1: \mu_X - \mu_Y \neq 0.$$

$$\overline{X} = 12, \; s_X = 1.5, \; n_X = 20; \; \overline{Y} = 10, \; s_Y = 2.0, \; n_Y = 5$$

$$t = \frac{(\overline{X} - \overline{Y}) - (\mu_X - \mu_Y)}{\sqrt{\dfrac{s_X^{\,2}}{n_X} + \dfrac{s_Y^{\,2}}{n_Y}}} = \frac{(12 - 10) - 0}{\sqrt{\dfrac{1.5^2}{20} + \dfrac{2.0^2}{5}}}$$

$$= \frac{2.0}{.95} = 2.1$$

$$D = \frac{\left(\dfrac{1.5^2}{20} + \dfrac{2.0^2}{5}\right)^2}{\left(\dfrac{1.5^2}{20}\right)^2 \dfrac{1}{21} + \left(\dfrac{2.0^2}{5}\right)^2 \dfrac{1}{6}} = \frac{.83}{.11} = 8$$

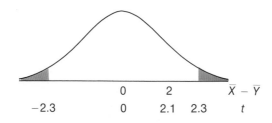

	0	2	$\overline{X} - \overline{Y}$
−2.3	0	2.1 2.3	t

$\alpha = .05$, 8 degrees of freedom, two-tail test, so $t = \pm 2.3$; H_0 is accepted. (In Example 2 it was rejected.)

12.8★   DERIVATION OF $\sigma_{(\overline{X}-\overline{Y})}$ WHEN POPULATION VARIANCES ARE EQUAL

If each score in a set has equal importance, the mean is called **unweighted.** If each score in a set is assigned a weight according to its relative importance, a **weighted** mean can be computed. To do this, first each score is multiplied by its assigned weight, then these products are summed, and finally the total is divided by the sum of the weights. If W is used for weight, then the weighted mean $\overline{X}_W = \dfrac{\Sigma(WX)}{\Sigma W}$.

Example 1

Sam Boggs has a batting average of .300 in the first 6 weeks of a season and .400 in the next 4 weeks. What is his season batting average? He was at bat 70 times in the first 6 weeks, and 30 times in the next 4 weeks.

The question cannot be answered without the information given in the last sentence. With this information, the question can be answered in two ways:

1. He had .300 * 70 = 21 hits in the first 6 weeks and .400 * 30 = 12 hits in the next 4 weeks. 21 + 12 = 33 hits in 70 + 30 = 100 times at bat gives a season batting average of .330.

2. If we weight his batting average by the number of times at bat, the weighted mean is

$$\frac{70 * .300 + 30 * .400}{70 + 30} = .330.$$

You have used a weighted mean in finding the mean of grouped scores, with weighting according to the frequency in each class.

X	f
1–3	9
4–6	18

$$\mu = \frac{2(9) + 5(18)}{9 + 18} = 4$$

The population variance σ^2 is estimated not by s_X^2 nor by s_Y^2, but by their weighted average — with weights assigned by the degrees of freedom; that is,

$$\sigma^2 \approx \frac{(n_X - 1)s_X^2 + (n_Y - 1)s_Y^2}{(n_X - 1) + (n_Y - 1)}.$$

Under the assumption that both populations have the same variance ($\sigma_X = \sigma_Y = \sigma$),

$$\sigma_{(\bar{X}-\bar{Y})}^2 = \frac{\sigma_X^2}{n_X} + \frac{\sigma_Y^2}{n_Y} = \sigma^2\left(\frac{1}{n_X} + \frac{1}{n_Y}\right) = \frac{(n_X - 1)s_X^2 + (n_Y - 1)s_Y^2}{n_X + n_Y - 2}\left(\frac{1}{n_X} + \frac{1}{n_Y}\right).$$

Now take square roots, and you arrive at formula (a) on page 211.

12.9 DIFFERENCES BETWEEN MEANS OF DEPENDENT SAMPLES

Sometimes you do not have independent samples: you want to take one sample and compare the people in it before and after a certain treatment, or you want to be sure certain factors (age, religion, or income, for example) do not affect the results by pairing people with the same age, religion, or income. The purpose in either case is to control variables other than the one being tested. A difference of means test cannot be used here, since you do not have independent samples. Instead, individuals in the two samples are paired, and a t test is made on the mean difference of the scores in each pair. [Note that the differences in scores of all pairs in the population(s) should be normally distributed.] Let d be the difference in each pair, $\bar{d}$ and s_d the mean and standard deviation, respectively, of these differences, and μ_d the mean of the population differences. If there are n pairs of scores, then $D = n - 1$.

Example 1

A random sample of 10 students take a calculus quiz, and receive the grades shown below (X). Then a review session in algebra is held and a similar quiz given (Y). At a .05 level of significance, are the grades better on the second quiz?

Student	Quiz 1 (X)	Quiz 2 (Y)	$Y - X = d$	$d - \bar{d}$	$(d - \bar{d})^2$
1	80	84	+4	1	1
2	50	56	+6	3	9
3	78	81	+3	0	0
4	90	92	−2	−5	25
5	75	76	−1	−4	16
6	70	75	+5	2	4
7	62	72	+10	7	49
8	87	90	+3	0	0
9	95	93	−2	−5	25
10	68	72	+4	1	1
			+30		130

$$H_0: \mu_d = 0$$

$$H_1: \mu_d > 0$$

Mean of differences $= \bar{d} = 30/10 = 3.$

Standard deviation of differences $= s_d = \sqrt{130/9} = 3.8.$

A t test is applied to 10 differences in test scores; $\mu_d = 0$ implies that the mean difference in test scores for the whole population is 0; H_1 says a one-tail (right) test should be made.

$n = 10$, so there are 9 degrees of freedom; the number of differences (10) is used rather than the number of test scores (20). $\alpha = .05$, $D = 9$, one-tail test; the critical value of t is 1.83.

$$t = \frac{\bar{d} - \mu_d}{s_d/\sqrt{n}} = \frac{3 - 0}{3.8/\sqrt{10}} = 2.5$$

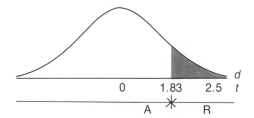

H_0 is rejected; there is a significant difference in scores on the two tests. This does not show, however, that this difference is caused by the review of algebra. It might be due to learning from mistakes on the first quiz, for example.

Example 2

From 240 students registered for a statistics course, 8 pairs of students are selected, well matched for number of years and grades of math courses in high school and college, college class (freshman, sophomore, and so on), and cumulative grade-point average up to the beginning of the statistics course. Then students in each pair are randomly assigned, one (X) to a section of 30 that meets 3 times a week, the other (Y) to 3 lectures each week for 210 students plus a recitation for 15 students once a week. The grades of the 8 pairs of students on the final examination common to both groups are shown below. At a .01 level of significance, do the mean grades on this examination differ?

Pair	X	Y	X − Y = d	d − 3	(d − 3)²
1	90	82	+ 8	5	25
2	85	95	−10	−13	169
3	75	79	− 4	− 7	49
4	78	81	− 3	− 6	36
5	95	88	+ 7	4	16
6	95	91	+ 4	1	1
7	60	50	+10	7	49
8	87	81	+ 6	3	9
			24		354

$$H_0: \mu_d = 0; \ H_1: \mu_d \neq 0$$

$$d = \frac{24}{8} = 3, \ s_d = \sqrt{354/7} = 7.22$$

$n = 8, D = 7, \alpha = .01$, two-tail test; the critical values of t are ± 3.5.

$$t = \frac{3 - 0}{7.22/\sqrt{8}} = 1.2$$

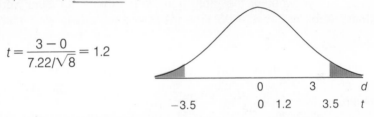

H_0 is accepted; there is no significant difference in final grades.

12.10 VOCABULARY AND SYMBOLS

t distribution	t	independent samples
(Student's t distribution)		paired differences $\qquad$ d
degrees of freedom	D	μ_d
$\sigma_{(\bar{X}-\bar{Y})}$		s_d
weighted mean		

12.11 EXERCISES

1.   What is the probability that the mean of a sample of size 17 drawn at random from a normal population will have a t score less than 2.583?

2.   In a t distribution with 4 degrees of freedom, 10 per cent of t scores are above what score?

3.   Nine workers, chosen at random from a work force of 250 workers in a factory, have a mean wage of $125 a week with a standard deviation of $12. Estimate the mean wages of all workers at a 95% confidence level. (Assume the distribution of all workers' wages is normal.)

4.   A metallurgist made 8 measurements of the melting point of an alloy. His

student's *t* distributions

measurements were 850°, 848°, 852°, 848°, 851°, 850°, 851°, 850°. What is his 90 per cent confidence interval for the melting point of the alloy? Assume his errors in measurement are normally distributed. *2 tail*

5.   Five patients with hypertension are given a new drug which lowers their blood pressure by 4, 25, 13, 18, and 20 points, respectively. At a .05 level of significance, does the new drug lower blood pressure by at least 20 points? Assume the changes in blood pressure of all patients with hypertension are normally distributed. *1 tail*

6.   The Speed Reading Institute claims that adults will read 400 words per minute faster after completing its course. A random sample of 12 adults are found to increase their reading speed by 350 words per minute, with a standard deviation of 50 words per minute. Is the Institute's claim substantiated? (.05 level of significance.)

7.   For a new fuse manufactured by the Portland Fuse Co., the mean time for the filament to melt is 5 seconds, according to specifications, with a standard deviation of 1 second. A random sample of 16 fuses has a mean time of 5.8 seconds with a standard deviation of .95 second. *don't need* Does the batch of fuses from which the sample was taken meet the specifications at a .01 level of significance? Assume the times of all fuses in the batch are normally distributed.

 # know σ ∴ Z-scores

8.   The following numbers are taken in order from a random number table: 3, 9, 5, 2, 7, 1, 7, 6, 2, 4.
   Test to see whether the table from which these numbers were taken is a random number table; $\alpha = .05$. (The mean of a table of random numbers is 4.5.)

9.   An inspector from the Department of Weights and Measures weighs 10 one-pound samples of peanut butter; he finds their mean weight is 15.8 oz. with a standard deviation of .4 oz.   *two tail test*
   (a) Do the weights of packages of peanut butter sold by the shop from which these samples were taken differ from the announced weight? ($\alpha = .05$)
   (b) Are they significantly lighter than they should be? (.05 level of significance.)

10.   A home owner needs to determine the pH of her soil; she will add lime if it is less than 6.5. She takes 16 samples from cores whose location is randomly chosen, and finds the average pH of the samples is 6.3. Must she add lime to the soil? ($\alpha = .10$).

11.   Ten journeyman typesetters in a printing office have an average of 2.3 mistakes per column set. A new apprentice is taken on; a random sample of 10 of his columns has a mean of 2.6 errors with a standard deviation of .5. Does the apprentice have significantly more mistakes than the experienced printers?

Can skip

12.   The water absorbency of 8 randomly chosen diapers is measured when brand new and again after washing 3 times, with the following results:

two sample testg

Diaper:	1	2	3	4	5	6	7	8
Absorbency when new (oz.):	3.4	3.3	3.5	3.2	3.5	3.4	3.6	3.6
Absorbency after washing (oz.):	3.7	3.5	3.5	3.4	3.9	3.6	4.0	4.0

After washing, does the absorbency of diapers increase significantly? ($\alpha = .01$).

13. In the fall semester, 5 students, chosen at random, in a statistics course, have a mean grade of 75 with a standard deviation of 5. In the spring semester, a random sample of 10 students from a statistics course have a mean grade of 77 with a standard deviation of 5. At a .05 level of significance, do students get better grades in statistics in the spring semester? Grades are normally distributed both semesters and have the same variance.

14. Eight white lab coats were washed in Blue Suds and 6 in Whitelite detergent. After washing, the coats were rated for whiteness on a 1–10 scale (10 is whitest); the mean for the Blue Suds coats was 7.5 with a standard deviation of .3; the mean for Whitelite was 6.8 with a standard deviation of .4. Assume the ratings of all lab coats are normally distributed for both kinds of detergents, but the variances are not necessarily equal. Do the two detergents differ, at a .05 level of significance?

15. A high school teachers' group is investigating summer work patterns. It finds that the mean monthly income of 20 randomly selected teachers who teach in the summer is $600 with a standard deviation of $100, while a random sample of 10 teachers who sell real estate during the summer is $700 with a standard deviation of $50. At a .02 level of significance, is there any difference in the earnings of the two groups? The teachers' group believes the pay for both kinds of work is normally distributed, but has no reason to assume the standard deviations are the same.

16. Two types of guns are tested at Picatinny Arsenal. Twelve of the first type and 10 of the second type, chosen at random, are tested. Those of the first type have a mean error of 3.0 inches with a standard deviation of .40 inch, while those of the second type have a mean error of 2.1 inches with a standard deviation of .30 inch. Assume the errors of both guns are normally distributed, with the same variance. At a .05 level of significance, is the mean error of the second type at least 1 inch less than that of the first type?

17. A random sample of 12 male freshmen who live in the dormitory are weighed when they enter college in September and again on November 1, with the following results:

Student	1	2	3	4	5	6	7	8	9	10	11	12
Sept.	150	210	160	150	180	170	200	240	190	193	200	185
Nov. 1	155	210	153	150	185	165	205	230	193	198	205	189

Have male freshmen students in general gained weight in 2 months, at a .05 level of significance? Assume weights of all male freshmen are normally distributed and have a constant standard deviation.

18. An economist is studying the differences in pay of workers in two mills operated by the same textile company. He takes a random sample of size 16 from each plant, pairs them for seniority with good matching, and then discovers the following hourly rates:

Pair:	1	2	3	4	5	6	7	8
X:	$2.50	3.40	3.20	2.60	3.00	2.75	4.00	3.75
Y:	$2.80	3.70	3.00	3.00	3.10	2.85	4.10	3.85
Pair:	9	10	11	12	13	14	15	16
X:	$5.20	4.50	3.00	4.05	4.20	4.15	4.30	4.60
Y:	$5.30	4.60	3.20	4.15	4.30	4.35	4.00	4.70

Test for differences in pay at a .01 level of significance. Assume the differences in pay are normally distributed.

19.   Two different machines are used for manufacturing automobile horns in a factory. The number produced in an hour on each machine is noted for hours chosen at random over a period of several months. Notes on the first machine, taken over 10 hours, show a mean production per hour of 100 horns with a standard deviation of 6.0. The second machine, noted also for 10 hours, has a mean production of 94 horns with a standard deviation of 5.0. Does the first machine produce at least 5 more horns per hour, at a .05 level of significance? Assume the production of both machines is normally distributed, and that the standard deviations of both are the same.

20.   A psychologist discovers that 10 orphans, viewing pictures (which have the same size) of friends and of unknown children, judge the friends to be 2 inches taller, on the average, with a standard deviation of .05 inch. Eight children living with both parents are given the same test and estimate friends to be 1.5 inches taller, with a standard deviation of .04 inch. Find a 95 per cent confidence interval for the difference in heights estimated by orphans and by children living with both parents.

ANSWERS

1. $D = 16$; $p(t > 2.583) = .01$; $p(t < 2.583) = .99$.   *one tail*

3. $\overline{X} = 125$, $s = 12$, $n = 9$, $D = 8$. At a 95 per cent confidence level with $D = 8$, $t = \pm 2.31$.

$$\pm 2.31 = \frac{125 - \mu}{12/\sqrt{9}}$$

$$\mu = 125 \pm 9.2 = 115.8 \text{ or } 134.2$$

The 90 per cent confidence interval is $115.80 to $134.20.

5. $\overline{X} = 16$, $s = 8.0$, $n = 5$, $D = 4$, $\alpha = .05$. $H_0 : \mu = 20$; $H_1 : \mu < 20$; one-tail test. H_0 is rejected if t score of sample mean < -2.132.
$$t = \frac{16 - 20}{8.0/\sqrt{5}} = -1.1; \; H_0 \text{ is accepted.}$$

see p. 2
p. 190

7. $\sigma \, (= 1)$ is known, so use z scores, not t scores! $H_0 : \mu = 5$; $H_1 : \mu \neq 5$; two-tail test; $\alpha = .01$. H_0 is rejected if z score of sample mean is > 2.58 or < -2.58.

$$\overline{X} = 5.8, \, n = 16, \, z = \frac{5.8 - 5}{1/\sqrt{16}} = 3.2$$

H_0 is rejected; the specifications are not met.

9.   (a) $H_0 : \mu = 16.0$, $H_1 : \mu \neq 16.0$. $\overline{X} = 15.8$, $s = .4$, $n = 10$, *two-tail* $t = \dfrac{15.8 - 16.0}{.4/\sqrt{10}} =$
-1.58; $\alpha = .05$, $D = 9$, two-tail test: the critical values of t are ± 2.26. H_0 is accepted.
     (b) $H_0 : \mu = 16.0$, $H_1 : \mu < 16.0$. Same computed value of t as in (a). The critical value of t for a one-tail test with $D = 9$, $\alpha = .05$, is -1.83. H_0 is still accepted.

11.   The mistakes of the new apprentice are unlikely to be normally distributed.

They are more likely to be skewed to the right, since the minimum number is 0, but there is no (theoretical) maximum.

13. $\overline{X} = 75$, $s_X = 5$, $n_X = 5$; $\overline{Y} = 77$, $s_Y = 5$, $n_Y = 10$.

$H_0: \mu_X - \mu_Y = 0$; $H_1: \mu_X - \mu_Y < 0$. One-tail test; $\alpha = .05$. $D = 13$; H_0 is rejected if $t < -1.77$.

$$t = \frac{(75 - 77) - 0}{\sqrt{\dfrac{4 * 5^2 + 9 * 5^2}{13}\left(\dfrac{1}{5} + \dfrac{1}{10}\right)}} = -.73; H_0 \text{ is accepted.}$$

15. $H_0: \mu_X - \mu_Y = 0$, $H_1: \mu_X - \mu_Y \neq 0$. $\overline{X} = 600$, $n_X = 20$, $s_X = 100$; $\overline{Y} = 700$, $n_Y = 10$, $s_Y = 50$.

$$t = \frac{(600 - 700) - 0}{\sqrt{\dfrac{100^2}{20} + \dfrac{50^2}{10}}} = 3.65$$

$$D = \frac{\left(\dfrac{100^2}{20} + \dfrac{50^2}{10}\right)^2}{\left(\dfrac{100^2}{20}\right)^2 \dfrac{1}{21} + \left(\dfrac{50^2}{10}\right)^2 \dfrac{1}{11}} - 2 = 30, \alpha = .02$$

So the critical values of t are ± 2.33. H_0 is rejected: there is a significant difference in mean monthly pay of teachers who teach and who sell real estate.

17. $H_0: \mu_d = 0$; $H_1: \mu_d > 0$. d: 5, 0, -7, 0, 5, -5, 5, -10, 3, 5, 5, 4.

$$\overline{d} = 10/12 = .833, s_d = \frac{\sqrt{324 - 10^2/12}}{11} = 5.36$$

$\alpha = .05$, one-tail test, $n = 12$, $D = 11$; critical value of t is 1.80.

$$t = \frac{.833 - 0}{5.36/\sqrt{12}} = .54$$

H_0 is accepted; there is no difference.

19. $H_0: \mu_X - \mu_Y = 5$; $H_1: \mu_X - \mu_Y < 5$. $D = 10 + 10 - 2 = 18$, $\alpha = .05$, critical value of $t = -1.73$.

$$\sigma_{(\overline{X} - \overline{Y})} = \sqrt{\frac{9 * 6^2 + 9 * 5^2}{18} * \left(\frac{1}{10} + \frac{1}{10}\right)} = 2.47, t = \frac{(100 - 96) - 5}{2.47} = -.40$$

H_0 is accepted.

chapter thirteen

χ^2 (chi square)

In the last four chapters, you learned to find a confidence interval or test a hypothesis about the population mean(s) or proportion(s). Inferences about the mean of course involved metric data (it is nonsense to talk about the mean of categorical data); inferences about the proportion can be made about categorical data, but then we are restricted to two categories. In this chapter you will learn to make statistical inferences about categorical data in which the number of categories may be larger than two.

A new family of probability distributions, χ^2, will be introduced; as in t *distributions, there is a different member of the family for each different number of degrees of freedom. Two types of problems will be discussed: "goodness of fit" problems, in which you will determine how well an observed frequency distribution agrees with a theoretical frequency distribution, and secondly, problems in which you decide whether two variables are independent or related ("Does income at age 40 depend on amount of education or not?")*

13.1 INTRODUCTION

First, become familiar with the Greek letter χ (chi, pronounced as the first two letters of "kite"). It is not equivalent to any English letter, but is similar to the German ch as in "ich." We shall see it **only** in the form χ^2 and shall refer to "chi-square distributions" or "chi-square tests."

You have met various families of probability distributions: the binomial, normal, and t distributions. Now you will meet another family of probability distributions; as with t distributions, there will be a different χ^2 distribution for each different number of degrees of freedom. But before becoming familiar with a χ^2 distribution, let us look at a problem in which there is a need for it.

Example 1

Studies made in 1950 showed that, among all men aged 30, 20 per cent were college graduates, 50 per cent were high school but not college graduates, and 30 per cent were not high school graduates. Is the present population distribution the same? A study of 1000 men, chosen at random from those now 30 years old, is carried out, and it is found that 250 are college graduates, 520 are high school but not college graduates, and 230 have not finished high school. Test at .01 level of significance.

H_0: There is no change in the distribution between 1950 and now. (Note that H_0 is an assumption not about the mean or proportion of a population, but about the whole distribution of a population.)

H_1: The present population (namely, all men now 30 years old) has a distribution different from that in 1950.

Let O—the letter O, not the numeral zero—be used as abbreviation for Observed frequency (at the present time) and E for Expected frequency. If there is no change in the distribution between 1950 and now, the given information can be presented as follows:

	"Observed" O	"Expected" E from 1950's
College graduates	250	200
High school graduates	520	500
Not high school graduates	230	300
	1000	1000

In 1950, 20 per cent were college graduates. In a sample of 1000, then, 20 per cent or 200 men would be expected to be college graduates, 50 per cent or 500 men to be high school graduates, and 30 per cent or 300 to be "not high school graduates." Note that we start off by scaling the E column so that its sum is the same as that of the O column. It is quite inappropriate to express the O column in numbers of men (total 1000) and the E column in percentages or proportions (total 100 or 1, respectively). Both must add up to the total number of scores in the sample.

A decision choice must now be made: What should be used as the criterion for accepting or rejecting H_0? Would $\Sigma(O - E)$ describe the difference between the two distributions? But $\Sigma(O - E) = \Sigma O - \Sigma E = n - n = 0$. Or, as shown in the $O - E$ column below, $50 + 20 - 70 = 0$.

	O	$p(x)$	E	$O - E$	$(O - E)^2$	$(O - E)^2/E$
College graduates	250	.2	200	50	2500	12.5
H.S. graduates	520	.5	500	20	400	.8
Not H.S. graduates	230	.3	300	−70	+ 4900	+ 16.3
Totals	1000		1000	= 0 (always)	7800	29.6 } χ^2_p

Do you remember that when we looked at measures of variability we considered $\frac{\Sigma(X - \overline{X})}{n}$, but it turned out that $\Sigma(X - \overline{X})$ always equals 0? Then we tried $\Sigma(X - \overline{X})^2$ and finally $\frac{\Sigma(X - \overline{X})^2}{n}$, the variance. So here we might see whether $\Sigma(O - E)^2$ reflects the difference between the two distributions. $\Sigma(O - E)^2 = 0$ only if there is a perfect fit between the observed and expected frequencies, and may be quite large if they are not in agreement.

Instead of simply using $\Sigma(O - E)^2$, however, we use $\Sigma(O - E)^2/E$. The reason for this is that a large entry in the $(O - E)^2$ column is more disturbing if it comes from a category with a small expected frequency than if the expected frequency is large, so the former case is weighted more heavily. An entry of 2500 in the $(O - E)^2$ column becomes 12.5 if the expected frequency E is 200, but only 5.0 if the expected frequency is 500, and 2.5 if the expected frequency in that category is 1000.

Now we define χ^2_P:

$$\chi^2_P = \Sigma \frac{(O - E)^2}{E}$$

The subscript P stands for Karl Pearson, the statistician who developed this formula. χ^2_P is a statistic (it is based on sample data). In the next section you will study new kinds of distributions, called χ^2 distributions. The distribution of χ^2_P for all samples of size n is approximately a χ^2 distribution. In many books, the subscript P for the statistic is dropped. It will be retained in this text, however, to remind you of the difference between the statistic and the distribution.

In the given example, we fill in a column for $(O - E)^2/E$ and add, finding that $\chi^2_P = 29.6$. But if we look at a χ^2 table (Table 6, Appendix C), we find, as in the case of t distributions, a column labeled D at the left. D again stands for **degrees of freedom.** To determine D, look at the column marked E. There are 3 entries here, but if any 2 are given, then the third is determined, *(already known)* since the sum of this column must be 1000. Therefore $D = 3 - 1 = 2$.

Confidence, (level of significance)

α	.10	.05	.02	.01
D_{N-1}		Values of χ^2		
1	2.71	3.84	5.41	6.63
$(3-1) = \to 2$	4.61	5.99	7.02	9.21
3	6.75	7.82	9.84	11.3

Probabilities (or areas under χ^2 curve above given values of χ^2)

$\alpha = .01$, $D = 2$, so we look on the $D = 2$ line under .01 and find $\chi^2 = 9.21$. This gives us the decision choice for the null hypothesis: H_0 is accepted if the computed value of χ^2_P is less than or equal to 9.21, and is rejected if the computed value is greater than 9.21. In our example, $\chi^2_P = 29.6$, so H_0 is rejected and H_1 is accepted. There is a significant difference in the distribution of 30-year-old men by education now as compared with 1950.

Ans:

We shall return later (Sections 13.3 and 13.4) to a general discussion of χ^2 tests and to further examples. But now that you have a faint notion of the direction in which we are heading, let us digress and see how a χ^2 table is made.

13.2 χ^2 DISTRIBUTIONS

From a normal distribution of X scores with mean μ and standard deviation σ, take all possible different samples of size 1 ($n = 1$). For each sample, determine

Squaring ea. side:

$$z^2 = \frac{(X - \mu)^2}{\sigma^2}.$$

The number of z^2's that must be computed equals the number of scores in the original population. All these z^2's form a new distribution, the χ^2 distribution with 1 degree of freedom:

$\chi^2 = z^2$ when $D = 1$.

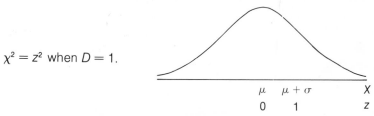

What does the χ^2 distribution look like when $D = 1$?

1. z values extend indefinitely to the left and to the right; z^2 is always greater than or equal to 0, so z^2 (and therefore χ^2) extends from 0 indefinitely to the right.

2. 68 per cent of z scores are between -1 and $+1$, so 68 per cent of χ^2 scores are between 0 and 1 when $D = 1$.

3. 95.4 per cent of z scores are between -2 and $+2$, so 95.4 per cent of χ^2 scores are between 0 and 4 when $D = 1$.

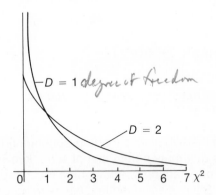

For χ^2 with 2 degrees of freedom, take all possible samples of size 2 from the population. If a sample consists of scores X_1 and X_2, this sample gives a χ^2 score of

$$\frac{(X_1 - \mu)^2}{\sigma^2} + \frac{(X_2 - \mu)^2}{\sigma^2} = z_1^2 + z_2^2.$$

Do this computation for all possible samples of size 2, and thus arrive at the χ^2 distribution for $D = 2$. Here one cannot easily arrive at areas under the curve by looking at probabilities. If $z_1 < 1$ and $z_2 < 1$, then $z_1^2 + z_2^2 < 2$. But $p(\chi^2 < 2) \neq p(z_1 < 1$ and $z_2 < 1)$; there are other possibilities. If, for example, z_1 is close to 0, then z_2 might be close to (but less than) $\sqrt{2}$ and still χ^2 would be less than 2.

Similarly, the χ^2 distribution with $D = 3$ is the distribution of scores $z_1^2 + z_2^2 + z_3^2$, where z_1, z_2, and z_3 are the z scores of the 3 members of a sample of size 3.

χ^2 distributions in general have the following characteristics:

1. Every χ^2 distribution extends indefinitely to the right from 0.

2. Every χ^2 distribution has only one (right) tail. We shall study **only one-tail tests** using χ^2.

3. As D increases, the χ^2 curves get more bell-shaped and approach the normal curve.

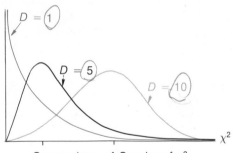

Comparison of Graphs of χ^2
for Different Degrees of Freedom

What is the relation between χ^2 as defined here and $\chi^2_p = \sum \frac{(O-E)^2}{E}$, used in Example 1, page 222? The answer is that χ^2_p is a statistic (sample data is used in computing it) whose distribution is approximately χ^2. The critical value of χ^2 for accepting or rejecting a null hypothesis is determined by consulting Table 6 for the appropriate value of α and degrees of freedom, and then a decision is made according to which region χ^2_p falls in. "The distribution of χ^2_p is **approximately** χ^2" implies some limitations: at least 3/4 of the entries in the E column should be 5 or larger, and all should be greater than 1. If this is not the case, increase the size of the sample.

13.3 TESTING "GOODNESS OF FIT" WITH χ^2

The comparison of distributions of 30-year-old men by education in 1950 and now, considered at the beginning of this chapter, is an example of a **"goodness of fit"** problem. In this use of χ^2 distributions, the observed frequencies in the distribution of a sample are used to test the hypothesis that the population from which the sample is taken does not differ in its distribution from that of some known population.

H_0 has the form "the population *(A)* from which the sample is taken has the same distribution as population *B*"; it is assumed that the distribution of *B* is known. H_1 states "populations A and B do not have the same distribution." A sample of size *n* is taken, and frequencies in various categories or of various scores are observed (*O* column). Then the frequencies expected for these categories or scores if the sample had the same proportion in each category or score as population *B* is computed (*E* column). Then $\sum \frac{(O-E)^2}{E}$ is determined.

For all χ^2 tests, the number of degrees of freedom *D* equals the number of entries in the *E* column minus the number of sample statistics used in determining the *E* column. In the examples of this section, only one sample statistic is used—namely, the total number of scores in the sample. Therefore **D = (number of E entries) − 1.**

The critical value of χ^2 for accepting or rejecting H_0 is determined from Table 6, Appendix C, for the given α and *D*. This critical value is compared with the computed statistic $\sum \frac{(O-E)^2}{E}$ and a decision is reached.

Example 1

meaning X = 200 B or hirls

In a survey of 400 infants, chosen at random, it is found that 185 are girls. Are boy and girl births equally likely, according to this survey? (.05 level of significance.)

$H_0: P = 1/2; H_1: P \neq 1/2.$

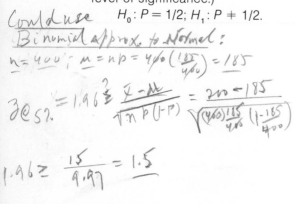

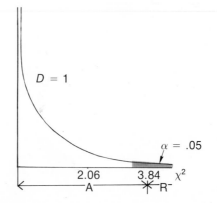

$\frac{1}{2}(400)$

	O	E	$O-E$	$(O-E)^2$	$(O-E)^2/E$
Girls	185	200	−15	225	1.03
Boys	215	200	15	225	1.03
	400	400			$2.06 = \chi^2_P$

There are two entries in the E column; one sample statistic (namely, n) is used in computing the E column, so $D = 2 - 1 = 1$. The critical value of χ^2, for $\alpha = .05$ and $D = 1$, is 3.84; H_0 is accepted: the proportion of girl births is 1/2, since $2.06 < 3.84$.

The question could also have been answered using the methods of Chapter 11: (BINOMIAL APPROX. to NORMAL)
$H_0: P = 1/2$; $H_1: P \neq 1/2$; $\alpha = .05$: The critical values of z are ±1.96.

$$p = 185/200 = .46, \quad n = 400, \quad \sigma_p = \sqrt{PQ/n} = \sqrt{.46 * .54/400} = .025$$

Population

$z = \frac{\overline{x} - \mu}{\sigma}$; $z = \frac{.46 - .50}{.025} = -1.6$

Area no diff. (small Hyp b)
no diff. (small Hyp) accepted

.025		.025

.5 p
−1.96   −1.6   1.96 z
R   A   R

Again, H_0 is accepted.

As this example illustrates, there are often several ways to go about testing the same hypothesis. Note that a two-tail test was used for z scores and a one-tail test for χ^2_P scores.

Example 2

An industrial engineer has constructed a model for the number of reams of paper used by typists in a large office per day. In his model there are 5 events: 1, 2, 3, 4, and 5 corresponding to the number of reams used; the probabilities of these events are .10, .30, .40, .10, .10, respectively. To check this model, the number of reams used is checked for 300 days. He finds that 1 ream is used on 20 days, 2 on 60 days, 3 on 110 days, 4 on 50 days, and 5 on 60 days. Is his model correct, at a .10 level of confidence?

$p(X)$ 300

X	O	$p(X)$	E	$O-E$	$(O-E)^2$	$(O-E)^2/E$
1	20	.1	30	−10	100	3.3
2	60	.3	90	−30	900	10.0
3	110	.4	120	−10	100	.9
4	50	.1	30	20	400	13.3
5	60	.1	30	30	900	30.
	300	1.0	300	0		$57.5 = \chi^2_P$

The E column is computed by multiplying $p(X)$ by 300. Thus, the probability that 1 ream is used is .1, so in 300 days it is expected that 1 ream is used on $.1 * 300 = 30$ days.

There are 5 entries in the E column; $D = 5 - 1 = 4$.
At a .10 level of significance with $D = 4$, $\chi^2 = 7.78$.
H_0 is rejected; the industrial engineer had better look for a better model.

In many experiments, you are pleased if H_0 is rejected since you have found a significant difference from the usual result. In testing goodness of fit with χ^2_p, however, you are often testing a hypothesis which you hope is true. If you are a careful investigator, you want the probability β of accepting H_0 when it is false to be small, and yet the convention of choosing α in reaching a decision is still used. Therefore many applied statisticians use $\alpha = .20$ or $\alpha = .25$ in such problems, since β is decreased as α is increased for fixed n.

13.4   RELATED AND INDEPENDENT VARIABLES; CONTINGENCY TABLES

In the last section, you learned how to use χ^2_p to test "goodness of fit"—that is, whether testing a distribution obtained from sample data leads to the conclusion that the distribution of the population from which the sample was taken fits some theoretical distribution.

χ^2_p can be used not only to test "goodness of fit" but also to test whether two variables are related or are independent. Does a higher proportion of men than of women vote Republican, or is there no relation between sex and political party? Is there a relation between years of schooling after high school and income at age 40? Do different strains of rats have differing abilities to "solve" a maze? A χ^2 test will help in each of these problems.

Example 1

The following table shows the relation between the number of accidents in one year and the age of the driver in a random sample of 500 drivers between 18 and 50. Test, at a .01 level of significance, the hypothesis that the number of accidents is independent of the driver's age.

Age of Driver

	18—25	26—40	41—50	
0	75	120	105	300
1	50	60	40	150
2	25	20	5	50
	150	200	150	500

(Number of Accidents)

There are 75 drivers between 18 and 25 who have no accidents, 120 between 26 and 40 with no accidents, and so on.

Such a table is called a **contingency table.** Each "box" containing a frequency is called a **cell.** This is a 3×3 table, since it has 3 rows and 3 columns. Note that the row and column totals are not counted in giving the size of a contingency table. The following is a 2×4 table:

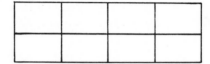

Rows are counted first, then columns: $R \times C$.

The null hypothesis for this type of problem always states that the classifications are independent (here, that there is no relation between age of driver and number of accidents). H_1: the variables are dependent (or related). $\alpha = .01$.

What are the expected frequencies? Concentrate first on the upper left-hand cell, and remember that if A and B are independent events, then $p(A \text{ and } B) = p(A)p(B)$. If one driver is chosen at random from a group of 500, the probability he is between 18 and 25 is 150/500 and the probability he has no accidents is 300/500. If H_0 is true (if age and number of accidents are independent), then $p(\text{age } 18-25 \text{ and no accidents}) = p(\text{age } 18-25)p(\text{no accidents}) = \dfrac{150}{500} * \dfrac{300}{500}$. This is the probability for one driver; out of 500 we would then expect $\dfrac{150}{500} * \dfrac{300}{500} * 500 = 90$ to be between 18 and 25 and have no accidents if H_0 is true. Note, however, that the same result is obtained by multiplying the column total (150) by the row total (300) and dividing by the grand total (500). Before you read on, find the expected number of drivers aged 26 to 40 who have had no accidents.

Answer

$$p(26-40 \text{ and } 0 \text{ accidents}) = \frac{200}{500} * \frac{300}{500},$$ so the expected number in 500 drivers is $\dfrac{200}{500} * \dfrac{300}{500} * 500 = 120$; **or** column total (200) * row total (300) divided by grand total (500) $= \dfrac{200 * 300}{500} = 120$.

The results can be summarized in a table:

Frequencies expected if H_0 is true

Age of Driver

Number of Accidents		18—25	26—40	41—50	Totals
	0	$\dfrac{150 * 300}{500}$ $= 90$	$\dfrac{200 * 300}{500}$ $= 120$	$\dfrac{150 * 300}{500}$ $= 90$	300
	1	$\dfrac{150 * 150}{500}$ $= 45$	$\dfrac{200 * 150}{500}$ $= 60$	$\dfrac{150 * 150}{500}$ $= 45$	150
	2	$\dfrac{150 * 50}{500}$ $= 15$	$\dfrac{200 * 50}{500}$ $= 20$	$\dfrac{150 * 50}{500}$ $= 15$	50
		150	200	150	500

χ^2 (chi square)

Note that the marginal totals in this table are the same as the marginal totals in the table of observed frequencies on page 228. Once these marginal totals are copied from the earlier table, then the expected frequency is entered in each cell **without** reference to the observed frequency in that cell: **Multiply the row total by the column total, and divide by the grand total.** Now the χ^2 test can be carried out with the same formula as before:

$$\chi^2{}_P = \sum \frac{(O - E)^2}{E}$$

Number of Accidents	Age of Driver	0	E	O − E	(O − E)²	(O − E)²/E
0	18–25	75	90	−15	225	2.5
0	26–40	120	120	0	0	0
0	41–50	105	90	15	225	2.5
1	18–25	50	45	5	25	.6
1	26–40	60	60	0	0	0
1	41–50	40	45	− 5	25	.6
2	18–25	25	15	10	100	6.7
2	26–40	20	20	0	0	0
2	41–50	5	15	−10	100	6.7
					0	19.6 = $\chi^2{}_P$

What about degrees of freedom, D? How many cells can be filled in freely before all are determined by the requirement that the row and column totals come out right? If any 2 cells in the first row are filled in, then the frequency in the third cell is determined by the requirement that the total for that row is 140. (Checks ✔ are used to indicate the two freely chosen cells, and an **X** to indicate the one which is then determined.)

✓	✓	X	140
✓	X	✓	252
X	X	X	108
100	300	100	

Similarly, if any 2 cells in the second row are filled in, then the third is determined by the requirement that the total for that row is 252. Finally, the frequency in each of the cells in the third row is now determined by the requirement that the column totals must be, respectively, 100, 300, and 100. You are free to fill in only 4 cells, in the whole pattern (and no 3 of these can be in the same row or column), so $D = 4$.

From the χ^2 table (Table 6, Appendix C), the critical value of χ^2 for $D = 4$, $\alpha = .01$ is 13.3. Since the computed value of $\chi^2{}_p$, 19.6, is greater than 13.3,

H_0 is rejected: There **is** a relationship between number of accidents and age of the driver.

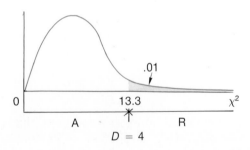

Degrees of Freedom in Contingency Tables

If the marginal totals in a 2×2 contingency are given, then choose a number for cell (1). In the example shown, your choice must be between 0 and 10, since the sum of the numbers in that row must be 10 and a frequency is never negative. As soon as you choose an entry for cell (1), the other three cell frequencies are determined. If your choice is 9, then in cell (2) the frequency is $10 - 9 = 1$, in cell (3) it is $20 - 9 = 11$, and in cell (4) it is $40 - 11 = 29$ (or $30 - 1 = 29$). A 2×2 contingency table has 1 degree of freedom.

(1) ✓	(2) X	10
(3) X	(4) X	40
20	30	50

9	1	10
11	29	40
20	30	50

Try a more complicated problem. See how many cells you can fill in freely in the 3×4 contingency table below.

(1)	(2)	(3)	(4)	25
(5)	(6)	(7)	(8)	50
(9)	(10)	(11)	(12)	25
10	20	30	40	100

If cells (1), (2), and (3) are filled in, (4) is determined, since the sum of these four numbers must be 25. If (5), (6) and (7) are filled in, then (8) is fixed. But now you can make no more choices, since the sums of the columns are given. By subtracting from the marginal totals, the frequencies in the last row and column are determined. In this example, then, the number of degrees of freedom is $(3 - 1) * (4 - 1)$. **In general, the number of degrees of freedom in an $n \times m$ contingency table** is

$$D = (n - 1)(m - 1)$$

Example 2

Is there a relationship between wearing of glasses and amount of education? It is found that, among 50 college graduates, 15 wear glasses and 35 do not. Seventy of 200 high school graduates wear glasses and 130 do not, while among 100 adults who attended only grade school, 28 wear glasses. Test at a .05 level of significance.

H_0: There is no relationship between wearing of glasses and amount of education.

H_1: There is a relationship between wearing of glasses and amount of education. (Note that, even if H_0 is rejected, no conclusion is drawn about the nature of the relationship between the two variables—not even whether people with more education are more or less apt to wear glasses, and certainly no cause-and-effect relation is shown.)

The observed frequencies give us the following 2×3 contingency table:

Amount of Education

	College	High School	Grade School	
Wear glasses	15	70	28	113
No glasses	35	130	72	237
	50	200	100	350

To find the expected frequencies if there is no relationship between wearing of glasses and amount of education, we use the same marginal totals; the expected frequency in a cell is the product of the marginal totals in the row and column to which the cell belongs divided by the grand total. The contingency table for expected frequencies is:

Amount of Education

	College	High School	Grade School	
Wear glasses	$\frac{50 \cdot 113}{350} = 16.1$	$\frac{200 \cdot 113}{350} = 64.6$	$\frac{100 \cdot 113}{350} = 32.3$	113
No glasses	$\frac{50 \cdot 237}{350} = 33.9$	$\frac{200 \cdot 237}{350} = 135.4$	$\frac{100 \cdot 237}{350} = 67.7$	237
	50	200	100	350

Now χ^2 can be computed:

Glasses	Education	O	E	$O-E$	$(O-E)^2$	$(O-E)^2/E$
Yes	College	15	16.1	−1.1	1.21	.08
Yes	H.S.	70	64.6	5.4	29.2	.45
Yes	Grade S.	28	32.3	−4.3	18.5	.57
No	College	35	33.9	1.1	1.21	.04
No	H.S.	130	135.4	−5.4	29.2	.22
No	Grade S.	72	67.7	4.3	18.5	.27
				0		$1.63 = \chi^2$

In a 2×3 table, $D = (2 - 1) * (3 - 1) = 1 * 2 = 2$. From Table 6, $D = 2$ and $\alpha = .05$ implies $\chi^2 = 5.99$. Since 1.63 is less than 5.99, H_0 is accepted at a .05 level of significance. There is no relationship between wearing of glasses and amount of education.

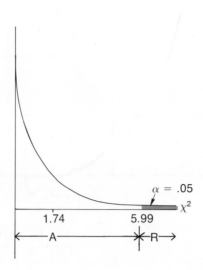

Example 3

A school psychologist finds that some students in his city are having problems because of their use of alcohol, glue, or marijuana. He carefully defines the term "drug user" and then collects information on a random sample. One hundred high school students who use drugs are classified by grade and type of drug as follows:

	Grade			
	Sophomore	Junior	Senior	
Alcohol	16	12	12	≃ 40
Glue	8	15	7	≃ 30
Marijuana	6	8	16	≃ 30
	≃ 30	≃ 35	≃ 35	≃ 100

Type of Drug (left axis label)

At a .01 level of significance, test whether there is a relation between grade in high school and the type of drug used.

H_0: There is no relation between grade in school and type of drug used.

H_1: There is such a relation.

Expected frequencies if there is no relation:

Grade

	Sophomore	Junior	Senior
Alcohol	$\dfrac{30 * 40}{100} = 12$	$\dfrac{35 * 40}{100} = 14$	$\dfrac{35 * 40}{100} = 14$
Glue	$\dfrac{30 * 30}{100} = 9$	$\dfrac{35 * 30}{100} = 10.5$	$\dfrac{35 * 30}{100} = 10.5$
Marijuana	$\dfrac{30 * 30}{100} = 9$	$\dfrac{35 * 30}{100} = 10.5$	$\dfrac{35 * 30}{100} = 10.5$

Type of Drug (row label, vertical)

Type of Drug	Grade	O	E	O − E	(O − E)²	(O − E)²/E
Alcohol	Sophomore	16	12	4	16	1.33
Alcohol	Junior	12	14	−2	4	.29
Alcohol	Senior	12	14	−2	4	.29
Glue	Sophomore	8	9	−1	1	.11
Glue	Junior	15	10.5	4.5	20.2	1.92
Glue	Senior	7	10.5	−3.5	12.2	1.16
Marijuana	Sophomore	6	9	−3	9	1.00
Marijuana	Junior	8	10.5	−2.5	6.25	.60
Marijuana	Senior	16	10.5	5.5	30.25	2.88
				0		$8.83 = \chi^2{}_P$

The contingency table is 3×3, so $D = (3 − 1) * (3 − 1) = 2 * 2 = 4$. If $\alpha = .10$ and $D = 4$, then $\chi^2 = 13.28$. H_0 is accepted; there is no relationship between grade in school and type of drug used.

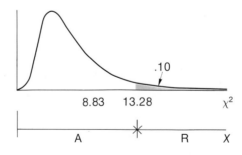

13.5 YATES' CORRECTION

When D = 1 the distribution of $\chi^2{}_P = \sum \dfrac{(O − E)^2}{E}$ is only approximately that of χ^2.

A statistician named Yates discovered that a statistic whose distribution is a better approximation to χ^2 when $D = 1$ is given by

$$\chi^2_P = \sum \frac{(|O - E| - .5)^2}{E}$$

Why is this correction needed? It is for the same reason that, when a normal distribution was used as an approximation to a binomial distribution, you added or subtracted 1/2 (see page 118). You are approximating a discrete distribution (here, the frequency in each category, which is an integer) by a continuous distribution (χ^2). Fortunately, this correction is unnecessary except when $D = 1$.

Example 1

Are the variables shown in the following contingency table independent? ($\alpha = .05$).

	C	D	
A	40	105	145
B	60	95	155
	100	200	300

O	E	$(O - E)$	$\|O - E\| - .5$	$(\|O - E\| - 5)^2$	$(\|O - E\| - .5)^2/E$
40	48.3	8.3	7.8	60.84	1.26
105	96.7	8.3	7.8	60.84	.63
60	51.7	8.3	7.8	60.84	1.18
95	107.3	8.3	7.8	60.84	.57
					$3.64 = \chi^2_P$

The critical value of χ^2 for $D = 1$ and $\alpha = .05$ is 3.84, so H_0 is accepted.

Yates' correction is used **only** when $D = 1$, but it **should** be used then, both for tests involving contingency tables and in "goodness of fit" problems.

13.6 VOCABULARY AND SYMBOLS

chi-square distribution	χ^2	goodness of fit	
χ^2_P		related variables	
observed frequency	O	contingency table	
expected frequency	E	cell	
degrees of freedom	D	Yates' correction	

13.7 EXERCISES

1. In a χ^2 distribution with $D = 10$, what is the probability that a score is greater than 18.31? *p.35? one tail @ .05*

2. In a χ^2 distribution with 8 degrees of freedom, (a) 1 per cent of scores are over what value? (b) 90 per cent are under what value? *(a) 20.1 (b) 13.4*

3. How many degrees of freedom does a χ^2 distribution have if the probability is .05 that a score in the distribution is greater than 26.3? *D = 16*

4.   A die is tossed 600 times with the following results:

Face	1	2	3	4	5	6
Frequency	95	110	90	85	112	108

Test the hypothesis that the die is fair at a .05 level of significance.

5.   In a survey of 160 families with 4 children, it is discovered that in 6 families there are no girls, while 34 families have 1 girl, 56 have 2, 46 have 3, and 18 families have 4 girls. Are boy and girl births equally probable, according to this survey? (.05 level of significance.)

6.   A box of after-dinner mints contains 40 green (sassafras), 60 white (peppermint), 50 yellow (lemon), 50 pink (wintergreen), and 40 orange (cinnamon). The packing apparatus in candy factory A uses mints in the following percentages: green 20, white 25, yellow 20, pink 15, orange 20, while in factory B the comparable percentages are 15, 30, 20, 20, 15. Compute χ^2_p for each factory to see if this particular box was more likely to have been packed at factory A or factory B.

7.   In a cross between tall, potato-leaf tomatoes and dwarf, cut-leaf tomatoes, the following plants are observed:

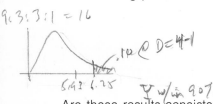

9:3:3:1 = 16

$D = 4-1$

	O	E	$(0-E)^2$	$(0-E)^2/E$
Tall, potato-leaf	280	(9/16 x 480)=270	100	.37
Tall, cut-leaf	100	3/16×480≥90	100	1.11
Dwarf, potato-leaf	80	3/16×480≥90	100	1.11
Dwarf, cut-leaf	+20	1/16×480=30	100	3.33
	470			$\chi^2 = 5.92$

Are these results consistent with the hypothesis that the ratios are 9:3:3:1? ($\alpha = .10$.)

8.   Upon observation of 1000 dented fenders, it is found that 300 are the left front, 240 the right front, 210 the left rear, and 250 the right rear. At a .01 level of significance, are all fenders equally likely to be dented?

9.   A manufacturer of television sets receives component transistors from 4 suppliers. Each month 1000 transistors from each supplier are chosen at random and carefully tested. In January the numbers of defective transistors are 46, 53, 42, 59, from suppliers A, B, C, D, respectively. Is the number of defective transistors the same from each supplier? ($\alpha = .05$)

10.   On a true-false exam, a playful professor includes 5 questions that he thinks can be answered only by guessing. The results on these five questions are as follows:

$p(x)$ $p(x) n = "E$

$5c_0(\frac{1}{2})^5 \cdot \frac{1}{2}^0 = .03 \times 75 = 2.34$

$5c_1 \frac{1}{2}^4 \frac{1}{2}^1 = .16 \times 75 = 11.72$

$5c_2 \frac{1}{2}^3 \frac{1}{2}^2 = .31 \times 75 = 23.44$

$5c_3 \frac{1}{2}^2 \frac{1}{2}^3 = .31 \times 75 = 23.44$

$5c_4 \frac{1}{2}^1 \frac{1}{2}^4 = .16 \times 75 = 11.72$

$5c_5 \frac{1}{2}^0 \frac{1}{2}^5 = .03 \times 75 = 2.34$

75

Number Correct	0-E	Number of Students	$(0-E)^2$	$(0-E)^2/E$
0	-1.34	1	1.8	.77
1	-1.72	10	2.96	.25
2	-3.44	20	11.83	.5
3	1.56	25	2.43	.1
4	3.28	15	10.76	.92
5	1.66	4	2.76	+1.18
	0			75

$\chi^2 = 3.72$ $D = 5-1$

Binom:

$\bar{m} = 10(1)+20(2)+25(3)+15(4)+4(5) = 208$

$n = 75(5) = 375$

$w = np = 375$

Binom

Is his belief correct at a .10 level of significance?

11.   The professor described in Exercise 10 tries the same experiment on a larger class; this time the numbers of students answering 0, 1, 2, 3, 4, 5 questions correctly are 2, 20, 40, 50, 30, 8, respectively. Is he correct, at a .01 level of significance?

$375 = \frac{x-\mu}{\sqrt{np(1-p)}} = \frac{705 - 5(75)}{\sqrt{(375)\frac{1}{2}(\frac{1}{2})}} = 1.91$; $3160 = \frac{2(187.5-205)}{\sqrt{2(375)\frac{1}{2}\frac{1}{2}}} = \frac{2}{\sqrt{2}} 375 = 1.41 \; 375$

higher level of signif.

12. A chemical society sends a questionnaire to its members with the following results:

Degree	Number Returned	Number Not Returned	Total
B.S.	50	40	90
B.Chem.E.	100	90	190
M.S.	50	40	90
M.Chem.E.	100	90	190
Ph.D	200	240	440

Is the proportion of questionnaires returned independent of the degree? ($\alpha = .10$.)

13. In a random sample of 100 blacks in Chicago, 50 have incomes less than $10,000, 40 earn between $10,000 and $15,000, and 10 earn over $15,000 per year. Among 100 whites, 30 earn under $10,000, 40 between $10,000 and $15,000, and 30 over $15,000 per year. At a .01 level of significance, are race and income related?

14. Test, on the basis of the following data and at a .05 level of significance, whether there is a relation between political affiliation and attitude towards a proposed law:

	Favor	Opposed	Undecided
Democrats	100	150	150
Republicans	150	100	150

15. Three brands of hair spray are tested. Among the blondes who use them, 20 prefer Brand A, 30 prefer Brand B, and 10 prefer Brand C. Among brunettes the number preferring the 3 brands are 30, 30, and 20, respectively, while among redheads who test the 3 brands A is preferred by 10, B by 5, and C by 5. At a .025 level of significance, find out whether brand preference is related to hair color.

16. Lipsticks are manufactured by three different processes (A, B, and C.) Defective lipsticks are classified as those having defects in color, having defects in consistency, or being chipped. At a .05 level of significance, do the following data show that cause of defect is independent of the process used?

		Defect in Color	Defect in Consistency	Chipped
	A	40	25	20
Process	B	10	30	20
	C	10	25	20

17. The fruit production in a state for 1970–73 is as follows, expressed in thousands of crates:

	Peaches	Cherries	Apples	Strawberries
1970	50	10	50	10
1971	55	8	50	12
1972	35	20	50	14
1973	60	12	50	14

At a .05 level of significance, is the type of fruit produced independent of the year?

18.   A professor claims that the grades in his course are normally distributed and that he gives A to those 1σ above the mean, B between .5 and 1σ above the mean, C if 1σ below the mean to $.5\sigma$ above, D if 1σ to 2σ below the mean, and F only if more than 2σ below the mean. Among 100 students in his class, there are 10 A's, 15 B's, 63 C's, 7 D's, and 5 F's. At a .05 level of significance, is his claim justified?

19.   Some people (tasters) find phenylthiourea to be very bitter when tasted, while some (non-tasters) find it tasteless or almost so; it is hypothesized that the ratio of tasters to non-tasters is 2:1. In a test of 300 randomly chosen subjects, 87 are found to be non-tasters and 213 are tasters. Is the hypothesis supported by this evidence? (.10 level of significance.)

20.   A maker of sleeping bags claims that 90 per cent of those using his bags will be comfortable sleeping in them when the temperature is 10°. Among 100 campers using them in winter, 15 are uncomfortably cold at 10°. Is the maker's claim correct? ($\alpha = .10$.)

21.   A geneticist is investigating the inheritability of a certain disease. He collects the following data on a random sample of 180 children:

		Child had disease?	
		Yes	No
Mother	Yes	10	70
had			
disease?	No	8	92

At a .01 level of significance, are children more likely to have the disease if their mothers have had it?

22.   A random sample of 50 people who hear the Crazy Eights perform in Wichita are queried about their reaction to the music, with the following results:

		Reaction	
		Like	Dislike
Age	Under 25	20	8
	Over 25	10	12

At a .05 level of significance, is there a relation between the age of the listener and attitude towards the music?

23.   A sociology major asks a random sample of 60 students to rate 40 statements from 1 ("I strongly agree") to 5 ("I strongly disagree"). On the basis of the answers, he classifies each of the students as showing or not showing racial prejudice. He also classifies students in 30 other two-way categories (upper or lower classman, father completed college or not, student belongs to a church or not, male or female, etc.) according to the answers to 30 other questions. Then he carries out 30 χ^2 tests at a .05 level of significance to see whether or not racial prejudice is related to any of these other categories.

(a) Assume that "H_0: Attitude toward race is independent of the sex of the student" is true; what is the probability that it will be rejected in the χ^2 test? What is the probability it is accepted?

(b) Assume that all thirty null hypotheses that he tests are true. What is the probability that all of them will be accepted, using χ^2 tests, if they are independent tests?

(c) Assume that all thirty null hypotheses are true, and that the tests are independent. What is the probability that at least one is rejected?

24. One hundred women and 200 men between ages 20 and 40 are surveyed. Fifteen of the women and 50 of the men favor capital punishment. Use a χ^2 test at a .05 level of significance to see whether women and men differ in their views of capital punishment.

25. In a random sample of 40 TV viewers in New York, 15 per cent were watching Channel 13 at 9:00 and 85 per cent were watching other channels. Is this consistent with the hypothesis that 20 per cent of all viewers were watching Channel 13? ($\alpha = .05$)

ANSWERS

1. .05.

3. $D = 16$.

5. E column: 10, 40, 60, 40, 10; $\chi^2_P = 1.6 + .9 + .3 + .9 + 6.4 = 10.1$; $D = 4$, $\alpha = .05$, $\chi^2 = 9.49$. H_0 is rejected; the distribution of girls in the families surveyed is not consistent with the distribution that would be expected if $P = 1/2$.

7. E column: 270, 90, 90, 30; $\chi^2_P = .37 + 1.11 + 1.11 + 3.33 = 5.93$. $D = 3$; $\chi^2 = 6.521$, > 5.93. Yes, the results are consistent with the hypothesis.

9. $\chi^2_P = .32 + .18 + 1.28 + 1.62 = 3.40$; $D = 3$, $\chi^2 = 7.82$; H_0 is accepted.

11. E column: 4.7, 23.4, 46.9, 46.9, 23.4, 4.7. $\chi^2_P = 7.45$. $D = 4$, $\alpha = .01$, $\chi^2 = 13.3$. His surmise is supported. Compare the answers to Exercises 10 and 11, and note what happens to χ^2_P when each frequency in the E column is doubled.

13. $\chi^2_P = 2.5 + 0 + 5.0 + 2.5 + 0 + 5.0 = 15.0$; $D = (2-1) \ast (3-1) = 2, \chi^2 = 9.21$. Reject H_0; there is a relation between income and race.

15. E column: 22.5, 24.4, 13.1, 30, 32.5, 17.5, 7.5, 8.1, 4.4; $\chi^2_P = 4.9$; $D = (3-1) \ast (3-1) = 4$, $\chi^2 = 11.1$. H_0 is accepted: Preference in hair spray is independent of hair color.

17. E column (peaches, then cherries and apples, and finally strawberries): 48, 50, 47.6, 54.4, 12, 12.5, 11.9, 13.6, 48.0, 50.0, 47.6, 54.4, 12.0, 12.5, 11.9, 13.6; $\chi^2_P = 13.5$; $D = 9$, $\chi^2 = 16.9$; the type of fruit produced is independent of the year.

19. (Remember Yates' correction.) $\chi^2_P = .78 + 1.56 = 2.34$; $D = 1$, $\chi^2 = 2.71$; the hypothesis is supported.

21. E column: 8, 72, 10, 90; $\chi^2_P = .28 + .03 + .23 + .03 = .56$; $D = 1$, $\chi^2 = 6.63$. H_0 cannot be rejected; children are not more likely to have the disease if their mothers have had it.

23. (a) Probability of rejection $= .05$ (this is measured by α, the level of significance). Probability of acceptance $= 1 - .05 = .95$.
    (b) $(.95)^{30} = .21$.
    (c) $1 - (.95)^{30} = 1 - .21 = .79$.

More advanced methods are available to treat this data correctly.

25. O column: 6, 34; E column: 8, 32; $(|O - E| - .5)^2/E$ column: .28, .07. $\chi^2_P = .35$; $D = 1$, $\alpha = .05$, $\chi^2 = 3.84$. The sample results are consistent with the hypothesis.

chapter fourteen

analysis of variance

In Chapter 11, we tested hypotheses about the differences between the means of two populations. An economist might use the methods discussed there to compare wages in a certain industry in New York and New Jersey; a psychologist might compare performance of an experimental group and a control group. It is often desirable, however, to compare many means rather than just two: the economist wishes to compare wages in more than two states, or the psychologist wishes to compare performance of groups treated in a variety of ways with a control group. The procedure that is used to compare more than two means is called **analysis of variance.** *An introduction to F distributions is a prerequisite for understanding this procedure. After becoming familiar with F distributions you will first use them to solve a simpler problem, that of comparing the variances of two populations, and then you will go on to the comparison of many means. Finally, if the hypothesis that all means are the same is rejected, you will learn how to find which pairs of means are significantly different.*

14.1 *F* DISTRIBUTIONS

Imagine that you have two populations, that **both** are normally distributed, and that they have the same variance. They may have different means. From the first population you take a sample of size 9, and from the second a sample of size 21. Suppose $s_X^2 = 5.0$ and $s_Y^2 = 2.0$. Then the ratio $F = s_X^2/s_Y^2$ is computed: for these two samples it is $5.0/2.0 = 2.5$. Then these samples are replaced, and new samples are taken from each population—again of size 9 and size 21 respectively. This time $s_X^2 = 4.0$, $s_Y^2 = 8.0$, and $F = 4/8 = .5$. This process is continued for all possible different pairs of samples of size 9 from the first population and of size 21 from the second population selected independently. The resulting F scores form a new distribution. Since neither s_X^2 nor s_Y^2 is negative, an F score is never negative. Since the two populations have the same variance, you would expect many of your samples to have the same variances, resulting in F scores near 1. Some ratios are very large, however (if the variance of the first sample is very large and that of the second sample is very small). Five per cent of the F scores will be above 2.91.

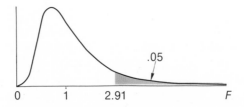

If the sizes of your samples are changed—say, 28 and 7 instead of 9 and 21—you will get a different F distribution. In general there will be a double family of F distributions; each member of the family will be determined by particular choices for the sizes of the samples from the two populations. Remember that an F distribution can be used **only** if both populations are normally distributed. This is true even if samples are large; there is no Central Limit Theorem which implies the assumption that the populations are normally distributed can be relaxed if sample sizes are large enough.

All F distributions have certain things in common:
1. An F score is the quotient of two independently estimated population variances:

$$F = \frac{s_X^2}{s_Y^2}$$

where s_X^2 is the estimate of the variance of one population based on a sample of size n_X, and s_Y^2 is the same for another population based on a sample of size n_Y.
2. Each F distribution is a probability distribution—that is, the area under the entire curve is 1 and the probability that an F score falls between a and b is given by the area under the curve between a and b.
3. In every F distribution the range of F values is from 0 to $+\infty$.
4. F tables are given not in terms of n_X and n_Y but rather in terms of degrees of freedom:

$$D_X = n_X - 1, \qquad D_Y = n_Y - 1 \qquad (n_X > n_Y)$$

The shape of any F distribution depends on the values of D_X and D_Y. All are skewed to the right, but become more symmetrical as the number of degrees of freedom gets larger.

14.2 HOW TO USE AN F TABLE

Since there are so many different F distributions, the values given for each are severely limited. Table 7, Appendix C, gives values of F when the right-hand tail of the distribution has an area of .05, .025, or .01. As a result, two-tail tests of hypotheses can only be found using these tables if $\alpha = .10$, .05, or .02.

Example 1

$n_X = 15$, $n_Y = 7$. Five per cent of F scores are above what value?

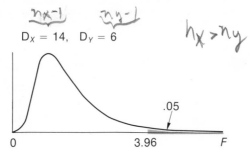

$n_X - 1 \qquad n_Y - 1$

$D_X = 14, \quad D_Y = 6$

$n_X > n_Y$

.05

0 3.96 F

$D_X = 14$, $D_Y = 6$. In Table 7(a), Appendix C, look across the first row until you find the number 14; look down the first column to the number 6. In the intersection of the 14th column and the 6th row is found the number 3.96.

analysis of variance

$p\,(F > 3.96) = .05$, or 5 per cent of F scores are above 3.96 if samples of size 15 and 7, respectively, are chosen at random from two normal populations with the same variance. Here is a useful notation: if $\alpha = .05$, $F(14,6) = 3.96$.

Example 2

If $n_X = 4$, $n_Y = 11$, 97.5 per cent of F scores are below what value?

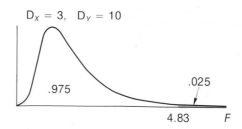

$D_X = 3$, $D_Y = 10$

.975

.025

4.83 F

(See Table 7b, Appendix C.) $D_X = 3$, $D_Y = 10$; 97.5 per cent of F scores are below 4.83.

Two-tail tests using F distributions or one-tail tests using the left-hand tail are quite possible. Table 7 gives only values of F for percentages in a right-hand tail, so an extra computation is required. It will always be possible to avoid use of a left-tail test and computation of the lower critical value when using a two-tail test. (See Examples 3 and 4, pages 244 and 245.) But the following rule makes it possible to find the critical value for a left-hand tail if you don't want to avoid it:

To find a left-hand critical value of F for D_X and D_Y degrees of freedom, find a right-hand critical value of F for D_Y and D_X degrees of freedom (note that the order of the D's is reversed!) and take its reciprocal.

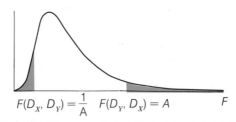

$F(D_X, D_Y) = \dfrac{1}{A}$ $F(D_Y, D_X) = A$ F

Example 3

If $n_X = 7$ and $n_Y = 15$, 5 per cent of F scores are below what value?
Here $D_X = 6$, $D_Y = 14$. From Example 1, we know that if $D_X = 14$, $D_Y = 6$, then 5 per cent of F scores are above 3.96, and therefore if $D_X = 6$, $D_Y = 14$, 5 per cent of F scores are below $1/3.96 = .25$.

Example 4

$H_0: \sigma_X^2 = \sigma_Y^2$; $H_1: \sigma_X^2 \neq \sigma_Y^2$; $\alpha = .02$. What are the critical values of F which determine the acceptance region for H_0 if $n_X = 21$, $n_Y = 25$?

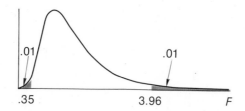

A two-tail test is used, with the area in each tail equal to .01. $D_X = 20$, $D_Y = 24$. For $\alpha = .01$, $F(20,24) = 2.74$. This is the upper critical value. Since $F(24,20) = 2.86$ when $\alpha = .01$, the lower critical value for F is $1/2.86 = .35$.

14.3 USE OF _F_ DISTRIBUTIONS IN TESTING HYPOTHESES ABOUT VARIANCE

Example 1

Two machines are used to fill boxes with crackers. In a random sample of 25 boxes filled by the first machine, the mean weight of crackers is 1.03 lb. with a variance of .008; a random sample of 31 boxes filled by the second machine has a mean weight of 1.06 lb. with a variance of .006. At a .05 level of significance, does the first machine show more variability than the second? Assume weights are normally distributed for both machines.

$H_0: \sigma_X^2 = \sigma_Y^2; H_1: \sigma_X^2 > \sigma_Y^2; \alpha = .05$; one-tail test.

$$n_X = 25, D_X = 24, s_X^2 = .008$$
$$n_Y = 31, D_Y = 30, s_Y^2 = .006$$

The mean weights are given, but are not relevant to the problem.

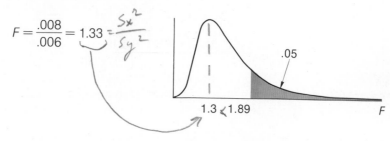

$$F = \frac{.008}{.006} = 1.33 = \frac{S_x^2}{S_y^2}$$

The critical value of F is $F(24,30) = 1.89$ for $\alpha = .05$.

Since 1.3 is less than 1.94, H_0 is accepted: The two types of machines have the same variability in filling boxes with crackers.

Example 2

A psychologist believes that scores of children who are under stress when they take an arithmetic test will be more variable because bright children will do better but dull children will panic and do worse than when not under stress. He finds that 20 children under stress have a standard deviation of 15.0, while a control group of 11 children not under stress have a standard deviation of 8.6. At a .025 level of significance, do children under stress show more variability in their grades on this test than children not under stress?

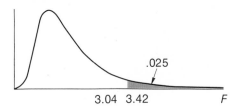

.025

3.04 3.42 F

$H_0: \sigma_X^2 = \sigma_Y^2; H_1: \sigma_X^2 > \sigma_Y^2.$ right- tail

$$n_X = 19, D_X = 18, s_X^2 = 15^2 = 225$$
$$n_Y = 11, D_Y = 10, s_Y^2 = 8.6^2 = 74$$
$$F = 225/74 = 3.04$$

For $\alpha = .025$, the critical value of $F(18,10) > F(20,10) = 3.42$. Therefore H_0 is accepted: Children under stress do not show more variability in their scores.

Example 3

Are the salaries of associate professors less variable than those of full professors at Great Western University? A random sample of 17 associate professors has a standard deviation of $1,000 in salaries, while 25 full professors chosen at random have a standard deviation of $2,500. Test at a .05 level of significance.

Solution 1

$H_0: \sigma_X^2 = \sigma_Y^2; H_1: \sigma_X^2 < \sigma_Y^2;$ left-tail test.

$$s_X^2 = 1000^2, s_Y^2 = 2,500^2; F = s_X^2/s_Y^2 = .16$$

$D_X = 16, D_Y = 24. F(24,16) = 2.24$ when $\alpha = .05$.

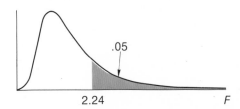

.05

2.24 F

Note that $F(24, 16) = F(D_Y, D_X)$. For a left-tail test, the critical value is $1/2.24 = .45$.

Since the computed F value, .16, is less than .45, H_0 is rejected: There is indeed less variability in the salaries of associate professors.

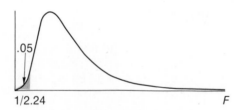

Solution 2

The necessity of finding $F(D_Y, D_X)$ and then taking its reciprocal can be avoided if the population whose sample has the larger variance is given the subscript *X* even if it is mentioned last in the statement of the problem. In this example,

$$H_0: \sigma_X^2 = \sigma_Y^2; \; H_1: \sigma_X^2 > \sigma_Y^2; \; \text{right-tail test.}$$
$$s_X^2 = 2500^2, \; s_Y^2 = 1000^2, \; F = s_X^2 / s_Y^2 = 6.25$$

$D_X = 24, D_Y = 16.$ $F(24, 16) = 2.24$ when $\alpha = .05$, so H_0 is rejected.

If this same strategem of naming the larger variance s_X^2 is used for two-tail tests, then the smaller critical value of *F* need not be ascertained: it will certainly be less than 1 since it is the reciprocal of a value in the table, and all tabular values are greater than 1; but the computed *F* value will be greater than 1.

Note: The computations in Example 3 were carefully done and the samples were random. Nevertheless, the result may not be reliable. Why? (Think before you read on.) Are the salaries of associate and full professors really normally distributed? It depends on the practices at Great Western University. Suppose there are definite ranges of salary at each rank. If Great Western has expanded recently and promoted a large number of new professors, the salary curve for professors may be skewed to the right. Or if 30 assistant professors are appointed associate each year and tend to stay at this rank for about 5 years, with an increment each year, before becoming full professors, the distribution for associate professors may be uniform (the "curve" is a horizontal line) rather than normal. **Use of the *F* distribution is based on the assumption of normality,** even in large samples. Tests for variances differ in this way from *z* tests for means.

Example 4

A psychologist wonders whether children under stress in a test show a different variance in their test scores than children not under stress. Bright children might do better but dull children panic and do worse, causing greater variability. On the other hand, perhaps stress has no effect on children who know the material well but causes those with less knowledge or ability to do better, so there is less variability in the results. He finds that a random sample of 15 children in a control group have a mean score of 77 with a standard deviation of 3.1, whereas 13 children under stress have a mean score of 82 with a standard deviation of 3.3. At a .05 level of significance, do the two groups have different variances?

$H_0: \sigma_X^2 = \sigma_Y^2$; $H_1: \sigma_X^2 \neq \sigma_Y^2$; two-tail test.

$$s_X^2 = 3.3^2 = 10.89, n_X = 13$$
$$s_Y^2 = 3.1^2 = 9.61, n_Y = 15$$
$$F = s_X^2/s_Y^2 = 1.13$$

Note that X scores are children under stress.

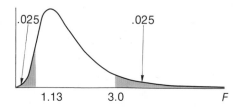

$D_X = 12, D_Y = 14$; $F(12, 14) = 3.0$ when $\alpha = .05$. It is not necessary to determine the lower critical value for acceptance of H_0; it is certainly less than 1, whereas the computed F, 1.13, is greater than 1. H_0 is accepted: the variances of the two groups of children do not differ.

Note: F curves differ somewhat in shape as D_X and D_Y vary. The same illustration has been used for all curves of this type, however, so use common sense and a grain of salt in interpreting the curves.

14.4 INTRODUCTION TO ANALYSIS OF VARIANCE

You have now learned how to use F distributions to answer the question, "Do two populations have the same variance?" If the answer is **yes,** then the F ratio for samples taken from the two populations is expected to be close to 1. If it is not close to 1, this difference may be due not just to sampling fluctuation but to different variances in the two populations.

Now F distributions will be used for quite a different kind of hypothesis testing: Do three or more populations have the same mean? Suppose tomato seeds can be treated with any one of three different chemicals or left untreated, and the yield of tomatoes under these four methods is to be compared. Four random samples of seeds are selected; the first three are each treated with a different chemical, and the fourth is left untreated. Then a field is divided into, say, 40 plots of the same size, and the 40 plots are randomly assigned to four groups; all the plots in a group are then planted with the same number of seeds from one of the samples.† The mean yield per plot of each of the four different samples is determined, with the following results:

Chemical 1	Chemical 2	Chemical 3	Untreated
$\overline{X}_1 = 96$	$\overline{X}_2 = 99$	$\overline{X}_3 = 92$	$\overline{X}_4 = 101$

Are the yields so different that it can be concluded that different chemical treatments affect yield? Should we accept or reject $H_0: \mu_1 = \mu_2 = \mu_3 = \mu_4$?

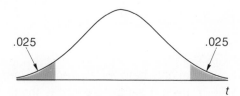

For any pair of means we could, of course, carry out a t test. But since there are four different means to be compared, $_4C_2 = 6$ t tests would have to be carried out. Remember that if $\alpha = .05$, the probability is .05 that given samples will have means whose difference falls in the rejection region even if $H_0: \mu_X = \mu_Y$ is true. If k independent t tests of differences of means are made, the probability is $1 - (.95)^k$ that in one of them the null hypothesis—that the samples come from populations with the same mean—is rejected even when it is true. If 6 independent tests are carried out, this probability is .26, even though $\alpha = .05$ for each individual test. If you are testing 4 samples a pair at a time (6 tests), you do not have independent t tests, so the probability computed above is only an approximation. This argument should persuade you, however, that t tests are not appropriate when more than 2 sample means are compared to see if the samples all come from the same population. A new technique is needed; this technique is called **analysis of variance**, often abbreviated **ANOVA**.

Suppose 3 samples, each of size 3, are taken from the same population and that there is no variability within the population. The sample scores then might be:

Sample 1	Sample 2	Sample 3
5	5	5
5	5	5
5	5	5

If, however, the samples come from populations which are affected by different treatments, but again there is no variability within a given population, the sample scores then might be:

---

†Both the seeds **and** the plots in which they are grown are selected at random. If this is not done, the effects of microclimate and different kinds of soil may have more influence on yield than the chemical treatment of the seed.

Sample 1	Sample 2	Sample 3
4	6	5
4	6	5
4	6	5

But there is, inevitably, variability in each population, so the sample scores would more likely be:

Sample 1	Sample 2	Sample 3
3	5	3
4	7	7
5	6	5
$\overline{X}_1 = 4$	$\overline{X}_2 = 6$	$\overline{X}_3 = 5$

Note that differences exist not only between scores in a sample but between scores in different samples. If we are given 3 samples such as those in the last table above and wish to decide whether or not they come from the same population, we must find out how to distinguish fluctuations in sampling from effects due to different treatments.

Let us start on this very simple problem with a much smaller sample size than is usually the case. Suppose that 3 groups of 3 students are taught a statistical technique by three different methods, and then are tested on one problem. The 9 students receive the following scores:

Group 1	Group 2	Group 3
3	5	3
4	7	7
5	6	5
$\overline{X}_1 = 4$	$\overline{X}_2 = 6$	$\overline{X}_3 = 5$

At a .05 level of significance, do the means of the populations taught by the three methods differ?

$H_0: \mu_1 = \mu_2 = \mu_3$

H_1: Differences exist between at least some of the means.

If the populations from which the three samples were chosen are identical — if they have the same mean and standard deviation, and if all three are normally distributed — then the three groups can be considered to be three samples from one large population. The sample means (4, 6, and 5, respectively) are not the same, but we hardly expect them to be identical. Is the variation between the sample means what is expected if the samples come from the same population, or are the differences between them so great that it is likely to occur not from sampling fluctuations but rather because the three populations are not in fact identical?

We shall make two different estimates of the population variance σ^2 under the assumption that the three populations are identical and therefore the three samples come from one large population. The first estimate will depend on the variations of the sample means about the grand mean that is obtained for all scores taken together;

this estimate, then, will depend on the differences between samples. The second estimate will depend on the variations of scores in a sample from the sample mean; this estimate, then, will depend upon the differences within each sample, but will not compare one sample with another. The first estimate will depend on sampling fluctuation and on real differences between samples; if the samples come from populations with different means (if H_0 is not true), then this estimate will be significantly larger than the second estimate, which depends on sampling fluctuation alone. The ratio of the two variances—their F score—will then be significantly greater than 1. But if H_0 is true and the three samples can be thought of as three slices of the same population, then the two estimates differ only because of sampling fluctuations and their ratio (their F score) should be approximately 1. A right-tail F test will be used, with $\alpha = .05$. Since this is the case, we will use the notation $F_\alpha(D_x D_y)$ [$F_{.05}(2,8)$, for example] instead of $\alpha = .05$, **one-tail test (right), $F(D_x D_y)$.** But be careful not to use this notation for two-tail tests!

The first estimate is based on the variability among the sample means: $\overline{X}_1 = 4, \overline{X}_2 = 6,$ $\overline{X}_3 = 5$. The mean of all 9 scores is $\overline{X}_T = (3 + 4 + 5 + 5 + 7 + 6 + 3 + 7 + 5)/9 = 5$. We find the variance of the three sample means about the mean of all sample scores $\overline{X}_T$ in the usual way; since we want an unbiased estimator we divide by one less than the number of samples:

$$s_X^2 = \frac{(4-5)^2 + (6-5)^2 + (5-5)^2}{2} = 1.$$

This quantity we have computed gives an estimate of $\sigma_X^2 = \dfrac{\sigma^2}{n}$, so $\sigma^2 \approx n s_X^2 = 3 * 1 = 3,$ since each sample is of size 3. This first estimate of σ^2 depends on variability between samples and will be referred to as $\hat{\sigma}^2$ **between** (read "σ hat squared between"; the hat is put on to remind you that this is an estimate based on sample quantities):

$$\hat{\sigma}^2 \text{ between} = 3.$$

It should be noted that we could have computed σ^2 directly by multiplying each squared deviation by the size of the sample involved:

$$\hat{\sigma}^2 \text{ between} = \frac{3(4-5)^2 + 3(6-5)^2 + 3(5-5)^2}{2} = 3.$$

When samples are of different sizes, each squared deviation will be weighted by the size of the sample from which it comes. Thus, if

$$n_1 = 9, \overline{X}_1 = 4, n_2 = 9, \overline{X}_2 = 6, n_3 = 7, \overline{X}_3 = 5, \overline{X}_T = 5$$

then

$$\hat{\sigma}^2 \text{ between} = \frac{9(4-5)^2 + 9(6-5)^2 + 7(5-5)^2}{2} = 9.$$

The mathematical justification requires more background than you possess. Be of great faith, please!

Now we need another and independent estimate of σ^2. We learned earlier that s^2 is an unbiased estimator of σ^2. If the three samples come from the same population (if H_0 is true), then we appear to have three ways of estimating σ^2: we could use s_1^2 or s_2^2 or

$s_3{}^2$. Instead, we use a weighted average of all three, assigning weights according to the degrees of freedom in each sample.

	Sample 1	Sample 2	Sample 3	Total
ΣX	12	18	15	45
ΣX^2	50	110	83	243
n	3	3	3	
D	2	2	2	
s^2	1	1	$4 = \dfrac{\Sigma X^2 - \dfrac{(\Sigma X)^2}{n}}{n-1}$	

$$\sigma^2 \approx \frac{2(1) + 2(1) + 2(4)}{2 + 2 + 2} = 2$$

This estimate of σ^2 depends on variability within a sample, and will be called $\hat{\sigma}^2$ **within:**

$$\hat{\sigma}^2 \; within = 2.$$

$$F = \frac{\hat{\sigma}^2 \; between}{\hat{\sigma}^2 \; within} = \frac{3}{2} = 1.5$$

What about degrees of freedom? For $\hat{\sigma}^2$ *between,* the degrees of freedom (D_x) is the number of samples minus 1. In this example, $D_x = 3 - 1 = 2$. For $\hat{\sigma}^2$ *within,* the degrees of freedom (D_y) is the total number of scores (in all samples) minus the number of samples. In this example, $D_y = 9 - 3 = 6$. For $\alpha = .05$, the critical value of $F(2,6)$ is 5.14 for a right-tail test. The computed F value, 1.5, falls in the acceptance region, so H_0 is accepted: $\mu_1 = \mu_2 = \mu_3$. The three populations from which the samples were taken have the same means. Since it was assumed that they have the same variances and that all are normally distributed, this means that they are identical populations.

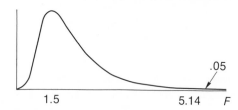

Now let's do the same problem again, and find computational formulas that will expedite the work when more or larger samples are used. First, some new notation is needed:

Let k = the number of samples.

$\quad n_T$ = the total number of scores in all samples.

The following steps are carried out (in Section 14.6, page 253, an explanation of why they work will be given):

1. List, in columns, the scores in each sample.

2. Add all the scores in the first sample, square the sum, and divide by the number of scores in the first sample.

3. Repeat step 2 for each of the other samples.

4. Add the results of steps 2 and 3; call this sum A.

5. Add all the scores in all samples; call this sum B.

6. Square each score and then add these square scores for all samples; call this sum C.

7. Add the number of scores in each sample; call this sum n_T.

8. Compute $\hat{\sigma}^2$ *between* $= \dfrac{A - B^2/n_T}{k - 1}$

     and $\hat{\sigma}^2$ *within* $= \dfrac{C - A}{n_T - k}$

9. $F = \dfrac{\hat{\sigma}^2 \ between}{\hat{\sigma}^2 \ within}$ is compared with the critical value $F_\alpha(k - 1, n_T - k)$ found in the appropriate F table.

Here is the work involved in our example:

	Sample 1	Sample 2	Sample 3	
	3	5	3	
	4	6	5	
	5	7	7	Total
Sum:	12	18	15	$45 = B$
Sum squared:	144	324	225	
Number in sample:	3	3	3	$9 = n_T$
$\dfrac{\text{Sum squared}}{\text{Number in sample}}$:	48	108	75	$231 = A$
Sum of squares:	$9 + 16 + 25 =$ 50	$25 + 36 + 49 =$ 110	$9 + 25 + 49 =$ 83	$243 = C$

	Numerator	Degrees of Freedom (Denominator)	Quotient
$\hat{\sigma}^2$ *between*	$A - B^2/n_T = 231 - 45^2/9 = 6$	$k - 1 = 2$	$6/2 = 3$
$\hat{\sigma}^2$ *within*	$C - A = 243 - 231 = 12$	$n_T - k = 6$	$12/6 = 2$

$F = 3/2 = 1.5$; $F_{.05}(2,6) = 5.14$, > 1.5, so H_0 is accepted, as before.

14.5 A SHORTCUT: SUBTRACTING A CONSTANT FROM SAMPLE SCORES

Sometimes the sample scores you must work with will be quite large. Computation with large figures can be time-consuming (and also increases the chance of errors in arithmetic).

Example 1

An economist is testing, at a .01 level of significance, the hypothesis that carpenters, masons, and electricians in Madison, N.J. have the same weekly wages. He assumes that the pay of all three groups is normally distributed and has the same variance. He determines the weekly wages of small random samples, with the following results:

Carpenters	Masons	Electricians
240	230	290
200	270	270
220	250	250
160		270
190		

Is the hypothesis correct?

$$H_0: \mu_1 = \mu_2 = \mu_3. \quad H_1: \text{It is not true that } \mu_1 = \mu_2 = \mu_3.$$

	Sample 1	Sample 2	Sample 3	Total
ΣX	1010	750	1080	$2840 = B$
$(\Sigma X)^2$	1,020,100	562,500	1,166,400	
n	5	3	4	$12 = n_T$
D	4	2	3	9
$(\Sigma X)^2/n$	204,020	187,500	291,600	$683,120 = A$
$\Sigma(X^2)$	207,700	188,300	292,400	$688,400 = C$

	Numerator	Degrees of Freedom (Denominator)	Quotient
$\hat{\sigma}^2$ between	$683,120 - 2,840^2/12 = 10,987$	$3 - 1 = 2$	5,493.5
$\hat{\sigma}^2$ within	$688,400 - 683,120 = 5,280$	9	587

$$F = \frac{5,493.5}{587} = 9.4. \quad F_{.01}(2, 9) = 8.02. \quad H_0 \text{ is rejected.}$$

If a constant is subtracted from each score in a set, the variance is not changed (see page 49). Often work can be simplified if the same constant is subtracted from every score in every sample. If 250 is subtracted from each score in Example 1, we shall obtain the same F score.

Example 2

Repeat Example 1, with 250 subtracted from each sample score.

Sample 1	Sample 2	Sample 3
−10	−20	40
−50	20	20
−30	0	0
−90		20
−60		

				Total
Sum:	−240	0	80	−160 = B
Sum squared:	57,600	0	6400	
Number in sample:	5	3	4	12 = n_T
$\dfrac{\text{Sum squared}}{\text{Number in sample}}$	11,520	0	1600	13,120 = A
Sum of squares:	15,200	800	2400	18,400 = C

	Numerator	Degrees of Freedom (Denominator)	Quotient
$\hat{\sigma}^2$ *between*	13,120 − (−160)²/12 = 10,987	2	5,493.5
$\hat{\sigma}^2$ *within*	18,400 − 13,120 = 5,280	9	587

$F = 5{,}493.5/587 = 9.4$, which is the same value obtained before. H_0 is rejected: There is a significant difference between the means of the populations from which the samples were taken.

Note, however, that this test does **not** disclose how many pairs of means differ significantly. This problem will be discussed in Section 14.7, page 255.

14.6★ WHERE DO THE COMPUTATIONAL FORMULAS COME FROM?

Let's take a closer look at the three samples of students in the discussion on page 248. Subscripts 1, 2, and 3 will refer to the three samples: $\overline{X}_1$ is the mean of the first sample, n_3 is the number of scores in the third sample, and so on. In this notation,

$$A = \frac{(\Sigma X_1)^2}{n_1} + \frac{(\Sigma X_2)^2}{n_2} + \frac{(\Sigma X_3)^2}{n_3}$$

$$B = \Sigma X_1 + \Sigma X_2 + \Sigma X_3$$

$$C = \Sigma(X_1^2) + \Sigma(X_2^2) + \Sigma(X_3^2)$$

We found (page 249) that $\hat{\sigma}^2$ **between** is the variance of the sample means about the mean $\overline{X}_T$ of all sample scores, with weights according to the size of each sample:

$$\hat{\sigma}^2 \; between = \frac{n_1(\overline{X}_1 - \overline{X}_T)^2 + n_2(\overline{X}_2 - \overline{X}_T)^2 + n_3(\overline{X}_3 - \overline{X}_T)^2}{k - 1}$$

(In this example, $k = 3$.) Let's look at the first term in the numerator.

$$n_1(\overline{X}_1 - \overline{X}_T)^2 = n_1\overline{X}_1^2 - 2n_1\overline{X}_1\overline{X}_T + n_1\overline{X}_T^2 \qquad \text{squaring } (\overline{X}_1 - \overline{Y}_T)^2$$

$$= n_1\left(\frac{\Sigma X_1}{n_1}\right)^2 = 2n_1\left(\frac{\Sigma X_1}{n_1}\right)\overline{X}_T + n_1\overline{X}_T^2 \quad \text{since } \overline{X}_1 = \frac{\Sigma X_1}{n_1}$$

$$= \frac{(\Sigma X_1)^2}{n_1} - 2(\Sigma X_1)\overline{X}_T + n_1\overline{X}_T^2$$

analysis of variance

The other two terms in the numerator of $\hat{\sigma}^2$ *between*, $n_2(\overline{X}_2 - \overline{X}_T)^2$ and $n_3(\overline{X}_3 - \overline{X}_T)^2$, can be expanded in a similar fashion. Adding all three, the numerator of $\hat{\sigma}^2$ *between* becomes

$$\left[\frac{(\Sigma X_1)^2}{n_1} \qquad - 2(\Sigma X_1)\overline{X}_T \qquad + n_1\overline{X}_T^2\right]$$

$$+ \left[\frac{(\Sigma X_2)^2}{n_2} \qquad - 2(\Sigma X_2)\overline{X}_T \qquad + n_2\overline{X}_T^2\right]$$

$$+ \left[\frac{(\Sigma X_3)^2}{n_3} \qquad - 2(\Sigma X_3)\overline{X}_T \qquad + n_3\overline{X}_T^2\right]$$

$$= \underbrace{\left[\frac{(\Sigma X_1)^2}{n_1} + \frac{(\Sigma X_2)^2}{n_2} + \frac{(\Sigma X_3)^2}{n_3}\right]}_{= A} \underbrace{- 2(\Sigma X_1 + \Sigma X_2 + \Sigma X_3)\overline{X}_T + (n_1 + n_2 + n_3)\overline{X}_T^2}_{= B^2/n_T?}$$

If we can show the last terms in the sum $= -B^2/n_T$, then we have succeeded in showing that $\hat{\sigma}^2$ *between* $= \dfrac{A - B^2/n_T}{k - 1}$. But this is easy to show, since $n_1 + n_2 + n_3 = n_T$ and

$$\overline{X}_T = \frac{\Sigma X_1 + \Sigma X_2 + \Sigma X_3}{n_1 + n_2 + n_3} = \frac{(\text{sum of all } X \text{ scores})}{n_T} = \frac{B}{n_T}.$$

$$-2(\Sigma X_1 + \Sigma X_2 + \Sigma X_3)\overline{X}_T + (n_1 + n_2 + n_3)\overline{X}_T^2 = -2B\left(\frac{B}{n_T}\right) + (n_T)\left(\frac{B}{n_T}\right)^2$$

$$= \frac{-2B^2}{n_T} + \frac{B^2}{n_T} = \frac{-B^2}{n_T}$$

Therefore $\hat{\sigma}^2$ *between* $= \dfrac{A - B^2/n_T}{k - 1}$.

Now what about $\hat{\sigma}^2$ *within*? It is the weighted average of the sample variances, with weights assigned according to the degrees of freedom in each sample:

$$\hat{\sigma}^2 \text{ within} = \frac{(n_1 - 1)s_1^2 + (n_2 - 1)s_2^2 + (n_3 - 1)s_3^2}{(n_1 - 1) + (n_2 - 1) + (n_3 - 1)}$$

Again, concentrate first on the first term in the numerator.

Using the computational formula for sample variance (see page 48),

$$(n_1 - 1)s_1^2 = (n_1 - 1)\left[\frac{\Sigma X_1^2 - (\Sigma X)^2/n_1}{n_1 - 1}\right] = \Sigma X_1^2 - (\Sigma X_1)^2/n_1$$

Weighting by degrees of freedom "cancels" the denominator. The same will be true for the other samples, so the numerator becomes

$$\Sigma X_1^2 - \frac{(\Sigma X_1)^2}{n_1} + \Sigma X_2^2 - \frac{(\Sigma X_2)^2}{n_2} + \Sigma X_3^2 - \frac{(\Sigma X_3)^2}{n_3}.$$

After rearrangement, this is

$$(\Sigma X_1{}^2 + \Sigma X_2{}^2 + \Sigma X_3{}^2) - \left[\frac{(\Sigma X_1)^2}{n_1} + \frac{(\Sigma X_2)^2}{n_2} + \frac{(\Sigma X_3)^2}{n_3}\right] = C - A.$$

That takes care of the numerator. Since $(n_1 - 1) + (n_2 - 1) + (n_3 - 1) = n_T - k$,

$$\hat{\sigma}^2 \; within = \frac{C - A}{n_T - k}.$$

14.7 WHICH MEANS ARE DIFFERENT?

Suppose that $H_0: \mu_1 = \mu_2 \ldots = \mu_k$ has been rejected after an F test. The experimenter then would like to know how to account for the difference: which pairs of means are significantly different? A number of statisticians have derived tests which can be applied after the analysis of variance test has shown a significant difference in the means. The test to be described here was devised by Scheffé.† It can be used for samples of different sizes, and is not much affected by populations which are not normal or which do not have equal variances. This test is valid for **all pairs of samples at once,** and thus differs considerably from repeated use of t tests.

Scheffé's Test

Scheffé's test includes the following steps:

1. Compute $S = \sqrt{(k - 1) \, F_\alpha(k - 1, n_T - k)(\hat{\sigma}^2 \; within)}$

2. If samples are of the same size n, see whether $|\overline{X}_i - \overline{X}_j|$ is less than or greater than $S\sqrt{2/n}$.
   If samples are of different sizes, see whether $|\overline{X}_i - \overline{X}_j|$ is less than or greater than
   $$S\sqrt{\frac{1}{n_i} + \frac{1}{n_j}}$$
   Carry out this test for each pair of sample means $\overline{X}_i, \overline{X}_j$.

3. If $|\overline{X}_i - \overline{X}_j|$ is larger, $H_0: \mu_i = \mu_j$ is rejected, and we conclude that that pair of means contributes significantly to the rejection of the analysis of variance test. If $|\overline{X}_i - \overline{X}_j|$ is smaller, however, then we are unable to reject the hypothesis that $\mu_i = \mu_j$.

Note: α in Scheffé's test is often the same as for the original analysis of variance test, but it doesn't need to be.

Example 1

A botanist plants random samples of each of 5 different strains of corn on 5 plots of land. The plots are of the same size and fertility and the same fertilizer is used on each. The mean yields of corn are 3.4, 7.4, 10, 10, and 11.8 bushels per plot. At a .05 level of significance, $H_0: \mu_1 = \mu_2 = \mu_3 = \mu_4$

---

†Scheffé, H., *The Analysis of Variance,* John Wiley, New York, 1959.

$= \mu_5$ is rejected, with $A = 2027.8$, $C = 2217$ (see Exercise 13, page 258). Which pairs of population means differ, at a .05 level of significance?

$$n_T = 25, k = 5, \hat{\sigma}^2 \text{ within} = \frac{2217 - 189.2}{25 - 5} = 9.46; F_{.05}(4, 20) = 2.87; S\sqrt{2/n}$$
$$= \sqrt{4 * 2.87 * 9.46} * \sqrt{2/5} = \sqrt{108.6} * \sqrt{.4} = 6.5.$$

The only sample means which differ by more than 6.5 are strains 1 and 5, so the statement "$\mu_1 = \mu_5$" can be rejected. We are unable, however, to reject the statement that any other pairs of means are equal.

Example 2

Consider the carpenters, masons, and electricians of Example 1, page 252. Here an analysis of variance test shows that $H_0 : \mu_1 = \mu_2 = \mu_3$ is rejected, if $\alpha = .01$. Apply Scheffé's test at the same level of significance, to see which pair or pairs of means differ.

$$k - 1 = 3 - 1 = 2, F_{.01}(2, 9) = 8.02, \hat{\sigma}^2 \text{ within} = 587, S = \sqrt{2 * 8.02 * 587}$$
$$= 97.1.$$

Since all pairs of means are to be examined, it is convenient to use a table such as the following:

		Differences of means		n_i	$\sqrt{\dfrac{1}{n_i} + \dfrac{1}{n_j}}$		$S\sqrt{\dfrac{1}{n_i} + \dfrac{1}{n_j}}$	
		$\overline{X}_2 = 250$	$\overline{X}_3 = 270$	n_i	n_2	n_3		
Carpenters	$\overline{X}_1 = 202$	-48	-68	5	.730	.671	70.9	65.1
Masons	$\overline{X}_2 = 250$		-20	3		.764		74.1
Electricians	$\overline{X}_3 = 270$			4				

The statement "$\mu_1 = \mu_3$" is rejected since $|-68| > 65.1$. We are unable to reject the statements "$\mu_1 = \mu_2$" (since $|-48| < 70.9$) or "$\mu_2 = \mu_3$" (since $|-20| < 74.1$).

14.8 VOCABULARY AND SYMBOLS

F score

$F(D_X, D_Y)$

$F_\alpha(D_X D_Y)$

F distribution

analysis of variance   ANOVA

$\hat{\sigma}^2$ between

$\hat{\sigma}^2$ within

k

n_T

A, B, C

Scheffé's test S

14.9 EXERCISES

1. Find the probability that $F(8, 10)$ is less than 5.06.

2. Find the probability that $F(7, 12)$ is less than .280.

3.   Ten ropes made of nylon are chosen at random; their mean breaking strength is 800 lbs. with a standard deviation of 50 pounds, whereas 12 ropes made of tufflon have a mean breaking strength of 900 pounds with a standard deviation of 28 pounds. At a .05 level of significance, test the hypotheses (a) that there is no difference in the variability of the two types of ropes in breaking strength, (b) that nylon ropes are more variable in their breaking strength than are ropes made of tufflon.

4.   On a final examination, 30 statistics students taught by TV tapes have a mean score of 78 with a variance of 25, while 21 students taught "live" have a mean of 80 and a variance of 30 on the same examination. At a .01 level of significance, do students taught by TV tapes show less variability?

5.   Twenty-five bottles of diet Cwik Cola chosen at random are tested for calorie content; the mean is 12.0 calories with a standard deviation of .60. Then 41 bottles of Pappy's Cola are found to have a mean calorie content of 20.0 calories with a standard deviation of .80. At a .02 level of significance, does Cwik Cola differ from Pappy's Cola in variability of calorie content?

6.   Refer to Exercise 14, page 219. At a .05 level of significance, do the two kinds of detergent have the same variance on tests of whiteness?

7.   Random samples of size 8 and 6, respectively, are chosen from among first graders in the Tacoma Elementary School and are given reading readiness tests. The first sample, of children who have had a year of kindergarten, has a mean of 104 and a standard deviation of 7 on the readiness test. For the second sample, of children who did not go to kindergarten, the mean on the reading readiness test is 95 with a standard deviation of 4. Assume that scores on the reading readiness tests of all children are normally distributed (both with and without kindergarten).

    (a) Do the test scores of children who have been to kindergarten have the same variance as those who have not? (.02 level of significance.)

    (b) Is there a difference in the reading readiness scores of children who have not been to kindergarten? ($\alpha = .05$.)

8.   At a .05 level of significance, do the populations from which the following samples are taken have the same means? Assume they are normally distributed and have the same variance.

Sample 1	Sample 2	Sample 3	Sample 4
0	−1	1	3
1	2	4	5
2	5	7	8
3			10

For Exercises 9–12, use analysis of variance to test whether the samples come from populations with the same mean ($\alpha = .05$). State assumptions that must be made; H_0 and H_1; and whether H_0 is accepted or rejected.

9.

Sample 1	Sample 2	Sample 3	Sample 4
0	4	4	8
−2	0	6	6
2	2	2	4

10.

Sample 1	Sample 2	Sample 3
2	3	−1
4	5	0
5	6	1
		3

11.

Sample 1	Sample 2
10	8
12	10
14	12
16	14
18	16
20	

(Suggestion: Subtract 12 from each score.)

12.

Sample 1	Sample 2
10	6
12	8
14	10
16	12
18	14
20	

13.   A botanist plants random samples of each of 5 different strains of corn on 5 plots of land. The plots are of the same size and fertility and the same fertilizer is used on each. The yields per plot are as follows:

Strain 1	Strain 2	Strain 3	Strain 4	Strain 5
4	7	10	16	10
3	8	14	14	13
6	9	12	10	12
2	8	9	7	10
2	5	5	3	14

At a .05 level of significance, do the different strains produce the same yield?

14.   Students are taught statistics by 3 different methods: (a) using TV tapes, (b) using audio tapes, (c) with a live instructor. A random sample of 5 students taught by each method shows the following grades on the final examination.

(a)	(b)	(c)
75	80	90
78	75	95
80	70	60
60	75	80
95	80	80

Is there a significant difference in grades of students taught by different methods? (.05 level of significance.) Hint: Subtract 80 from each score.

15.   The data in this table show the relative performance in German of random samples chosen from those using a traditional language lab and those using computer-assisted instruction:

Language lab	Computer-assisted
1	2
2	3
3	4
3	4
3	4
4	5
5	6
6	6
27	6
	6
	46

(a) Use a t test to test $H_0: \mu_X = \mu_Y$ at a .05 level of significance; first find out whether the populations have the same variance.

(b) Use analysis of variance to test $H_0: \mu_1 = \mu_2$ at a .05 level of significance.

16. The tensile strength of three different brands of cotton thread is compared by measurements on three spools of each brand, chosen at random. Assume the tensile strength of all three bands is normally distributed, and that they have the same variance. Test at a .01 level of significance, using the following data:

Brand A	Brand B	Brand C
98	105	110
100	103	98
102	101	106

17. An agricultural chemist is measuring the effectiveness of three different formulations of an insecticide. He tests random samples of each formulation under standard conditions on colonies of fire ants, and collects the following data:

Percentage Kill, using Formulation

#1	#2	#3
40.2	50.4	71.2
35.1	60.2	60.3
50.3	53.8	75.4
	40.2	80.3

At a .05 level of significance, is there a difference in the mean percentage kill of the three formulations? Assume normality and equal variances.

18. The same agricultural chemist as in Exercise 17 is carrying out the same experiment with three new formulations of insecticide. This time he collects the following data:

Percentage Kill, using Formulation

#1	#2	#3
40.2	50.4	71.2
39.6	51.0	72.0
43.0	52.0	73.0
	50.6	70.0

Test again at a .05 level of significance.

analysis of variance

19. The following table gives the grades of 5 students on 4 quizzes in statistics:

Student

		1	2	3	4	5
	1	85	70	90	95	70
Quiz	2	90	80	95	90	90
	3	70	75	80	85	70
	4	80	65	75	80	75

Test, at a .05 level of significance, the hypotheses that (a) the students have the same mean grades, and (b) the quizzes are equally difficult. Hint: Subtract 80 from each grade.

20. The output of four workers on the same type of machine was measured in 4 randomly chosen time periods. At a .05 level of significance, do they all have the same mean output?

Worker

		A	B	C	D
	1	22	26	20	28
Time	2	20	28	16	26
Period	3	21	24	18	30
	4	24	22	20	26

Hint: Subtract 25 from each score.

21. Refer to exercise 9. With $\alpha = .05$, $H_0: \mu_1 = \mu_2 = \mu_3 = \mu_4$ is rejected. Use Scheffé's test with $\alpha = .05$ to see which pairs of means are significantly different. ($\hat{\sigma}^2$ *within* = 4.)

22. Refer to Exercise 10. Use Scheffé's test with $\alpha = .01$ to see which pairs of means differ significantly. ($\hat{\sigma}^2$ *within* = 2.59.)

23. Refer to Exercise 12. Use Scheffé's test with $\alpha = .05$ to see which pairs of means differ significantly. ($\hat{\sigma}^2$ *within* = 12.22.)

24. Refer to Exercise 20. In this problem $H_0: \mu_1 = \mu_2 = \mu_3$ is rejected when $\alpha = .05$. Find out which pairs of means are different (a) at a .05 level of significance, (b) at a .01 level of significance.

25. An analysis of variance test is made on the following data:

X_1	X_2	X_3	X_4
0	2	4	6
1	3	5	7
1	3	5	7
2	4	6	8
2	4	6	8

$H_0: \mu_1 = \mu_2 = \mu_3 = \mu_4$ is rejected at a .05 level of significance. Use Scheffé's test to see which pairs of means are different at a .05 level of significance. ($A = 452.8$, $C = 464$.)

26. See Exercise 13. The 5 different strains of corn did not give the same mean yield, using an analysis of variance test at a .05 level of significance. Test, again with $\alpha = .05$, to determine which pairs of means are different ($\hat{\sigma}^2$ *within* = 9.46).

27.   The amount of cola given out by 3 soda machines is measured as follows:

Machine A: 6.1, 6.4, 6.2, 6.3 oz.
Machine B: 5.7, 5.8, 5.9, 5.7, 5.3 oz.
Machine C: 6.5, 6.4, 6.2 oz.

At a .01 level of significance, do the machines give drinks of the same size?

28.   Refer to Exercise 27. Use Scheffé's test, at a .05 level of significance, to see which means differ. ($\hat{\sigma}^2$ *within* = .034.)

ANSWERS

1. $F(8, 10) = 5.06$ when $\alpha = .01$. Therefore the probability that $F(8, 10)$ is less than 5.06 is .99.

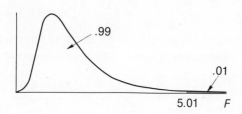

3. $H_0: \sigma_X^2 = \sigma_Y^2; H_1: \sigma_X^2 \neq \sigma_Y^2$ (two-tail test).

$n_X = 10, s_X^2 = 50^2, n_Y = 12, s_Y^2 = 28^2$.

$F = s_X^2/s_Y^2 = 3.2. D_X = 9, D_Y = 11, F(9, 11) = 3.59$ when the area in the right tail is .025.

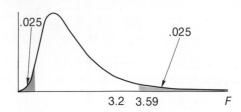

H_0 is accepted; there is no difference in variability.
(b) $H_0: \sigma_X^2 = \sigma_Y^2; H_1: \sigma_X^2 > \sigma_Y^2$ (one-tail test, right-hand tail).
Computed F score is the same as in (a). $F(9, 11) = 2.90$ when $\alpha = .05$.

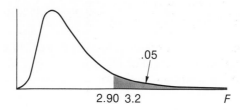

H_0 is rejected.

5. $H_0: \sigma_X^2 = \sigma_Y^2$; $H_1: \sigma_X^2 \neq \sigma_Y^2$; two-tail test. $n_X = 41$, $s_X = .80$, $n_Y = .60$; $F = .80^2/.60^2 = 1.78$. $\alpha = .02$, $F(40, 24) = 2.49$.

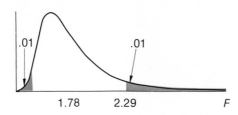

H_0 is accepted; Cwik and Pappy's have the same variability.

7.   (a) $s_X = 7$, $n_Y = 8$, $s_Y = 4$, $n_Y = 5$, $F = \dfrac{7^2}{4^2} = 3.06$; if $\alpha = .02$, 2-tail test, $F(7, 5) = 10.46$. The variances are the same.

(b) $\sigma_{(\bar{X}-\bar{Y})} = \sqrt{\dfrac{7 * 7^2 + 4 * 4^2}{8 + 5 - 2} \left(\dfrac{1}{8} + \dfrac{1}{5}\right)} = 3.47$; $t = \dfrac{(104 - 95) - 0}{3.47} = 2.59$;

$\alpha = .05$, $D = 11$, so the critical values of t are ± 2.20. H_0 is rejected; there is a difference in scores of children who have gone to kindergarten and those who have not.

9. $F = \dfrac{(168 - 36^2/12)/3}{(200 - 168)/8} = \dfrac{20}{4} = 5.00$; $F_{.05}(3, 8) = 4.07$. H_0 is rejected; the population means are not all the same.

11. $F = \dfrac{(54 - 18^2/11)/1}{(164 - 54)/9} = \dfrac{24.55}{12.22} = 2.01$; $F_{.05}(1, 9) = 5.12$. Accept H_0.

13.   (Subtract 10 from each score.) $F = \dfrac{213.0/4}{189.2/20} = \dfrac{53.3}{9.46} = 5.6$; $F_{.05}(4, 20) = 2.87$. H_0 is rejected; the yields are different.

15.   (a) $H_0: \mu_X - \mu_Y = 0$, $H_1: \mu_X - \mu_Y \neq 0$, $\alpha = .05$.
$\bar{X} = 27/8 = 3.38$, $s_X^2 = (109 - 27^2/8)/7 = 2.55$
$\bar{Y} = 46/10 = 4.60$; $s_Y^2 = (230 - 46^2/10)/9 = 2.04$

$s_X^2/s_Y^2 = 2.55/2.04 = 1.25$; with $\alpha = .05$ and a two-tail test, $F(9, 7) = 4.82$, so it can be assumed that the populations have the same variance.

$$t = \dfrac{(3.38 - 4.6) - 0}{\sqrt{\dfrac{7 * 2.55 + 9 * 2.04}{(8 - 1) + (10 - 1)} * \left(\dfrac{1}{8} + \dfrac{1}{10}\right)}} = \dfrac{-1.22}{.71} = -1.72$$

$D = (8 - 1) + (10 - 1) = 16$, $\alpha = .05$, critical value of $t = 2.12$. H_0 is accepted; there is no significant difference between the two methods of instruction.

(b) $F = \dfrac{(302.7 - 73^2/18)/1}{(339 - 302.7)/16} = \dfrac{6.7}{2.26} = 2.96$; $F_{.05}(1, 16) = 4.49$. Accept H_0.

Notice that the computed and critical values of F are, within roundoff, the square of the computed and critical values of t. For $H_0: \mu_1 = \mu_2$, $H_1: \mu_1 \neq \mu_2$, the t test and the analysis of variance test are exactly equivalent.

17.   (Subtract 50 from each score.) $F = \dfrac{(2104.70 - 67.4^2/11)/2}{(2651.6 - 2104.7)/8} = \dfrac{845.9}{68.4} = 12.4$; $F_{.05}(2, 8) = 4.46$. H_0 is rejected; the three formulations do not have the same mean kill.

19. (a) $F = \dfrac{(612.50 - 10^2/20)/4}{(1600 - 612.50)/15} = 2.31$; $F_{.05}(4, 15) = 3.06$. The students' mean grades do not differ.

(b) $F = \dfrac{661.25 - 10^2/20)3}{(1600 - 661.25)/16} = 2.80$; $F_{.05}(3, 16) = 3.24$. The means of the quizzes do not differ.

21. $\overline{X}_1 = 0, \overline{X}_2 = 2, \overline{X}_3 = 4, \overline{X}_4 = 6$; $S\sqrt{2/n} = \sqrt{3 \star 4.07 \star 4} \sqrt{2/3} = 5.71$; $\mu_1 \neq \mu_4$, since $|6 - 0| > 5.71$. We are unable to conclude that any of the other pairs of means differ significantly.

23. $F_{.05}(1, 9) = 5.12$; $S\sqrt{(1/n_1) + (1/n_2)} = \sqrt{1 \star 5.12 \star 12.22} \star \sqrt{(1/6 + (1/5)} = 4.79$. $\overline{X}_1 = 5, \overline{X}_2 = 0$; $\mu_1 \neq \mu_2$ since $|5 - 0| > 4.79$.

25. σ^2 within $= (464 - 452.8)/16 = .70$, $F_{.05}(3, 16) = 3.24$, $S\sqrt{2/n} = \sqrt{5 \star 3.24 \star .70}$ $\star \sqrt{2/5} = 2.13$. $\overline{X}_1 = 1.20, \overline{X}_2 = 3.20, \overline{X}_3 = 5.20, \overline{X}_4 = 7.2$. $\mu_1 \neq \mu_3, \mu_1 \neq \mu_4, \mu_2 \neq \mu_4$; the other pairs of means are not found to differ significantly.

27. (Subtract 6.0 from each score.) $F = \dfrac{(1.165 - .50^2/12)/2}{(1.470 - 1.165)/9} = \dfrac{.572}{.034} = 16.9$. $F_{.01}(2, 9)$ $= 8.02$. The machines do not give drinks of the same size.

chapter fifteen

AKA "Least Squares"

linear regression and correlation

Many problems in statistics are concerned with the relation be-tween two variables which can be paired together — height and weight of students, age and take-home pay of plumbers, IQ and score on a mechanical aptitude test, for example. There are two different questions that need to be answered: Are the variables related? If they are related, can knowledge of one be used to predict the other? (Is there a relation between a student's rank in his class at Lynn High School and his grade-point average at Boonton College? And if the Dean of Admis-sions at Boonton knows a particular student's rank at Lynn, can he predict how well the student will do at Boonton if admitted there?)

In this chapter you will learn how to plot the pairs of data, find the equation of the straight line which best fits the plotted points, and find a coefficient which measures how well the points fit the straight line. For this it will not be necessary to make any assumptions about the population from which the samples are taken. If the second variable is to be predicted when a value of the first is known, or if the closeness with which all paired scores in the populations fit a straight line is to be measured, then it is essential to make assumptions about the dis-tributions in the population. These assumptions and inferences about the populations will be taken up in Chapter 16.

15.1 INTRODUCTION

Suppose you know the years of school completed by and the income of each of a number of people. In each case observations are paired: for the first individual both years of schooling (or X) and income (or Y) are known; the pair (X_1, Y_1) represents the first individual; the second individual provides the pair of observations (X_2, Y_2), and so on. We shall not be interested in a list of years of schooling and a separate list of in-comes, but only in measurements that are paired together. (Can X scores be the number of births per year in the United States, 1930–1950, and Y scores the number of U.S. college freshmen, 1948–1968? The answer is "yes" if each X score is paired with a Y score 18 years later.) The pair (X, Y) is called an **ordered pair** since order matters: (12, 10,000) implies 12 years of school and an income of $10,000, while (10,000, 12) implies 10,000 years of school and an income of $12.

A warning about notation: You are probably accustomed to x- and y-coordinates. Here we shall use X and Y instead, because capital letters have been used consistently for scores. Also the equation $y = mx + b$, where m is the slope and b the y-intercept of the line for which this is the equation, is probably familiar to you. Here, however, the equation of a straight line will be $Y = a + bX$, so that you can refer to other statistics texts which are pretty uniform in this respect; now the slope is b and the Y-intercept is a.

A certain familiarity with straight lines is required. If you can answer the following questions, skip to Section 15.5, page 271. If not, refer to the section noted. The answers are at the bottom of page 265.

1. Can you graph the line $Y = 1 + 2X$? (If not, see Section 15.2.)

2. What is the slope of the line $Y = 4 + 2X$? What is the Y-intercept? (Section 15.3.)

3. What is the equation of the line through $P: (1,4)$ and $Q: (2,3)$? (Section 15.4.)

15.2 GRAPH OF A STRAIGHT LINE

Equations of the form $Y = a + bX$ have graphs which are straight lines. (It is also true that any straight line which is not a vertical line has the equation $Y = a + bX$, and therefore this is called a linear equation.) Upon replacing X by 2 and Y by 4, the equation $Y = 2 + X$ becomes an identity: $4 = 2 + 2$. We say that the point P whose coordinates are $(2,4)$ lies on the line $Y = 2 + X$. $Q: (0,2)$ is another such point, since $2 = 2 + 0$. Since two points determine a straight line, we can now sketch the graph. To check arithmetic, it is a good idea to find a third point, such as $R: (1,3)$, whose coordinates satisfy the equation, and see that it does lie on the line determined by P and Q. It is customary and efficient to put these results in a table, as shown below. Choose any convenient X values, and find the corresponding Y values by substituting the chosen X values in the equation.

X	Y
0	2
1	3
2	4

Example 1

Graph the line whose equation is $Y = 3 - 2X$.
Choose three convenient X values: 0, 1, 4, say.

X	Y
0	
1	
4	

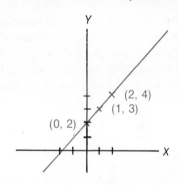

Just Algebra

skip to p.271

ANSWERS:

1.

$$y = \frac{4-3}{1-2} x + b; \; m = -1$$
$$y = -x + b$$
$$4 = -1 + b; \; b = 5$$
$$y = 5 - x$$

2. Slope $= 2$, Y-intercept $= 4$.
3. $Y = 5 - X$.

Then find the corresponding Y values. If $X = 0$, then $Y = 3 - 2 * 0 = 3$. Similarly, $Y = 1$ when $X = 1$; $Y = -5$ when $X = 4$. Then plot the points and draw the straight line.

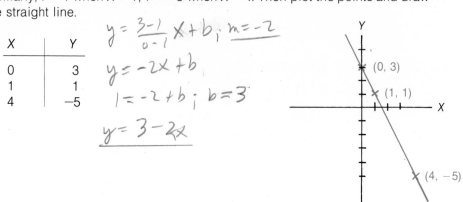

X	Y
0	3
1	1
4	−5

$y = \frac{3-1}{0-1}X + b; \ m = -2$

$y = -2x + b$

$1 = -2 + b; \ b = 3$

$y = 3 - 2x$

15.3   THE SLOPE OF A LINE

The slope of a line is a number that tells you not only whether Y values of points on the line are increasing or decreasing as X values increase but also how quickly they are changing. The slope of a vertical line is not defined; the slope of any horizontal line is 0. For any other line, we can take any two points on the line and find the ratio of the change in Y values to the change in X values:

$$\text{slope} = \frac{\text{change in } Y \text{ values}}{\text{change in } X \text{ values}}.$$

Example 1

What is the slope of the line $Y = 1 + 2X$?

Choose any two points P and Q on the line. Suppose P has coordinates $(1,3)$ and Q has coordinates $(3,7)$. Then the slope of this line is $\frac{7-3}{3-1} = \frac{4}{2} = 2$.

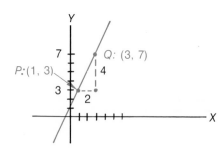

It doesn't matter in which order you take the two points:

$$\text{slope} = \frac{3-7}{1-3} = \frac{-4}{-2} = 2.$$

But if you take the Y-coordinate of Q first in the numerator, take the X-coordinate of Q first in the denominator $\left(\textbf{not} \ \frac{7-3}{1-3} = \frac{4}{-2} = -2\right)$. Nor does it

matter which two points on the line you take. If instead of P and Q, you choose
R: (0,1) and S: (4,9), the slope is $\dfrac{9-1}{4-0} = \dfrac{8}{4} = 2$.

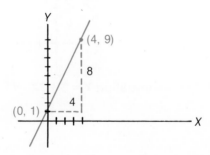

Why is the slope of a vertical line not defined? Choose, for example,
the points (2,1) and (2,3) on the vertical line X = 2; slope $= \dfrac{3-1}{2-2} = \dfrac{2}{0}$ = ? The
slope is not defined because division by 0 is not defined.

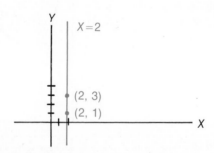

Choose two points on the horizontal line Y = 2 and check that its slope
is 0.

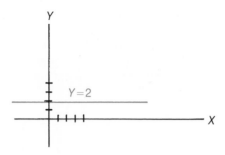

Example 2

What is the slope of the line whose equation is Y = 2 + 3X?
    (0,2) and (1,5) are two points on the line.

$$\text{slope} = \frac{5-2}{1-0} = 3 \left(\text{or } \frac{2-5}{0-1} = \frac{-3}{-1} = 3 \right).$$

X	Y
0	2
1	5

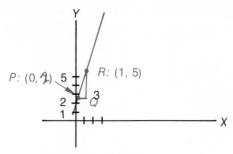

But **3 is the coefficient of X in the equation Y = 2 + 3X.** Will this always be the case?

Example 3

What is the slope of the line $Y = a + bX$, where a and b are any constants? Assume for graphing that a and b are positive.

If $X = 0$, $Y = a$; if $X = 1$, $Y = a + b$. So two points on the line are $P: (0,a)$ and $Q: (1,a+b)$.

$$\text{slope} = \frac{(a + b) - a}{1 - 0} = \frac{b}{1} = b$$

X	Y
0	a
1	$a + b$

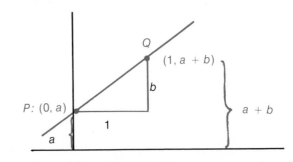

> The slope of the line $Y = a + bX$ is b.

In the examples shown so far, a point moving along the line has been rising as its X-coordinate increases. But what if the point **falls** as it moves to the right?

Example 4

What is the slope of the line $Y = 3 - 2X$?
$P: (0,3)$ and $Q: (1,1)$ lie on the line.

$$\text{slope} = \frac{3 - 1}{0 - 1} = \frac{2}{-1} = -2 \left(\textbf{or} = \frac{1 - 3}{1 - 0} = \frac{-2}{1} = -2 \right)$$

Lines *(a)* through *(c)* below have positive slope; lines *(d)* through *(f)* have negative slope; *(g)* has slope 0; the slope of *(h)* is not defined.

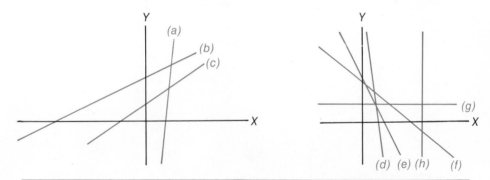

> The Y-intercept of a line is the Y-coordinate of the point at which the line crosses the X axis; it is found by letting $X = 0$ in the equation of the line.

Equation	Slope	Y-intercept
$Y = 1 + 2X$	2	1
$Y = 3 - 2X$	-2	3
$Y = -4 + X$	1	-4
$Y = a + bX$	b	a

If the slope and Y-intercept of a line are known, it is easy to graph the line, so note the last line above with care: The constant term is the Y-intercept and the co-efficient of X (the number that multiplies X) is the slope of the line $Y = a + bX$.

Example 5

Graph the line $Y = 4 - 2X$.

The Y-intercept is 4; that is, the line crosses the Y axis at (0,4). The slope is −2; as X increases by 1, Y decreases by 2. The points *P:* (0,4) and *Q:* (1,2) determine the line.

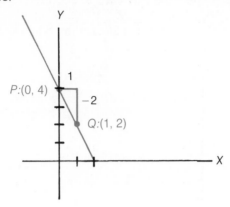

15.4 THE EQUATION OF A LINE THROUGH TWO GIVEN POINTS

The graph of the equation $Y = a + bX$ is a straight line. To find the equation of a line through two given points, you must determine the values of *a* and *b*. Finding *b* is easy — it is the slope of the line and is determined by the given points. To find *a*, concentrate on one of the given points. It lies on the line, so its coordinates satisfy the equation of the line. Using this information, we can find *a*. Study the following examples.

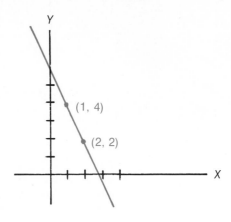

Example 1

Find the equation of the line through (1,4) and (2,2).
The equation is $Y = a + bX$. $a = ?$ $b = ?$

The slope of the line is $\dfrac{4-2}{1-2} = -2$, so $b = -2$ and the equation is $Y = a$ − 2X. The point *P*: (1,4) lies on the line, so these coordinates satisfy the equation. $4 = a - 2 * 1$, so $a = 6$; $Y = 6 - 2X$ is the equation. (Check that the line passes through *Q*: (2,2).)

Example 2

Find (a) the equation and (b) the *Y*-intercept (that is, the point where the line crosses the *Y* axis) of the line through (3,5) and (4,7).

(a) Slope $= \dfrac{5-7}{3-4} = 2 = b$; $Y = a + 2X$.

P: (3,5) lies on the line, so $5 = a + 2 * 3$, and therefore $a = -1$.
The equation of the line is $Y = -1 + 2X$.

(b) When $X = 0$, $Y = -1$. The *Y*-intercept is −1.

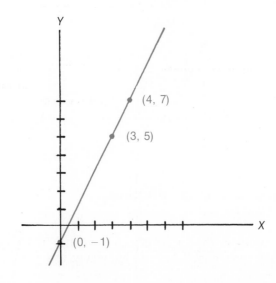

15.5 THE REGRESSION EQUATION

Suppose the number of hours spent in studying for a final exam in statistics and the number of questions answered correctly by each student are as follows:

Hours of study (X)	5	8	6	9	10	8	5	4	10	4	10	7	9
Correct answers (Y)	5	8	7	9	10	7	4	4	8	2	9	6	8

These data can be shown pictorially by plotting the corresponding pairs of scores as points on a graph. The resulting picture is called a **scatter diagram** or **scatter plot:**

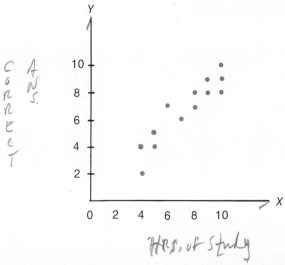

We shall learn how to draw the line which best fits a given set of points, and how to find its equation. The line is called a **regression line,** the equation a **regression equation.** The word "regression" is used because a man named Sir Francis Galton developed the techniques for finding the equation of the line while comparing the heights of parents and children. He discovered that short parents have children who are shorter than average, but not as short as their parents, while tall parents have children who are taller than average but not as tall as their parents—there is a "regression to the mean."

It will always be possible to find a straight line which best fits a given set of points. Sometimes it is not wise to do so, however. Later we shall study correlation, and then we shall have a measure of how close to a straight line the given points lie. Until then, always make a scatter diagram and see if a straight line seems reasonable as in (a) below; don't bother to find the regression line if the scatter diagram looks like (b) or (c):

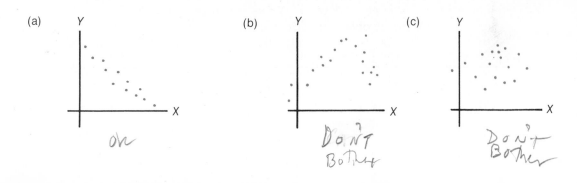

In more advanced texts, it is shown how to fit a regression curve of the form $Y = a + bX + cX^2$ to a scatter plot such as (b). We shall work only with straight lines, however.

The purpose of a regression equation is to use one variable to predict another. If you count cricket chirps per minute, can you tell the temperature? If you know an entering freshman's standing in his high school class, can the Dean of Admissions predict his grade-point average at the end of his freshman year? Can a company which manufactures small appliances give finger dexterity tests to job applicants and predict which ones will assemble parts most quickly? The variable whose value is known (cricket chirps per minute or standing in high school class or score on finger dexterity test) is called the independent variable and is (usually) represented by X; the variable whose value is being predicted (temperature or GPA or number of parts assembled) is called the dependent variable and is (usually) represented by Y.

Suppose we are given the following pairs of X and Y scores:

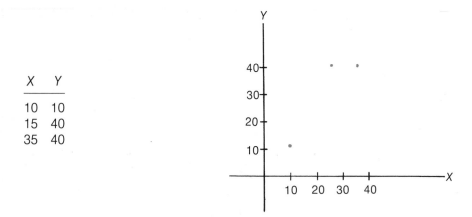

X	Y
10	10
15	40
35	40

We shall try to find an equation $Y_c = a + bX$ so that we can predict a Y score by substituting in a value of X. For example, if $X = 20$, what corresponding Y score would be predicted? If a and b can be determined, the answer is $a + 20b$. The subscript c (for *computed* Y) is used to distinguish Y values computed or predicted from the regression equation from Y values given in the original pairs.

The difference between the given score Y and the predicted score Y_c is known as the **error of estimation. The regression line, or the line which best fits the given pairs of scores, is the line for which the sum of the squares of these errors of estimation is minimized.**

Example 1

Which of the following equations describes the line which best fits the three points (i.e., the three pairs of X and Y values) given above?

(a) $Y_c = 30$. This is obtained by letting $a = Y, b = 0$.

(b) $Y_c = -50 + 6X$. This is the equation of the line determined by the first two points: (10,10) and (15,40).

(c) $Y_c = -2 + 1.2X$. This equation is satisfied by the first and third pairs: (10,10) and (35,40).

(d) $Y_c = 12.8 + .86X$. This is the true regression equation; the method by which it is derived will be shown shortly.

Answer

Least Squares

		(a) $Y_c = 30$			(b) $Y_c = -50 + 6X$			(c) $Y_c = -2 + 1.2X$			(d) $Y_c = 12.8 + .86X$		
X	Y	Y_c	$Y-Y_c$	$(Y-Y_c)^2$	Y_c	$Y-Y_c$	$(Y-Y_c)^2$	Y_c	$Y-Y_c$	$(Y-Y_c)^2$	Y_c	$Y-Y_c$	$(Y-Y_c)^2$
10	10	30	−20	400	10	0	0	10	0	0	21.4	−11.4	130.0
15	40	30	10	100	40	0	0	16	24	576	25.7	14.3	204.5
35	40	30	10	100	160	−120	14,400	40	0	0	42.9	2.9	8.4
				600			14,400			576			342.9

$\Sigma(Errors)^2 = 600$

the least of the four eqns.

(a)   (b)   (c)   (d)

The sum of the squares of errors is the least for the equation $Y_c = 12.8 + .86X$; therefore this is the best fit of the four equations given.

How is the regression equation determined? By more advanced mathematics, we can show that the values of b and a may be computed using these formulas:

m' slope =
$$b = \frac{n\Sigma XY - (\Sigma X)(\Sigma Y)}{n\Sigma X^2 - (\Sigma X)^2}$$

'b' y intercept =
$$a = \frac{\Sigma Y}{n} - b\frac{\Sigma X}{n} = \overline{Y} - b\overline{X}$$

Warnings

(1) n is the number of **pairs** of X and Y scores which are used in determining the regression line. In Example 1 below, $n = 3$, **not** 6.

(2) Be careful to distinguish between $(\Sigma X)^2$ and ΣX^2; $(\Sigma X)^2$ directs you to add first and then square the sum, while ΣX^2 directs you to square each X score and then add the squares.

(3) The denominator of b is always greater than 0. It is $n(n-1)$ times the variance of the X scores, and the variance is always greater than 0 unless all X's are the same.

But be of good cheer: it is easy to apply these formulas.

Example 2

Find the regression line for the pairs of X and Y scores shown below.

X^2	X	Y	XY
100	10	10	100
225	15	40	600
1225	35	40	1400
1550	60	90	2100

Notice that the X^2 column is written to the left of the X column, so that your eye doesn't have to travel far.

Answer

$$b = \frac{n\Sigma XY - (\Sigma X)(\Sigma Y)}{n\Sigma X^2 - (\Sigma X)^2} = \frac{3 * 2100 - 60 * 90}{3 * 1550 - 60^2} = \frac{6300 - 5400}{4650 - 3600} = \frac{900}{1050} = .857$$

$$a = \frac{90}{3} - \frac{60}{3} * .86 = 30 - 20 * .86 = 12.8$$

The equation of the regression line is, then,

$$Y_c = 12.8 + .86X.$$

Example 3

$n = 6$

Find the regression line for the six pairs of X and Y scores shown below.

X^2	X	Y	XY
4	2	0	0
9	3	3	9
16	4	4	16
25	5	4	20
36	6	6	36
49	7	11	77
139	27	28	158

Slope: $b = \dfrac{n(\Sigma xy) - (\Sigma x)\Sigma y}{n(\Sigma x^2) - (\Sigma x)^2}$

y-intercept: $a = \dfrac{\Sigma y}{n} - b\left(\dfrac{\Sigma x}{n}\right) = \bar{y} - b\bar{x}$ obvious!

$$b = \frac{(6 * 158) - (27 * 28)}{(6 * 139) - 27^2} = \frac{948 - 756}{834 - 729} = \frac{192}{105} = 1.829$$

$$a = \frac{28}{6} - 1.829 * \left(\frac{27}{6}\right) = -3.56$$

$$Y_c = -3.56 + 1.83X$$

Again it should be emphasized that n is 6 here, not 12.

Sometimes $n\Sigma XY$ is approximately equal to $(\Sigma X)(\Sigma Y)$ and it is necessary to avoid rounding off before subtraction. Here is an example to illustrate this.

Example 4

(a) What is the equation of the regression line for the pairs (X,Y) shown at the top of page 275?

(b)   What Y is predicted when X = 72?

X^2	X	Y	XY
4900	70	127	8890
4356	66	112	7392
5625	75	146	10950
5041	71	131	9301
4489	67	116	7772
3844	62	97	6014
28255	411	729	50319

(a) $$b = \frac{6 * 50319 - 411 * 729}{6 * 28255 - 411^2} = \frac{301,914 - 299,619}{169,530 - 168,921} = \frac{2,295}{609} = 3.68$$

$$a = \frac{729}{6} - \frac{411}{6} * 3.77 = -137$$

$$Y_c = -137 + 3.77X$$

(b)   When X = 72, Y_c = −137 + 3.77 * 72 = 134. (This gives you practice in using the regression equation for prediction, but note that the prediction is meaningless without assumptions about the distribution of the Y population that will be taken up in Chapter 16. This is similar to the point estimate of a population mean; see the discussion on page 138.)

The values of a and b are determined to three significant figures, but the results would have been very different if all the numbers in the numerator and denominator of b had been rounded to three or even four figures before subtraction.

In the examples given so far, b has been positive. The slope of the regression line may be negative.

Example 5

What is the predicted value of Y when X = 6 for the data shown below?

X^2	X	Y	XY
1	1	6	6
4	2	3	6
16	4	2	8
64	8	1	8
85	15	12	28

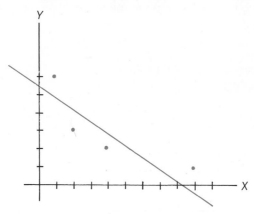

Answer

$$b = \frac{n(\Sigma xy) - \Sigma x(\Sigma y)}{n(\Sigma x^2) - \Sigma x^2}$$

$$a = \frac{\Sigma y}{n} - b\left(\frac{\Sigma x}{n}\right)$$

$$b = \frac{(4 * 28) - (15 * 12)}{(4 * 85) - 15^2} = \frac{112 - 180}{340 - 225} = \frac{-68}{115} = \underset{\text{(Slope)}}{\underline{.591}}$$

$$a = \frac{12}{4} - (\underset{b}{-.591}) * \frac{15}{4} = 3 + 2.21 = 5.22 \qquad \text{(y-intercept)}$$

$$\underset{a}{Y_c} = 5.22 - \underset{b}{.59}X$$

When $X = 6$, $Y_c = 5.22 - .59 * 6 = 1.67$.

A regression equation is used for prediction (in the last example, $Y_c = 1.67$ when $X = 6$). But a prediction is meaningless unless you can estimate how valid it is, and this can't be done unless certain assumptions about the population distribution can be made. This whole project will be discussed in the next chapter. For the moment, be aware of the following:

1. A regression equation $Y_c = a + bX$ can be found even when the data are far from linear; for example:

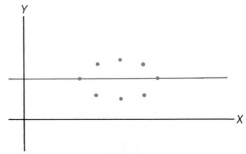

Finding the regression equation in such a case is a waste of time, however, so always make a scatter diagram first and see whether your data lie at least approximately on a straight line.

2. We have found only the best (in the sense of "least squares") linear equation $Y_c = a + bX$ to fit data; it may be preferable to use the least squares method but to fit an equation of another type: $Y_c = a + b \log X$ or $Y_c = a + bX + cX^2$, for example. But these other types are not considered in this book.

3. Finding a regression equation is possible when there is no cause and effect relationship between X and Y, and **never** proves that such a relationship exists. The number of cigarettes sold and the size of the police force in selected cities under 100,000 may show a linear relation, but smoking more cigarettes does not cause the hiring of more police; rather, both are related to the size of the city and thus increase as the population increases.

15.6 PEARSON CORRELATION COEFFICIENT: ROUGH ESTIMATES

A correlation coefficient, like a regression equation, is determined for X and Y scores that are paired. There are a number of different correlation coefficients; we shall concentrate now on the Pearson correlation coefficient, r. This correlation coefficient will give a measure of the closeness to linearity of a relationship between two variables, and will determine the precision with which the regression equation discussed in

Section 15.5 can be used for prediction. Note, however, that in correlation problems there is no necessity for considering one variable as independent and the other as dependent or predicted. For example, we may consider the relationship between weight and height of coal miners. Tall people tend to be heavier and short people lighter, but height could be used to predict weight or vice versa.

When all pairs of X and Y values are on the regression line, the correlation coefficient is either +1 or −1: it is +1 if the slope of the regression line is positive (in which case we say the correlation is **perfect and positive**); it is −1 if the slope of the regression line is negative (when the correlation is said to be **perfect and negative**).

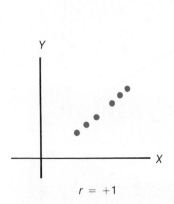

$r = +1$

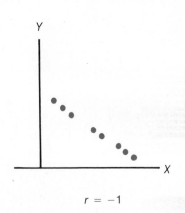

$r = -1$

If some pairs of X and Y scores do not lie on the regression line, the correlation coefficient r is between −1 and +1. (Warning: If in determining a correlation coefficient you come up with results such as $r = 2.3$ or $r = -1.5$, check your work! Something is wrong.) The value of r can be estimated **very roughly** by enclosing the cloud of scores in a lining, and measuring the length L and width W of the lining. Then

$$r = \pm\left(1 - \frac{W}{L}\right)$$

with the sign of r determined by the sign of the slope of L. Use any units of measurement you like, but it may be easiest to let the width W be 1 unit and estimate L as a multiple of W. Here are some examples:

Estimating r'

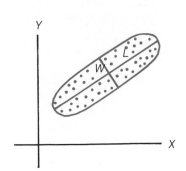

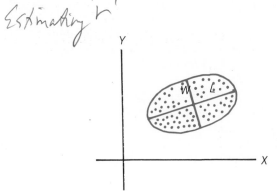

$W = 1, L = 4, r \approx 1 - \dfrac{1}{4} = \dfrac{3}{4}$

$W = 1, L = 2, r \approx 1 - 1/2 = 1/2$

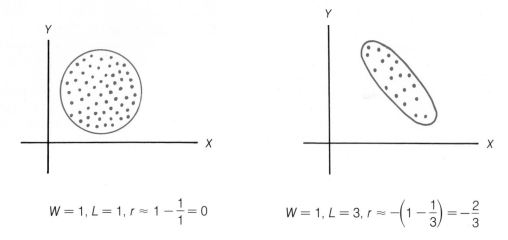

$$W = 1, L = 1, r \approx 1 - \frac{1}{1} = 0$$ $$W = 1, L = 3, r \approx -\left(1 - \frac{1}{3}\right) = -\frac{2}{3}$$

If only the points were fairly dense inside an elliptical cloud, this method would work pretty well.

$$1 - \frac{W}{L} = 1 - \frac{1}{4} = \frac{3}{4}; \ r \approx .7.$$

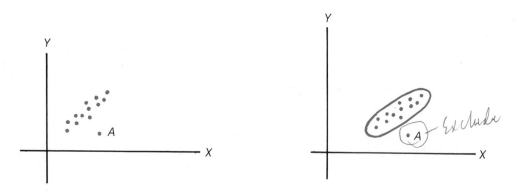

Unfortunately, in practice, scores don't tend to behave nicely, and isolated points make it hard to determine exactly how the lining should be drawn. Inclusion of the point A would lower the estimate of r. It's probably better to exclude, it, and then lower the estimate of r a little.

15.7 ERROR, EXPLAINED, AND TOTAL VARIATIONS

The fact that some observed points do not lie on a regression line, the line of "best fit," does not deny the hypothesis that the line represents a relation between X and Y scores. If these scores are obtained by a psychologist, say, in carrying out an experiment, he expects experimental errors in his observations. The term "experimental errors" lumps together possible differences in factors not controlled by the experiment as well as inaccuracies in measurement; there may also be round-off errors. The **standard error of estimate** (symbol: s_e) is a measure of these errors:

$$s_e = \sqrt{\frac{\Sigma(Y - Y_c)^2}{n - 2}}$$

(There are $n-2$ degrees of freedom because a and b of the regression equation are used in determining Y_c.)

Error Variation. The numerator $\Sigma(Y-Y_c)^2$ of the equation above is called the **error variation.** (In the remainder of this section we shall concentrate on variations rather than variances or standard deviations.†)

Recall that in finding the regression line, the criterion of "best fit" was that the sum of the squares of the errors $\Sigma(Y-Y_c)^2$ should be made as small as possible. Thus if the Y_c's are computed using any other a and/or b in the regression equation, the resulting standard error of estimate (and error variation) would be larger.

Example 1

Find (a) the error variation, (b) the standard error of estimate for the pairs of X and Y scores below. The regression equation is $Y_c=-3.564+1.829X$ (determined in Example 3, page 274.)

X	Y	Y_c	$Y-Y_c$	$(Y-Y_c)^2$
2	0	.094	− .094	.009
3	3	1.923	1.077	1.160
4	4	3.752	.248	.062
5	4	5.581	−1.581	2.500
6	6	7.410	−1.410	1.988
7	11	9.239	1.761	3.101
	28	28.00	.001	8.820

(a) Error variation $=\Sigma(Y-Y_c)^2=8.82$.
(b) $s_e=\sqrt{8.83/4}=1.49$.

In computing the quantities in the table, observe that $\Sigma Y=\Sigma Y_c$ and $\Sigma(Y-Y_c)=0$ (within round-off). Thus two totals for checking arithmetic are available.

Explained Variation. If knowing an X score is of no help whatsoever in predicting the corresponding Y score with a linear regression equation—that is, if there is no linear relationship between X and Y scores—then the best estimate of Y_c for any X is $Y_c=\bar{Y}$. In this case, $\Sigma(Y_c-\bar{Y})^2=0$. But if there is some linear relation between X and Y, $\Sigma(Y_c-\bar{Y})^2$ measures what is gained by using Y_c instead of $\bar{Y}$ for prediction; $\Sigma(Y_c-\bar{Y})^2$ is called the **explained variation.**

Example 2

For the same pairs of scores as in Example 1, compute the explained variation.

X	Y	Y_c	$Y_c-\bar{Y}$	$(Y_c-\bar{Y})^2$
2	0	.094	−4.572	20.90
3	3	1.923	−2.746	7.54
4	4	3.752	− .914	.84
5	4	5.581	.915	.84
6	6	7.410	2.744	7.53
7	11	9.239	4.573	20.91
	28		.000	58.56

†But we shall return to the standard error of estimate in Chapter 16.

Explained variation $= \Sigma(Y_c - \overline{Y})^2 = 58.56$.

Observe that the sum of the $Y_c - Y$ column is 0. This should always be the case (within round-off); again, you have a useful check on arithmetic.

Total Variation. The variation of the Y's about their mean can also be computed. The quantity $\Sigma(Y - \overline{Y})^2$ is called the **total variation.**

Example 3

For the same pairs of X and Y scores used in Examples 1 and 2, compute the total variation.

X	Y	$Y - \overline{Y}$	$(Y - \overline{Y})^2$
2	0	−4.667	21.81
3	3	−1.667	2.79
4	4	− .667	.45
5	4	− .667	.45
6	6	1.333	1.77
7	11	6.333	40.11
	28	− .002	67.38

$$\overline{Y} = \frac{28}{6} = 4.67$$

Total variation $= \Sigma(Y - \overline{Y})^2 = 67.5$.

Again there is a check on arithmetic: the sum of the $Y - \overline{Y}$ column must be 0 (except for round-off).

Study the figures below carefully so that you realize what distances are being squared to find the variations:

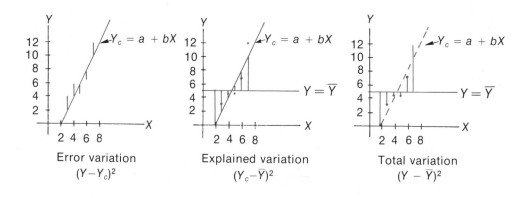

Error variation
$(Y - Y_c)^2$

Explained variation
$(Y_c - \overline{Y})^2$

Total variation
$(Y - \overline{Y})^2$

It can be shown that $\Sigma(Y - Y_c)^2 + \Sigma(Y_c - \overline{Y})^2 = \Sigma(Y - \overline{Y})^2$:

Error variation + Explained variation = Total variation

The same equation holds for corresponding variances:

Error variance + Explained variance = Total variance

No higher mathematics is needed to show that this will always be the case, but the algebra involved is so long and tedious that a proof is not included here.

Example 4

Check that the sum of the error variation and the explained variation equals the total variation, within round-off, for the scores in Examples 1, 2, and 3.

	Variation	Formula
Error	8.82	$\Sigma(Y - Y_c)^2$
Explained	58.56	$\Sigma(Y_c - \overline{Y})^2$
Total	67.33	$\Sigma(Y - \overline{Y})^2$

$$8.82 + 58.56 = 67.38$$
$$\approx 67.33$$

Within round-off, the formula checks. $Y = -3.564 + 1.83x$

If all the data points lie on the regression line, each $Y - Y_c = 0$ so $\Sigma(Y - Y_c)^2 = 0$; the explained variation equals the total variation. If, on the other hand, knowing an X score is of no help in predicting a Y score, then the best choice for the equation of the regression line is $Y_c = \overline{Y}$; therefore the explained variation $\Sigma(Y_c - \overline{Y})^2$ equals 0.

The ratio of the explained variation to the total variation measures how well the linear regression line fits the given pairs of scores. It is called the **coefficient of determination,** and is denoted by r^2.

$$r^2 = \frac{\text{Explained variation}}{\text{Total variation}}$$

In the examples above,

1, 2 & 3

$$r^2 = \frac{58.4}{67.5} = .86.$$

The explained variation is never negative (it is the sum of squares, which are never negative), and is never larger than the total variation (since the error variation, also the sum of squares, is never negative), so r^2 is always between 0 and 1. If the explained variation equals 0, $r^2 = 0$; this occurs when knowing the X value of a pair is of no help in predicting the Y value and the best choice of Y_c, for any X, is $\overline{Y}$.

If r^2 is known, then $r = \pm\sqrt{r^2}$. The sign of r is **the same as the sign of b from the regression equation.** Now at last the coefficient of correlation r is defined:

$$r = (\text{sign of } b)\ \sqrt{r^2} = (\text{sign of } b)\ \sqrt{\frac{\text{Explained variation}}{\text{Total variation}}}$$

Since r^2 is between 0 and 1, r is between -1 and $+1$.

In the above examples, the equation of the regression line was $Y_c = -3.57 + 1.83X$; $b = +1.83$, so $r = +\sqrt{.86} = .93$.

An r close to 1 or to -1 indicates a good fit, because then the explained variation is a large part of the total variation and the error variation is a small part; an r close to 0 indicates a poor fit, since in this case the explained variation is a small part of the total variation and the error variation is a large part of it.

15.8   COMPUTATIONAL FORMULAS FOR *r*

If you understand the definition of the correlation coefficient *r* given in the last section, you should be quite clear on what *r* does: it measures how closely the (X, Y) pairs of data are linearly related, and it tells you whether the slope of a regression line is positive or negative. But to find *r*, it's necessary to find first the regression equation $Y_c = a + bX$, and sometimes you are interested only in *r*. Also, the computations for explained and total variation are staggering if *n* is large. Fortunately, there is a computational formula for *r* which is very much simpler:

$$r = \frac{n\Sigma XY - (\Sigma X)(\Sigma Y)}{\sqrt{n\Sigma X^2 - (\Sigma X)^2}\ \sqrt{n\Sigma Y^2 - (\Sigma Y)^2}}$$

Again, the derivation is long and tedious, but not difficult—that is, it uses algebraic manipulations but not advanced mathematics—and is omitted here.

Example 1

Use the computational formula to determine *r* for the scores in Examples 1–3 of the previous section.

X^2	X	Y	Y^2	XY
4	2	0	0	0
9	3	3	9	9
16	4	4	16	16
25	5	4	16	20
36	6	6	36	36
49	7	11	121	77
139	27	28	198	158

$$n = 6;\ r = \frac{6 * 158 - 27 * 28}{\sqrt{6 * 139 - 27^2} * \sqrt{6 * 198 - 28^2}} = \frac{948 - 756}{\sqrt{105} * \sqrt{404}} = \frac{192}{206} = .93$$

as before.

Example 2

Find *r* for the pairs of *X* and *Y* scores below.

X^2	X	Y	XY	Y^2
1	1	6	6	36
4	2	3	6	9
16	4	2	8	4
64	8	1	8	1
85	15	12	28	50

$$r = \frac{4 * 28 - 15 * 12}{\sqrt{4 * 85 - 15^2}\ \sqrt{4 * 50 - 12^2}} = \frac{112 - 180}{\sqrt{340 - 225}\ \sqrt{200 - 144}} = \frac{-68}{\sqrt{115 * 56}} = \frac{-68}{80.3} = -.85$$

Since $n\Sigma X^2 - (\Sigma X)^2 = n(n - 1)\left[\dfrac{\Sigma X^2 - (\Sigma X)^2/n}{n - 1}\right]$ and $\dfrac{\Sigma X^2 - (\Sigma X)^2/n}{n - 1}$ is the variance of the X scores, the quantity under the square root sign **must** be positive. (If it were not, you couldn't find the square root anyway.) The same comment holds for $n\Sigma Y^2 - (\Sigma Y)^2$: check your arithmetic if it ever appears to be negative!

As with regression lines, n is the number of X scores or the number of Y scores, not the total number of scores.

Before reading on, try this next one yourself.

Example 3

Find r.

X^2	X	Y	XY	Y^2
1	1	4	4	16
4	2	5	10	25
9	3	6	18	36
6		15	32	77

$$r = \frac{n\Sigma XY - (\Sigma X)(\Sigma Y)}{\sqrt{n\Sigma X^2 - (\Sigma X)^2}\ \sqrt{n\Sigma Y^2 - (\Sigma Y)^2}} = ?$$

Answer

$$r = \frac{3 * 32 - 6 * 15}{\sqrt{3 * 14 - 6^2}\ \sqrt{3 * 77 - 15^2}} = \frac{6}{6} = 1$$

Are you surprised? The three points lie on the same line.

There are other formulas for r that sometimes prove useful. The similarity between the formula above for r and the equation for b in the regression equation $Y_c = a + bX$ should have struck you:

$$b = \frac{n\Sigma XY - \Sigma X \Sigma Y}{n\Sigma X^2 - (\Sigma X)^2}$$

$$r = \frac{n\Sigma XY - \Sigma X \Sigma Y}{n\Sigma X^2 - (\Sigma X)^2} * \frac{\sqrt{n\Sigma X^2 - (\Sigma X)^2}}{\sqrt{n\Sigma Y^2 - (\Sigma Y)^2}} = b * \frac{\sqrt{n\Sigma X^2 - (\Sigma X)^2}}{\sqrt{n\Sigma Y^2 - (\Sigma Y)^2}}$$

If the numerator and denominator of the fraction at the right are divided by $\sqrt{n(n-1)}$, these become, respectively, the standard deviations of the X and Y scores. The symbols s_X and s_Y are used; the subscripts simply refer to the scores; don't confuse these with standard errors of the mean $\sigma_{\bar{x}}$ and $\sigma_{\bar{y}}$—no sampling distribution is involved here.

Substituting, then,

$$r = b * \frac{s_X}{s_Y}$$

This formula for r is convenient if you have already found the regression equation and the standard deviations. Or, if you first determine r, then this equation can be solved for b:

$$b = r * \frac{s_Y}{s_X}$$

Example 4

In Example 5, page 275, the regression equation for the following scores was found to be $Y_c = 5.22 - .59X$. Determine r.

X	Y	X²	Y²
1	6	1	36
2	3	4	9
4	2	16	4
8	1	64	1
15	12	85	50

$$s_X = \sqrt{\frac{85 - 15^2/4}{3}} = 3.10$$

$$s_Y = \sqrt{\frac{50 - 12^2/4}{3}} = 2.16$$

$$b = -.59$$

$$r = -.59 * \frac{3.10}{2.16} = -.85$$

15.9 SOME WARNINGS

1. Even though two variables show a high correlation (whether positive or negative), this does not show any cause-and-effect relationship between them.

There is a high correlation between boys' length of pants and ability to read in grades K–6: kindergarteners wear shorts and read poorly, while sixth graders wear long pants and read pretty well. Would you save money on teachers' salaries by passing a law that all boys must wear long pants? In 100 cities chosen at random, a study shows a positive correlation between the number of churches and the number of bars. Would you conclude that religion causes drinking or vice versa? And if you wanted to encourage church attendance, would you increase the number of liquor licenses?

This is, of course, the source of the controversy about smoking and lung cancer. Even the tobacco companies admit there is a positive correlation between the two. Proving there is not a common cause of both — urban atmosphere or nervous tension, for example — is a more difficult matter.

2. Sometimes a sample will show a correlation of two attributes, whereas sub-samples do not show the same correlation.

Some people think there is a negative correlation between grades received and amount of time spent studying by college students. (It is probably not true that bright students study less than dummies — in fact, in challenging courses they are likely to study more.) But even if the correlation were negative, this wouldn't mean that you would get better grades by studying less! In a particular course, bright students might study less but get higher grades, while dull students were getting lower grades even though studying more.

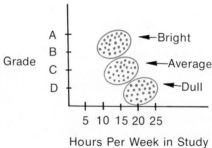

In general, if a population consists of several groups with different means, an apparent correlation may be produced or lost, as these diagrams indicate:

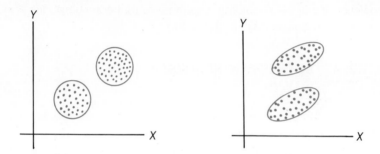

Your best remedies to these pitfalls are (a) always draw a scatter diagram, and (b) take a careful look at the source of your sample.

3. Be careful in interpreting r, once you have computed it.

An apparently helpful but very dubious statement is the following: "If r is greater than .7, there is a very strong association between X and Y scores; if r is between .3 and .7 there is some association; if r is between 0 and .3, there is negligible association." Of course the same statement should be made for negative values of r: $r = +0.8$ and $r = -0.8$ show exactly the same degree of correlation but in opposite directions.

Why is this statement so dubious? First, because r measures only linear relationships. In the scatter diagram shown below, r will be close to 0 even though there is a very close relationship between X and Y, since it is not a linear relationship. Usually you can tell whether this is the reason for a low correlation by looking at the scatter diagram.

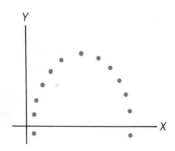

Secondly, recall that r^2 measures the proportion of the variation in Y scores explained by a linear relation between Y and X. If $r = .3$, then $r^2 = .09$; only 9 per cent of the variation in Y scores is explained. If $r = .7$, then 49 per cent is explained, and so on for other r values.

Finally, when can we say "There is a high correlation between these variables"? The answer depends on how r is to be used. A sociologist studying delinquency in teen-agers might correlate number of arrests with a variety of other scores: attitudes towards authority, education of parents, IQ, participation in religious services, and so on. Any correlation greater than .2 might indicate a factor worthy of further study. On the other hand, if the College Entrance Examination Board found a correlation as low as .7 between the scores of students on two different verbal aptitude tests, I'm sure they would be alarmed. In general, that which is a low correlation to a physicist or a chemist is a very high correlation to a sociologist.

4. So far, no assumptions have been made about the distributions of the populations from which the X and Y scores are drawn. Even without these assumptions, we can determine the regression equation and the correlation coefficient for a sample, as part of descriptive statistics. As soon as we are to make inferences about the populations, however, it will be absolutely essential to make assumptions about the populations. In the next chapter we shall do so.

15.10 THE RELATION BETWEEN REGRESSION AND CORRELATION

Correlation measures the strength of a linear relation between two variables, while the regression line actually describes the relation. There are other differences.

In a regression equation, one variable (X) is the independent variable and is the predictor, while the other variable (Y) is to be predicted. In correlation, on the other hand, there is no distinction between the two variables, and either one may be designated as X. (It should be emphasized again that in neither case is a cause-and-effect relationship assumed or shown.)

A sociologist is more apt to use correlation: He collects data on a great many variables, and wants to know whether there is a linear relationship between any of them. An economist, on the other hand, is much more apt to use regression: He has more exact data, and more prior theory to tell him a linear relationship does exist. Correlation is used when the relation between variables is first being studied, and the regression equation is determined in a later part of the investigation.

In correlation both variables may take on any values in a wide range. This is sometimes, but not always, true for prediction by a regression line. If, for example, the effect of diluting a chemical with water on freezing point is to be studied, one might add to a fixed amount of the chemical no water, then 10 cc., 20 cc., . . . of water. The X value is not chosen randomly, but is given fixed values by the experimenter.

15.11 VOCABULARY AND SYMBOLS

graph of a straight line	predicted value of Y
slope of a straight line	Pearson correlation coefficient $\quad r$
equation of a straight line	standard error of estimate $\quad s_e$
linear relationship	error variation
regression line	explained variation
error of estimation	total variation
"least squares" fit	coefficient of determination $\quad r^2$
$Y_c = a + bX$	

15.12 EXERCISES

1. For each of the following, determine the slope of the line and the Y-intercept, and sketch the graph.

    (a) $Y = -2 + 3X$.
    (b) $Y = 5 - 2X$.
    (c) $X = 4$.
    (d) $X + 2Y = 4$.

2. Find the equation of the line through the given points, and sketch its graph.

    (a) P: (0,4), Q: (4,6)
    (b) P: (1,3), Q: (2,1)
    (c) P: (3,0), Q: (3,4)
    (d) P: (-2,5), Q: (4,-3)

3. By computing $\Sigma(Y - Y_c)^2$ for each, find which of the following equations best fits (in the sense of "least squares") the pairs of scores (1,2), (2,2), (2,3), (3,4):

    (a) $Y_c = 1 + X$.
    (b) $Y_c = 1 + .5X$.
    (c) $Y_c = 3/4 + X$.

4. On a multiple-choice statistics exam, the professor made a mistake in computing the regression equation for the points (1,3), (1,4), (2,5), (3,7), and offered the following choice of answers: (a) $Y_c = 1$, (b) $Y_c = 1 + x$, (c) $Y_c = 1 + 3x$, and (d) $Y_c = 3x$. Unfortunately, none of these is the true regression equation. Which of these four choices gives the best fit in the sense of least squares?

5. Draw a scatter plot for each of the following:

(a)

X	3	5	6	2	3	8	6	4	7	1	2	4
Y	5	1	1	6	4	6	2	2	4	9	5	3

(b)

X	32	36	30	36	34	35	30	32	38	36	37	34	33
Y	5	10	4	8	7	7	2	7	12	9	10	6	5

For which of these sets does it make sense to find a linear regression equation? Explain.

6. Use the method of least squares to fit a straight line to the given data. Plot the data and the fitted line.

(a)

X	Y
1	2
2	2
2	3
3	4

(b)

X	Y
1	14
3	12
6	9
7	8
13	2

In each case, what is the predicted value of Y for $X = 4$?

7. In the following scores the X variable is IQ and Y is the grade on a statistics quiz.

X^2	X	Y	XY
10000	100	10	1000
12100	110	12	1320
14400	120	15	1800
16900	130	18	2340
19600	140	20	2800
73000	600	75	9260

(a) Make a scatter diagram.
(b) Find the equation of the straight line which best fits the given X and Y scores.
(c) What is the Y-intercept of the regression line? its slope?
(d) What is the predicted grade for a student whose IQ is 125? 105?
(e) Add the regression line to the scatter diagram.

8. In determining a gravitational force, a physicist plotted the velocity of a falling object versus the time interval between readings.

X^2	Time (sec) (X)	Velocity (Y)	XY	Y^2
-0625	1/4	3	.75	9
.25	1/2	7	3.5	49
.5625	3/4	10	7.5	100
1.0	1	11	11	121
1.5625	5/4 3.75	17 48	21.25	289
			44	568

3.4375

(a) Using the method of "least squares," fit a straight line to the above data.
(b) Plot the data and the regression line.
(c) What is the predicted Y value if $X = 7/8$?

9. Upon graduation, the scores of six chemistry majors on the mathematical aptitude test (MAT) were compared with their cumulative grade-point averages (GPA):

MAT	GPA
600	2.8
720	3.8
640	3.6
590	2.9
620	3.2
680	3.6

(a) Make a scatter diagram.
(b) Find the equation of the straight line which best fits the given values.
(c) Give the Y-intercept of the regression line and its slope.
(d) What is the predicted GPA for a student whose math score was 700?

10.

X	0	1	5
Y	3	4	5

For these scores, $\bar{X} = 2$, $\bar{Y} = 4$. Each of the following lines passes through $(\bar{X}, \bar{Y}) = (2,4)$:
(a) $Y_c = 4$; (b) $Y_c = 2X$; (c) $Y_c = 4 - 2b + bX$ (if $Y_c = a + bX$ passes through $(2,4)$, then $a = 4 - 2b$).
For each line given, show that $\Sigma(Y - Y_c) = 0$.

11. The heights and weights of 5 men are as follows:

Height (inches)	64	68	70	72	74
Weight (pounds)	160	170	180	190	195

(a) What weight would you predict for a man 69 inches tall?

(b) What height would you predict for a man who weighs 185 pounds?

12. An economist gives the following estimates of sales price and demand for a product:

Price (in dollars)	Demand (in tons)
1	9
2	7
3	6
4	3
5	1

(a) What demand would he predict if the sales price is $1.50?

(b) What price should be set for the product if the demand is to be 8?

13. List 3 sets of paired scores between which you would expect to find a high positive correlation.

14. List 3 sets of paired scores between which you would expect to find no correlation.

15. List 3 sets of paired scores between which you would expect to find a high negative correlation.

16. For the following sets of scores, plot a scatter diagram, estimate r from the scatter diagram, and compute r.

(a)

X	Y
1	1
3	3
5	4
7	8

(b)

X	Y
1	2
2	5
3	3
4	5
5	2

(c)

X	Y
1	7
3	5
2	6
1	6
7	1
2	5
6	3
6	2

17. Estimate r:

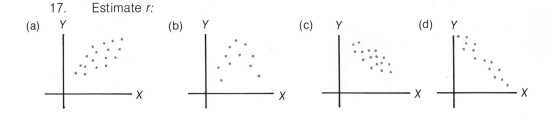

(a) Y ... X   (b) Y ... X   (c) Y ... X   (d) Y ... X

18. For which of the following problems would you find the equation for the regression line and for which would you find r?

(a) Is there a linear relationship between the exchange rate of pounds sterling and the price of Scotch in the United States?

(b) What is the relationship between the length of a boy's pants and his grade in school?

(c) Do factory workers who are heavy drinkers have more absenteeism?

(d) Can the attendance at a baseball game be predicted if the afternoon temperature forecast at 8 A.M. is known?

(e) How many people will attend a baseball game if the afternoon temperature forecast at 8 A.M. is 70°?

(f) What effect does the age of dishwashers have on the number of repairs needed?

(g) Can a student's grade-point average upon graduation be estimated if his freshman grades are known?

19. For the following scores, compute:
(a) the regression equation.
(b) the error variation $\Sigma(Y - Y_c)^2$.
(c) the explained variation $\Sigma(Y_c - \overline{Y})^2$.
(d) the total variation $\Sigma(Y - \overline{Y})^2$.
(e) r^2, using the results of (a) and (d).
(f) r.

Finally,
(g) draw a scatter diagram.

Does the result of (f) surprise you?

X	−1	0	1	−1	1
Y	1	0	1	−1	−1

20. Follow the same directions as in Exercise 19 for the following scores:

X	−1	0	2	−2	1
Y	1	0	−2	2	−1

21. (a) The regression equation for the points $(2,9)$, $(4,6)$, $(6,0)$ is $Y_c = 14 - 2.25X$. Compute (i) $\Sigma(Y - Y_c)^2$, (ii) $\Sigma(Y - \overline{Y})^2$, (iii) $\Sigma(Y - Y_c)^2/\Sigma(Y - \overline{Y})^2$, (iv) r.

(b) If each Y value in (a) is multiplied by 10, the new regression equation is $Y_c = 140 - 22.5X$. Compute the new values of (i) $\Sigma(Y - Y_c)^2$, (ii) $\Sigma(Y - \overline{Y})^2$, (iii) $\Sigma(Y - Y_c)^2/\Sigma(Y - \overline{Y})^2$, (iv) r.

(c) Compare the results of (a) and (b).

22. An account executive for an advertising firm presents the following figures to a client:

(X) Percentage of sales spent on advertising	(Y) Sales (thousands of dollars)
1	100
2	107
3	115
4	125
5	135

(a) Find the regression line which fits these data.
(b) Find $\Sigma(Y - Y_c)^2$.

(c) Find $\Sigma(Y - \bar{Y})^2$.

(d) Use (b) and (c) to find r.

(e) Use the computational formula to find r.

23.   The numbers of years in school (X) and the salaries (Y) in thousands of dollars of eight working women chosen at random in Denver are as follows:

X	12	16	19	10	13	16	12	12
Y	10	9	20	8	6	18	9	7

(a) Make a scatter diagram.

(b) What salary would you predict for a woman with a master's degree? (She has had 17 years of school.)

(c) Find the coefficient of correlation between the number of years in school and the salary earned by these women.

24.   A doctor collects the following information on mean calories per day and weight lost per week:

Calories:	2000	1200	1500	500	800
Weight lost:	1	3	2	5	4

(a) Find the correlation between calories per day and weight loss per week.

(b) What weight loss would you predict for a dieter who is restricted to 700 calories per day?

25.   Ministers' salaries (in thousands of dollars) and sales of alcohol (in billions of dollars) are as follows for the years 1970–1974:

	1970	1971	1972	1973	1974
Salaries	8.3	9.0	10.0	10.5	11.1
Alcohol consumption	7.9	8.3	8.7	9.2	9.6

(a) Find r.

(b) Can you explain why the coefficient of correlation is so high?

26.   Don Smith follows several stocks daily and compares their activity to that of the market as a whole. X values are average gain or loss in points each day in his selected stocks compared to Y values that represent point gain or loss in the market for an average share. For one work week, the data read:

	X	Y
Mon.	+1 1/2	−1/8
Tues.	−1/4	−3/4
Wed.	+1/4	−3/4
Thurs.	+1 1/8	+1/2
Fri.	−1/2	−3/8

(a) Compute r.

(b) Did Don Smith pick good stocks?

27.   The equation of the regression line through the points (1,2), (2,2), (2,3), and (3,4) is $Y_c = .75 + X$. Find the correlation between X and Y scores:

(a) using $r = b * \dfrac{s_X}{s_Y}$.

(b) using $r = \dfrac{n\Sigma XY - \Sigma X \Sigma Y}{\sqrt{n\Sigma X^2 - (\Sigma X)^2} \; \sqrt{n\Sigma Y^2 - (\Sigma Y)^2}}$.

ANSWERS

1.

	Slope	Y-intercept
(a)	3	−2
(b)	−2	5
(c)	Not defined	None
(d)	$-\dfrac{1}{2}$	2 $\quad \left(Y = 2 - \dfrac{1}{2}X\right)$

3.

			(a) $Y_c = 1 + X$		(b) $Y_c = 1 + .5X$		(c) $Y_c = 3/4 + X$	
X	Y		Y_c	$(Y - Y_c)^2$	Y_c	$(Y - Y_c)^2$	Y_c	$(Y - Y_c)^2$
1	2		2	0	1.5	.25	1.75	1/16
2	2		3	1	2	0	2.75	9/16
2	3		3	0	2	1	2.75	1/16
3	4		4	0	2.5	2.25	3.75	1/16
				1		3.50		12/16 = .75

Of the three, (c) gives the best fit, since .75 is less than 1 and also less than 3.50; (c) is the regression line.

5. (a)   (b)

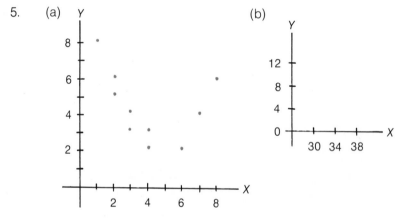

It makes sense for (b). The points in (a) will fit a parabola better than a straight line.

7. (b) $b = \dfrac{5 * 9260 - 600 * 75}{5 * 73000 - 600^2} = .26$, $a = 15 - .26 * 120 = -16.2$, $Y_c = -16.2 + .26X$.

(c) Y-intercept $= a = -16.2$; slope $= b = .26$.

(d) $Y_c = -16.2 + .26 * 125 = 16.3$ when $X = 125$.
$Y_c = -16.2 + .26 * 105 = 11.1$ when $X = 105$.

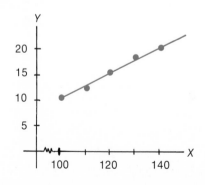

9. (a) Note that different scales are used on the X and Y axes.

(b) $\Sigma XY = 12863$, $\Sigma X = 3850$, $\Sigma Y = 19.9$, $\Sigma X^2 = 2482900$; $b = .00752$, $a = -1.51$; $Y_c = -1.51 + .00752X$.

(c) Y-intercept $= -1.51$, slope $= .00752$.

(d) 3.75.

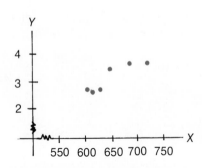

11. (a) If X is height and Y is weight, the regression equation is $Y_c = -77.3 + 3.68X$; if $X = 69$ inches, $Y = 176.6$ pounds.

(b) If X is weight and Y is height, the regression equation is $Y_c = 22 + .27X$; if $X = 185$ pounds, $Y = 72.0$ inches.

17. (a) $r \approx .5$; (b) $r \approx 0$; (c) $r \approx -.6$; (d) $r \approx -.9$.

19. (a) $b = \dfrac{5 * 0 - 0 * 0}{5 * 5 - 0^2} = 0$; $a = 0 - 0 * \dfrac{0}{5} = 0$, $Y_c = 0$. (b) 4. (c) 0. (d) 4. (e) $\dfrac{0}{4} = 0$. (f) 0.

21. (a) (i) $\Sigma(Y - Y_c)^2 = (-.5)^2 + 1^2 + (-.5)^2 = 1.50$; (ii) $4^2 + 1^2 + (-5)^2 = 42$; (iii) $1.50/42 = .04$; (iv) .96.

(b) (i) $(-5)^2 + 10^2 + (-5)^2 = 150$; (ii) $40^2 + 10^2 + (-50)^2 = 4200$; (iii) $150/4200 = .04$; (iv) .96.

(c) When Y values are multiplied by 10, $\Sigma(Y - Y_c)^2$ and $\Sigma(Y - \bar{Y})^2$ are both multiplied by 100; their quotient is, unchanged, and therefore r is not changed.

23. (a)

(b) $b = \dfrac{8 * 1282 - 110 * 87}{8 * 1574 - 110^2} = 1.394$

$a = \dfrac{87}{8} - 1.394 * \dfrac{110}{8} = -8.29$

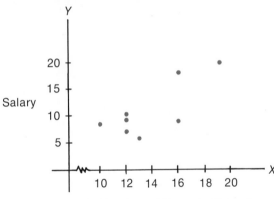

Number of Years in School

$Y_c = -8.29 + 1.39X$; if $X = 17$, $Y_c = 15.2$; her predicted salary is $15,200.

$$\text{(c)} \quad r = \frac{686}{\sqrt{492}\ \sqrt{1511}} = .80.$$

25.   (a) $r = \dfrac{5 * 430.43 - 48.9 * 43.7}{\sqrt{25.54}\ \sqrt{9.26}} = .99.$

         (b) Perhaps salaries increased because of inflation, and alcohol consumption because of increased population as well as inflation.

27.   (a) $r = 1 * \dfrac{.82}{.96} = .85.$ (b) $r = \dfrac{4 * 24 - 8 * 11}{\sqrt{4 * 18 - 8^2}\ \sqrt{4 * 33 - 11^2}} = .85.$

chapter sixteen

linear regression and correlation: statistical inference

In the last chapter, you learned how to find the Pearson correlation coefficient and the linear regression equation for a sample of paired X and Y values. It was not necessary to make any assumptions about the distribution of either X or Y values in the populations. Now it is time to answer questions about the relationship of X and Y scores in the populations, and about the predicted value of Y for a given X in the population. For these, it is absolutely essential to make some assumptions about the population distributions. There is an intimate relation between regression and correlation. But as you learn to make statistical inferences about regression and correlation in the population, you must also study the differences between them, in particular the different assumptions about the population distributions.

16.1 INFERENCES ABOUT REGRESSION: ASSUMPTIONS ABOUT THE POPULATIONS

Suppose that you have made a scatter diagram for n pairs of X and Y scores and decided that they do approximately fit a straight line, and that you have then determined the regression equation $Y_c = a + bX$. Now you are concerned about the population from which the X and Y pairs were taken. Three questions need to be answered: First, is there a linear relationship between **all** X and Y pairs? Second, if there is, can you estimate the mean of all Y scores for any X in the population? Last, if in the future another X score is chosen from the population, can you predict what the corresponding Y score will be?

The following assumptions must be made about the population:

1. If a psychologist, in carrying out an experiment, could measure **all** Y scores in the population for each X score, he expects experimental errors in his observations, due to innaccuracies in measurement, round-off errors, or differences in factors not controlled by the experiment. But it is assumed that the **mean** of the Y's for any given X falls on a straight line $Y = A + BX$. This means that there is a true linear relationship between pairs of scores, and that A and B for the population equation are approximated by a and b of the sample regression equation $Y_c = a + bX$.

2. For each X, the corresponding values of Y in the population are assumed to be normally distributed, and the different normal distributions for different values of X all have the same variance. A normal distribution is, of course, continuous and

therefore infinite, but the accompanying diagram with a finite number of Y's for each X should help you picture these first two assumptions.

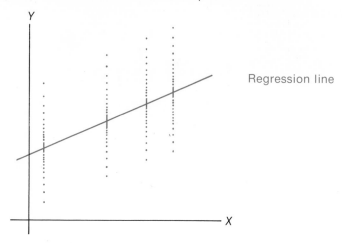

Regression line

3. X scores used in determining a regression line commonly do not form a random sample from the population, and no assumption about the X population need be made; rather, particular X scores may be chosen by the experimenter. For example, the chemist studying the boiling point of solutions may first add 5 grams of chemical to 100 cc. of water, then 10, 15, 20, . . . grams. A sociologist might study social adjustment scores of children with 0, 1, 2, 3, 4, . . . siblings. (When we study the accuracy of prediction of Y for a given X score, we shall find that it is greater for X values near the mean $\overline{X}$, so you may not want to have a wide spread in X values unless your sample size is large. But do not confuse this practical restriction on X with an assumption about the distribution of X scores in the population.)

16.2   TESTING H_0: $B = 0$

Is there indeed a linear regression equation $Y = A + BX$ for the population that will help in prediction? If there is not, then choosing $\overline{Y}$ for any X is the best that can be done. But this means the best population equation is $Y = \overline{Y}$—that is, $B = 0$. Therefore the hypothesis to be tested is H_0: $B = 0$. Only if this hypothesis is rejected is it worth going on to consider how the sample regression equation can be used for prediction of Y in the population, and with what accuracy this can be done.

What is the alternative hypothesis? Sometimes you will suspect from the nature of the experiment that the slope is indeed positive (or negative); in this case, use H_1: $B > 0$ (or $B < 0$) and a one-tail test; otherwise, use H_1: $B \neq 0$ and a two-tail test.

Before testing the null hypothesis, we must, as always, look at the sampling distribution of slopes. Suppose that, from a population which meets all the assumptions given in the preceding section, we take all different samples of size n and compute $\dfrac{b - B}{s_b}$ for each of these, where

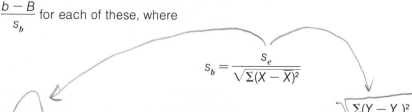

$$s_b = \frac{s_e}{\sqrt{\Sigma(X - \overline{X})^2}}$$

and s_e is your old acquaintance the standard error of estimate $\sqrt{\dfrac{\Sigma(Y - Y_c)^2}{n - 2}}$. (See page 278.)

The sampling distribution we end up with is a t distribution with $n - 2$ degrees of freedom.

Fortunately, computational formulas for $\Sigma(Y - Y_c)^2$ and $\Sigma(X - \bar{X})^2$ can be found:

$$\Sigma(Y - Y_c)^2 = \Sigma Y^2 - a\Sigma Y - b\Sigma XY$$

$$\text{and } \Sigma(X - \bar{X})^2 = \Sigma X^2 - (\Sigma X)^2/n$$

(The last one is the numerator of the formula for variance, with which you have long been familiar.) With a little algebra, the computational formula for s_b is seen to be

$$s_b = \sqrt{\frac{n(\Sigma Y^2 - a\Sigma Y - b\Sigma XY)}{(n-2)[n\Sigma X^2 - (\Sigma X)^2]}}$$

Note that $n\Sigma X^2 - (\Sigma X)^2$ is the denominator of b.

To test $H_0: B = 0$, then, we compute $t = \dfrac{b - B}{s_b}$ and compare this t value with the critical value determined for the given α and $n - 2$ degrees of freedom, using a one- or two-tail test as determined by H_1.

Example 1

At a .05 level of significance, test $H_0: B = 0$, $H_1: B \neq 0$, for the (X,Y) scores shown below.

X^2	X	Y	XY	Y^2
4	2	0	0	0
9	3	3	9	9
16	4	4	16	16
25	5	4	20	16
36	6	6	36	36
49	7	11	77	121
139	27	28	158	198

$$b = \frac{6 * 158 - 27 * 28}{6 * 139 - 27^2} = \frac{192}{105} = 1.8286\dagger$$

$$a = \frac{28}{6} - 1.8286 * \frac{27}{6} = -3.5619\dagger$$

$$\Sigma Y^2 - a\Sigma Y - b\Sigma XY = 198 - (-3.5619) * 28 - 1.8286 * 158 = 8.82$$

Now turn to page 279 and see that the error variation computed there for the same scores by using $\Sigma(Y - Y_c)^2$ has the same value, within round-off.

$$s_b = \sqrt{\frac{6 * 8.82}{4 * 205}} = .254$$

$$t = \frac{1.83 - 0}{.254} = 7.20$$

$$\approx \sqrt{\frac{n(\Sigma y^2 - \partial \Sigma y - b\Sigma xy)}{n-2(n\Sigma x^2 - (\Sigma x)^2)}}$$

$$= sb \ (\text{stdev}_{ony})$$

For $\alpha = .05$, $n = 6$, $D = 4$, the critical values of t for a two-tail test are ±2.78. Since 7.20 falls outside these limits, H_0 is rejected and H_1 accepted; the slope

---

†To prevent round-off problems, b and a are carried to four decimal places in demonstrating that indeed $\Sigma(Y - Y_c)^2 = \Sigma Y^2 - a\Sigma Y - b\Sigma XY$.

of the regression line in the population from which this sample was taken is
not 0.

16.3 CONFIDENCE AND PREDICTION INTERVALS FOR THE REGRESSION EQUATION

You have determined the regression equation $Y_c = a + bX$ for a sample of n pairs
of X and Y scores. Then you tested $H_0 : B = 0$ and rejected it, so you can assume there
is a meaningful linear relation between all pairs of X and Y scores in the population.
Can you then estimate the mean Y score for any X in the population? And, if so, <u>what is
the accuracy of your estimate?</u> Intuition perhaps suggests that we should in some
fashion find two straight lines parallel to the regression equation and make a claim such
as "a 95 per cent confidence interval for the mean Y corresponding to X_0 is $Y_c \pm$ (a constant)"

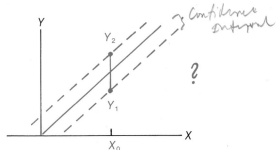

This implies that the confidence interval is the same for all values of X. Unfortunately,
this is not the case: It is smaller for values of X scores near the sample mean $\overline{X}$. Let us
see why this is so:

The sample regression line $Y_e = a + bX$ always passes through the point whose
coordinates are $(\overline{X}, \overline{Y})$: Since $a = \overline{Y} - b\overline{X}$, $Y_c = a + bX = \overline{Y} - b\overline{X} + bX$, and thus $Y_c = \overline{Y}$ if $X = \overline{X}$.
Then we should expect the population regression line to pass through (μ_X, μ_Y). There are
several sources of error. First, the slope of the population regression equation may not
be the same as that of the sample regression equation, so the population line is in some
region pivoted around the point (μ_X, μ_Y).

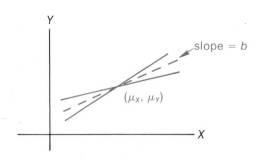

Secondly, we can determine confidence intervals for μ_X and μ_Y, but we do not know
them exactly; this is another source of difficulty in placing the population regression
line. The result is that the confidence interval for this mean Y score will be narrower
when X_0 is close to $\overline{X}$ than when it is far away.

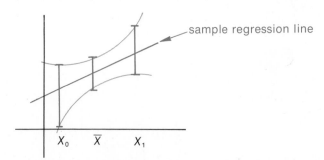

sample regression line

X_0 $\overline{X}$ X_1

Suppose X_0 is a particular X score, and remember the assumption that, for a given X_0, the corresponding Y scores in the population are normally distributed. Two questions can now be asked:

1. What is the mean of the Y scores corresponding to X_0?

2. If an observation is made in the future with $X = X_0$, can we give limits within which, with given probability, the observed Y value will fall?

Consider the first question. It may seem plausible to use the sample regression line and choose $a + bX_0$ as our estimate of the mean of the corresponding Y scores. But here we have the same problem as in finding μ when $\overline{X}$ is known for a sample: A point estimate is of no use unless we have some notion of its accuracy, and it becomes an interval estimate when we do (see the discussion on page 138). Let μ_{Y/X_0} be the mean value of Y corresponding to X_0. A confidence interval for μ_{Y/Y_0} is found by using a t-distribution with $n - 2$ degrees of freedom, with

$$t = \frac{Y_c - \mu_{Y/X_0}}{s_{Y/X_0}}$$

where

$$s_{Y/X_0} = s_e \sqrt{\frac{1}{n} + \frac{n(X_0 - \overline{X})^2}{n\Sigma X^2 - (\Sigma X)^2}}.$$ { Confidence Interval for Slope

s_e reflects the spread of Y values about the sample regression line, common for all values of X (see Assumption 2, Section 16.1); the square root reflects the increasing spread of the Y scores as X_0 gets further from $\overline{X}$. (Note that this factor has its smallest value when $X_0 = \overline{X}$ and the last term under the square root becomes zero.)

$$s_b = \frac{s_e}{\sqrt{\Sigma(x - x^2)}}$$

Example 1

For the scores given on the next page, the regression equation is

$$Y_c = -3.56 + 1.83X, \; s_e = \sqrt{\frac{8.82}{4}} = 1.485 \quad = \sqrt{\frac{\Sigma(Y - Y_c)^2}{n - 2}}$$

and $n\Sigma X^2 - (\Sigma X)^2 = 205$. (See Example 1, page 279.)

Find a 95 per cent confidence interval for μ_{Y/X_0} when X_0 is (a) 2.0, (b) 4.5.

X	Y
2	0
3	3
4	4
5	4
6	6
7	11

(a)   At a 95 per cent confidence level, with $n = 6$ and $D = 4$, $t = \pm 2.78$; $\overline{X} = 4.5$.

$$s_{Y/X_0} = 1.485\sqrt{\frac{1}{6} + \frac{6(2 - 4.5)^2}{205}} = .878$$

(annotation over 1.485: S_e)

When $X = 2$, $Y_c = -3.56 + 1.83 * 2 = -.100$.

$$\pm 2.78 = \frac{-.100 - \mu_{Y/X_0}}{.878}$$

$$\mu_{Y/X_0} = -.10 \pm 2.44 = -2.5 \text{ or } 2.3$$

When $X = 2$, a 95 per cent confidence interval for the mean of corresponding Y scores is -2.5 to 2.3.

(b)   Again $t = \pm 2.78$, $\overline{X} = 4.5$; $s_{Y/X_0} = 1.485 * \sqrt{1/6} = .606$; when $X = 4.5$, $Y_c = -3.56 + 1.83 * 4.5 = 4.68$.

$$\pm 2.78 = \frac{4.68 - \mu_{Y/X_0}}{.606}$$

$$\mu_{Y/X_0} = 4.68 \pm 1.68 = 3.0 \text{ or } 6.4$$

When $X = 4.5$, a 95 per cent confidence interval for the mean of corresponding Y scores is 3.0 to 6.4. Note that when $X = \overline{X}$, the width of the confidence interval is $6.4 - 3.0 = 3.4$, while it is $2.3 - (-2.5) = 4.8$ when $X = 2.5$.

Example 2

For the X and Y pairs of scores shown below,
(a)   Find the equation of the regression line.
(b)   Find 90 per cent confidence intervals for μ_{Y/X_0} when X_0 is (i) 1, (ii) 2, (iii) 2.5, (iv) 3, (v) 4.
(c)   Sketch the sample regression line and the limits of the confidence intervals for μ_{Y/X_0}.

X^2	X	Y	XY	Y^2	Y_c
1	1	0	0	0	.3
4	2	2	4	4	1.1
9	3	1	3	1	1.9
16	4	3	12	9	2.7
30	10	6	19	14	

Answer

(a) $b = \dfrac{4 * 19 - 10 * 6}{4 * 30 - 10^2} = \dfrac{16}{20} = .80$

$a = \dfrac{6}{4} - .8 * \dfrac{10}{4} = -.50$

$Y_c = -.50 + .80X.$

(b)   At a 90 per cent confidence level, with $n = 4$ and $D = 2$, $t = \pm 2.92$.

$$\mu_{Y/X_0} = Y_c \pm 2.92\, s_{Y/X_0}$$

$$s_{Y/X_0} = \sqrt{\dfrac{4 * 14 - 6^2 - .80 * 16}{2 * 4}} \sqrt{\dfrac{1}{4} + \dfrac{4(X_0 - 2.5)^2}{20}}$$

$$= .95\sqrt{.25 + .20(X_0 - 2.5)^2}$$

	X_0	Y_c	s_{Y/X_0}	Confidence interval for μ_{Y/X_0}	Width of interval
(i)	1	.3	.79	− .5 to 1.1	1.6
(ii)	2	1.1	.52	.6 to 1.6	1.0
(iii)	2.5	1.5	.48	1.0 to 2.0	1.0
(iv)	3	1.9	.52	1.4 to 2.4	1.0
(v)	4	2.7	.79	1.9 to 3.5	1.6

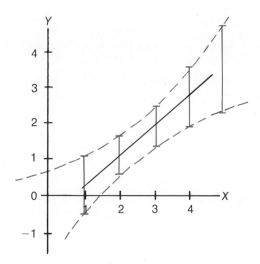

Now let's look at the second question asked on page 299: If an observation is made in the future with $X = X_0$, can we give limits within which the observed Y value will fall, with given probability? The computations here will be very similar to those for the first question, but the point of view is very different. We are no longer trying to estimate a population parameter (as in answering the first question); we are simply predicting a single observation to be made in the future.

The notation Y_p will be used for the Y score in a population predicted for the particular value X_0. A prediction interval for Y_p is found by using the t distribution

$$t = \frac{Y_c - Y_p}{s_{Y_p}}$$

with $n - 2$ degrees of freedom, where

$$s_{Y_p} = s_e \sqrt{1 + \frac{1}{n} + \frac{(X_0 - \bar{X})^2}{\Sigma(X - \bar{X})^2}}$$

This is almost the same formula as for s_{Y/X_0}; it differs only in 1 added under the square root sign. But adding 1 under the square root increases s_{Y/X_0} and therefore the prediction interval for Y_p is wider than the confidence interval for μ_{Y/X_0}. This should be expected, since you should expect more variability in 1 score picked from a normal distribution than in the mean of many scores.

Example 3

For the scores in Example 1, predict the interval within which the Y score corresponding to (a) $X_0 = 2$, (b) $X_0 = 4.5$, will fall, with probability .95 that your interval is correct.

(a)   As before, $s_e = 1.485$, $\bar{X} = 4.5$, $t = \pm 2.78$, and $Y_c = -.100$; when $X = 2$,

$$s_{Y_p} = 1.485 \sqrt{1 + \frac{1}{6} + \frac{6(2 - 4.5)^2}{205}} = 1.72.$$

$$\pm 2.78 = \frac{-.100 - Y_p}{1.72}$$

$$Y_p = -.100 \pm 4.78 = -4.9 \text{ or } 4.7$$

A 95 per cent prediction interval for Y when $X = 2$ is -4.9 to 4.7.

(b)   When $X = 4.5$, $s_{Y_p} = 1.485 \sqrt{1 + \frac{1}{6}} = 1.60$ and $Y_c = 4.68$.

$$\pm 2.78 = \frac{4.68 - Y_p}{1.60}$$

$$Y_p = 4.68 \pm 4.45 = .2 \text{ or } 9.1$$

A 95 per cent prediction interval for Y when $X = 4.5 (= \bar{X})$ is .2 to 9.1. Compare the confidence and prediction intervals for Y:

X_0	Confidence interval for μ_{Y/X_0}	Width	Prediction interval for Y_p	Width
2	−2.5 to 2.3	4.8	−4.8 to 4.7	9.5
4.5(= $\bar{X}$)	3.0 to 6.4	3.4	.2 to 9.1	8.9

Example 4

For the scores in Example 2, (a) find a 90 per cent prediction interval for Y when X_0 is (i) 1, (ii) 2, (iii) 2.5, (iv) 3, (v) 4; (b) sketch the regression line and the limits of the prediction intervals for Y_p.

Answer

(a)   As in Example 2, $t = \pm 2.92$ and $s_e = .95$. $Y_p = Y_c \pm 2.92\, s_{Y_p}$, and

$$s_{Y_p} = \sqrt{1 + \frac{1}{4} + \frac{4(X_0 - 2.5)^2}{20}}$$

	X_0	s_{Y_p}	Y_c	Prediction interval for Y_p	Width of interval
(i)	1	1.24	.3	−3.3 to 3.9	7.2
(ii)	2	1.08	1.1	−2.1 to 4.3	6.4
(iii)	2.5	1.06	1.5	−1.7 to 4.6	6.3
(iv)	3	1.08	1.9	−1.3 to 5.1	6.4
(v)	4	1.24	2.7	− .9 to 6.3	7.2

(b)

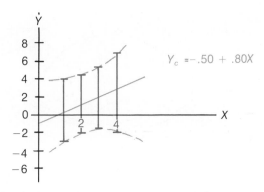

$Y_c = -.50 + .80X$

Both the confidence intervals for μ_{Y/X_0} and the prediction intervals for Y_p obtained in these examples have been very wide. How can they be narrowed? One way is to increase the sample size, since n and $n - 2$ occur in denominators. A second way is to decrease the spread of the X scores, since $\Sigma(X - \bar{X})^2$ also occurs in a denominator. If you have a random sample, nothing can be done about this. But a regression equation is commonly used in a planned experiment where X scores can be controlled. Do not be too ambitious about the spread of your X scores unless you are able to have a large sample size.

16.4   THE SAMPLING DISTRIBUTION OF *r*

If a population consists of the following pairs of scores:
$$X:\ \ 1\ \ 1\ \ 3\ \ 5\ \ 5$$

$$Y:\ \ 1\ \ 5\ \ 3\ \ 1\ \ 5$$
the population coefficient of correlation is 0. But if a sample of size 3 consists of the first, third, and fifth pairs, then the sample correlation coefficient is +1. This is, admittedly, a manufactured and unusual example, but it should make clear that it is necessary to test hypotheses about the population correlation after the sample *r* has been determined. We must look at the sampling distribution of *r*.

We need now another Greek letter: ρ is the Greek *r* (**rho,** pronounced to rhyme with **no** — the h is silent); ρ is the symbol for the population coefficient of correlation.

If all possible samples of given size *n* are taken from a population, and the correlation between pairs of X and Y scores is computed for each sample, what is the sampling distribution of the r's? The question cannot be answered unless it is assumed that **the population is normally distributed in both X and Y scores.** In more detail, this means that for each X the corresponding Y's must be normally distributed, and for each Y the corresponding X's must be normally distributed. The accompanying figure, representing three dimensions, should help you see what the distribution looks like. (The solid curve in color illustrates the distribution of Y scores when X = 2; it is a normal curve. The dotted curve in color illustrates the distribution of X scores when Y = 6; it is also a normal curve.) A distribution such as this is called a **bivariate normal distribution.**

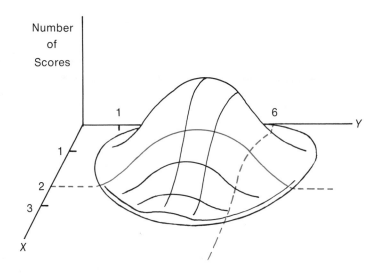

Unfortunately, the sampling distribution of r depends not only on the sample size n but also on the population correlation. When $\rho = 0$, the sampling distribution of r is symmetric about 0. The distribution is very skewed, however, if ρ is close to +1 or to −1. If $\rho = .80$, for example, samples taken from the population will have different r's. However, no r can be greater than +1, so the sample r's can not be more than .20 above the population correlation ρ, but may be considerably more below it. Therefore the sampling distribution of r is negatively skewed if ρ is close to 1. Similarly, it will be positively skewed if ρ is close to −1.

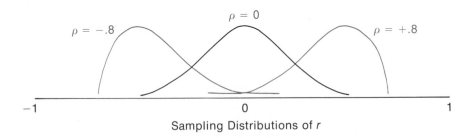

Sampling Distributions of r

For this reason, it will be necessary to use different methods for testing the null hypothesis $\rho = A$, depending upon whether $A = 0$ or is different from zero.

16.5   TESTING H_0: $\rho = 0$.

Suppose r is computed for each sample of size n in a bivariate normal population with $\rho = 0$. The distribution of r scores is neither normal nor a t distribution. But if for each sample the quantity $\dfrac{r\sqrt{n-2}}{\sqrt{1-r^2}}$ is computed, then these quantities have a t distribution with $n - 2$ degrees of freedom.

> Test H_0: $\rho = 0$ by computing $t = \dfrac{r\sqrt{n-2}}{\sqrt{1-r^2}}$ for a sample; compare with the critical values of t for $n - 2$ degrees of freedom and the given α.
>
> Use a one-tail test if H_1: $\rho < 0$ or H_1: $\rho > 0$; use a two-tail test if H_1: $\rho \neq 0$.

Example 1

A study of 66 students chosen at random from Johnson Junior High School shows there is a correlation of .3 between grades in art and in mathematics. Is there a significant correlation between grades in these two subjects for all students in the school? (.05 level of significance.) Assume that the grades of all students in both art and mathematics are normally distributed.

H_0: $\rho = 0$, H_1: $\rho \neq 0$; $\alpha = .05$, $D = 66 - 2 = 64$, two-tail test, so $t = \pm1.96$ (use the z table since $t > 40$); $r = .3$, $n = 66$.

$$t = \frac{r\sqrt{n-2}}{\sqrt{1-r^2}}$$

$$t = \frac{.3\sqrt{66-2}}{\sqrt{1-.3^2}} = \frac{2.4}{\sqrt{.91}} = 2.52$$

	0	.3 r
-1.96	0	1.96 2.52 t

Since 2.52 falls in the rejection region, H_0 is rejected: There is a significant correlation between grades in the two subjects.

Increase the size of your samples if you suspect that one variable or both are only approximately normal. If you think either one is **not** normally distributed, don't try to test a hypothesis about ρ.

Example 2

Eighteen students chosen at random from third grades in Morristown public schools show a correlation of $-.4$ between family income and hours per week spent watching television. At a .01 level of significance, is there a significant negative correlation for all Morristown third graders?

H_0: $\rho = 0$, H_1: $\rho < 0$; one-tail test, $\alpha = .01$; $r = 0.4$, $n = 18$.

$$t = \frac{r\sqrt{n-2}}{\sqrt{1-r^2}} = \frac{-.4\sqrt{18-2}}{\sqrt{1-.4^2}} = \frac{-1.6}{.916} = -1.74$$

The critical value of t for $n - 2 = 16$ degrees of freedom with $\alpha = .01$ is -2.58.

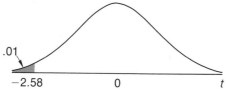

H_0 is accepted: There is no correlation between family income and time spent watching TV among Morristown third graders. This conclusion is not very trustworthy, however, for both family income and hours per week watching television are probably skewed to the right rather than normal distributions.

16.6   TESTING $H_0: \rho = A, A \neq 0.$

If $\rho \neq 0$, the sampling distribution of correlation coefficients r is skewed. But a famous statistician, R. A. Fisher, devised a transformation of r scores into Z scores so that the Z distribution is approximately normal for large n. (Do not confuse Z scores with standard z scores.) Fortunately, you do not need to carry out computations for this transformation; Table 8, Appendix C, gives corresponding r and Z values.

Example 1

What is the Z value corresponding to $r = .15$? to $r = -.70$?
If $r = .15$, then $Z = .151$.
If $r = -.70$, then $Z = -.867$.

If the Z score is computed for each pair of samples of size n in the populations, the distribution is approximately normal for large n ($n > 30$, say). Its mean μ_z is the Z value corresponding to ρ, and its standard deviation is

$$\sigma_z = \frac{1}{\sqrt{n-3}}$$

A hypothesis is then tested with the usual z test for a normal distribution:

$$z = \frac{Z - \mu_z}{\sigma_z}$$

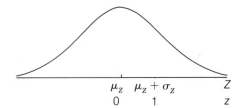

Example 2

Studies in past years have shown there is a correlation of .40 between grades in introductory economics and history courses at Drew University.

This year, a random sample of 39 students taking both courses shows a correlation of .34. Is there a significantly different correlation this year in the grades of all students taking both introductory courses? (.05 level of significance.)

$H_0: \rho = .40, H_1: \rho \neq .40; \alpha = .05.$
If $r = .34$, then $Z = .354$.
If $\rho = .40$, then $\mu_z = .424$; the transformation from ρ to μ_z has the same form as the transformation from r to Z.

$$\sigma_z = \frac{1}{\sqrt{n-3}} = \frac{1}{\sqrt{36}} = .167$$

$$z = \frac{Z - \mu_z}{\sigma_z} = \frac{.354 - .424}{.167} = -.42$$

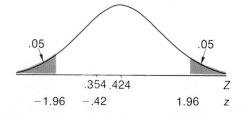

For a two-tail test with $\alpha = .05$, $z = \pm 1.96$; $-.42$ falls in the acceptance region: The correlation is still .40 for grades of all students taking introductory economics and history courses, and has not changed significantly.

The Fisher r-to-Z transformation is also used to find confidence intervals for ρ.

Example 3
Find a 95 per cent confidence interval for ρ if a sample of 103 paired X and Y scores has a correlation of .58.
With a .95 confidence coefficient, $z = \pm 1.96$.
If $r = .58$, then $Z = .663$; $n = 103$ so $\sigma_z = \dfrac{1}{10} = .10$.

$$z = \frac{Z - \mu_z}{\sigma_z} \text{ or } \pm 1.96 = \frac{.663 - \mu_z}{.1}; \; \mu_z = .467 \text{ or } .859$$

Table 8, Appendix C, is again used—this time looking up .467 and .859 in the Z column, and finding the corresponding ρ in the r column.
If $\mu_z = .467$, then $\rho = .44$.
If $\mu_z = .859$, then $\rho = .70$.
The 95 per cent confidence interval for ρ is .44 to .70.

Compare the results of the same problem if the sample size is considerably smaller:

Example 4
Find the 95 per cent confidence interval for ρ if $r = .58$, $n = 39$.
As in Example 1, $z = \pm 1.96$ and $Z = .663$.

$$\sigma_z = \frac{1}{\sqrt{39 - 3}} = \frac{1}{\sqrt{36}} = .167$$

$$z = \frac{Z - \mu_z}{\sigma_z}; \; \pm 1.96 = \frac{.663 - \mu_z}{.167}; \; \mu_z = .336 \text{ or } .990$$

If $\mu_z = .336$, then $\rho = .32$.
If $\mu_z = .990$, then $\rho = .76$.
The 95 per cent confidence interval for ρ is .32 to .76. A smaller n gives a wider confidence interval.

16.7 VOCABULARY AND SYMBOLS

$Y = A + BX$

sampling distribution of slopes

Y_p

$\sigma_{Y/X}$

ρ

sampling distribution of r

normal distribution in two variables

bivariate normal distribution

Z distribution $\quad \mu_z, \sigma_z$

Fisher r-to-Z transformation

16.8 EXERCISES

1.   The regression equation for the following pairs of scores is $Y_c = 5.21 - .59X$. At a .05 level of significance, find out whether or not $B = 0$.

X	1	2	4	8
Y	6	3	2	1

2.   Upon graduation, the scores of six chemistry majors on the Mathematical Aptitude Test (MAT) were compared with their cumulative grade-point average (GPA):

	1	2	3	4	5	6
(X) MAT	600	720	640	590	620	680
(Y) GPA	2.8	3.8	3.6	2.9	3.2	3.6

The regression equation is $Y_c = -1.51 + .0075X$ (see Exercise 9, page 288.) With $\alpha = .01$, test $H_0: B = 0, H_1: B \neq 0$.

3.

X	0	0	0	0	1	1	1	1	2	2	2	2	3	3	3	3
Y	−1	1	0	2	4	3	1	2	6	4	5	6	7	8	6	8

  (a) Test $H_0: B = 0, H_1: B \neq 0$ ($\alpha = .05$).
  (b) Test $H_0: B = 2.5, H_1: B \neq 2.5$ ($\alpha = .05$).
  (c) Find a 95 per cent confidence interval for B.

4.   Refer to Exercise 23, page 291. The regression equation is $Y_c = -8.29 + 1.39X$, $\Sigma XY = 1282$, $\Sigma X = 110$, $\Sigma X^2 = 1574$, $\Sigma Y = 87$, $\Sigma Y^2 = 1135$.
  (a) Test (with $\alpha = .05$) $H_0: B = 0, H_1: B \neq 0$.
  (b) Find a 95 per cent confidence interval for B.

5.

X	−6	−5	−4	−3	−2	−1	0	1	2	3	4	5	6
Y	8	6	6	4	4	2	2	0	0	−2	−2	−4	−4

$\Sigma X = 0$, $\Sigma X^2 = 182$, $\Sigma Y = 20$, $\Sigma Y^2 = 216$, $\Sigma XY = -182$, $n = 13$.
  (a) Find the equation of the regression line.
  (b) Test $H_0: B = 0, H_1: B < 0$ ($\alpha = .01$).
  (c) Test $H_0: B = -1.1, H_1: B \neq -1.1$ ($\alpha = .01$).
  (d) Find a 99 per cent confidence interval for B.

6.   Refer to Exercise 3. Suppose that a new set of 64 pairs of scores is made up by repeating each of the 16 pairs in Exercise 3 four times. Find a 95 per cent confidence interval for B, and compare your answer with the result of Exercise 3(c).
     Note: ΣX, ΣY, ΣX^2, ΣY^2, ΣXY, and n are each multiplied by 4.

7.   Suppose that a new set of 1600 pairs of scores is made up by repeating each of the 16 pairs in Exercise 3 100 times. Find a 95 per cent confidence interval for B, and compare your answer with the results of Exercises 3(c) and 6.

8.   Refer to Exercise 4, and (a) give a 95 per cent confidence interval for the mean of all Y scores in the population if (i) X = 0.0, (ii) X = 1.0, (iii) X = 1.5; (b) give a 95 per cent prediction interval for Y if (i) X = 0, (ii) X = 1, (iii) X = 1.5.

9.   A sociologist studying upward mobility gives 80 children a test which measures their social attitudes (Y). He thinks that one factor which can be used to predict the test score is the number of children in the family (X). He discovers that $\Sigma X = 200$, $\Sigma X^2 = 600$, $\Sigma Y = 440$, $\Sigma Y^2 = 4000$, $\Sigma XY = 1300$, $a = .500$, $b = 2.00$.
    (a) Find a 95 per cent confidence interval for the mean test score if there are 2 children in the family.
    (b) In addition to the 80 children, another child is to take the test; he has one brother and no sisters. What test score would the sociologist predict for this child, with .95 probability of being correct?

10.   A farmhand sorts apples into 2, 3, or 4 grades. The length of time he takes for sorting depends on the number of grades as follows:

Time (seconds)	3,1,2,1,1	2,3,2,1,2	2,3,3,1,4	3,4,4,5,2
Number of grades	2	3	4	5

(a) With .90 probability, what is the predicted time for 6 grades?

(b) Can you explain why it may not be wise to extrapolate prediction for X values outside the range in the sample?

11. Fifty-one grade school students are given a reading test and the height of each child is measured. The computed value of r between score on the test and height is .80.

(a) Test the hypothesis that there is no correlation between reading score and height for all grade school students. ($\alpha = .05$.)

(b) Test the hypothesis that the correlation is at least .85 for all grade school students. ($\alpha = .05$.)

(c) Can you explain the apparently high correlation?

(d) Can you explain why the tests suggested in (a) and (b) should not be used?

12. Test $H_0: \rho = 0$, $H_1: \rho \neq 0$ for the scores in Exercise 3. ($\alpha = .01$.)

13. Test $H_0: \rho = .90$, $H_1: \rho > .90$ for the scores in Exercise 3. ($\alpha = .05$.)

14. Test $H_0: \rho = .85$, $H_1: \rho < .85$ for the scores in Exercise 3. ($\alpha = .05$.)

15. The correlation between two samples is .30. Test $H_0: \rho = 0$, $H_1: \rho \neq 0$ if (a) $n = 11$, (b) $n = 102$. ($\alpha = .05$.)

16. The correlation between two samples is .30. Test $H_0: \rho = .50$, $H_1: \rho < .50$ if (a) $n = 11$, (b) $n = 102$. ($\alpha = .05$.)

17. The correlation between age and systolic blood pressure among 228 men chosen at random is .2. Find a 95 per cent confidence interval for this correlation among all men.

18. The correlation between the area and population of 10 counties in Texas, chosen at random, is $-.2$. At a .10 level of significance: (a) is there a positive correlation between area and population for all counties in Texas?; (b) is the correlation $-.3$ or less for all counties? (c) Find a 95 per cent confidence interval for ρ. Does this explain the apparent contradiction in (a) and (b)?

ANSWERS

1. $s_b = \sqrt{\dfrac{4(50 - 5.22 * 12 + .59 * 28)}{2(4 * 85 - 15^2)}} = .26$, $t = \dfrac{-.59 - 0}{.26} = -2.27$; $\alpha = .05$,

$D = 2$, two-tail test, $t = \pm 4.3$. H_0 is accepted: $B = 0$.

3. (a) $\Sigma X = 24$, $\Sigma X^2 = 56$, $\Sigma Y = 62$, $\Sigma Y^2 = 362$, $\Sigma XY = 139$, $n = 16$, $b = 2.30$,

$a = .43$, $s_b = \sqrt{\dfrac{16(362 - .43 * 62 - 2.30 * 139)}{14(16 * 56 - 24^2)}} = .236$, $t = \dfrac{2.30 - 0}{.236} = 9.75$; for $\alpha = .05$,

two-tail test, $D = 14$, the critical values of t are ± 2.14. H_0 is rejected.

(b) $t = \dfrac{2.30 - 2.50}{.236} = -.85$; H_0 is accepted.

(c) $\pm 2.14 = \dfrac{2.30 - B}{.236}$, $B = 2.30 \pm .51 = 1.8$ or 2.8. A 95 per cent confidence interval for B is 1.8 to 2.8.

5. (a) $b = \dfrac{13 * (-182) - 0 * 20}{13 * 182 - 0^2} = -1.00$, $a = \dfrac{20}{13} - 0 = 1.54$, $Y_c = 1.54 - 1.00X$.

(b) $s_b = \sqrt{\dfrac{13[216 - 1.54 * 20 + 1.00 * (-182)]}{11 * 13 * 182}} = .040, t = \dfrac{-1.00 - 0}{.040} = -25;$

the critical value of t is -2.72, so H_0 is rejected.

(c) $t = \dfrac{-1.00 - (-1.10)}{.040} = 2.50;$ the critical values of t are ±3.11, H_0 is accepted.

(d) $\pm3.11 = \dfrac{-1.00 - B}{.040}$, $B = -1.00 \pm .12 = -1.12$ or $-.88$; a 95 per cent confidence interval for B is -1.12 to $-.88$.

7. The regression equation is the same as in Exercise 3; $a = .43$, $b = 2.30$;

$s_b = \sqrt{\dfrac{1600(100 * 362 - .43 * 100 * 62 - 2.30 * 100 * 139)}{1598(1600 * 100 * 56 - 100^2 * 24^2)}} = .021; \pm1.96 = \dfrac{2.30 - B}{.021}$, B

$= 2.30 \pm .041 = 2.253$ or 2.341; a 95 per cent confidence interval for B is 2.25 to 2.34.

The width of the interval is $.09$; for 64 scores it was $.48$, for 16 scores it was 1.00. One hundred times as many scores gives a confidence interval approximately one-tenth as wide.

9. (a) $s = \sqrt{\dfrac{80(4000 - .500 * 440 - 2.00 * 1300)}{78 * 8000}} = .389, \overline{X} = 2.5;$ if $X_0 = 2, Y_c =$

$.50 + 2.00 * 2 = 4.50, s_{Y/X_0} = .389\sqrt{\dfrac{1}{80} + \dfrac{80(2 - 2.5)^2}{8000}} = .0476, D = 78, t = \pm1.99 =$ *from normal*

$\dfrac{4.50 - \mu_{Y/X_0}}{.0476}$, $\mu_{Y/X_0} = 4.41$ or $4.59.$

(b) $s_{Y_p} = .389\sqrt{1 + \dfrac{1}{80} + \dfrac{80(2 - 2.5)^2}{8000}} = .392; \pm1.96 = \dfrac{4.50 - Y_p}{.392}$, $Y_p = 3.73$ to $5.26.$

11. (a) $t = \dfrac{.80\sqrt{49}}{\sqrt{1 - .80^2}} = 9.33;$ $\alpha = .05$, $D = 49$, critical values of t are $\pm1.96;$ H_0 is rejected.

(b) $H_0: \rho = .85$, $H_1: \rho < .85$, $\sigma_z = 1/\sqrt{48}$, $z = \dfrac{1.10 - 1.26}{.144} = -1.11;$ the critical value of z is -1.65, so H_0 is accepted.

(c) Children's reading ability and height both increase as they grow older.

(d) Heights of children in any one grade may approximate a normal distribution, but for all grades they probably are not; the same is true of reading ability.

13. $r = \dfrac{16 * 139 - 24 * 62}{\sqrt{16 * 56 - 24^2}\sqrt{16 * 362 - 62^2}} = .932, Z = 1.67, \mu_z = 1.47, z = \dfrac{1.67 - 1.47}{1/\sqrt{13}}$

$= 1.09.$ The critical value of z is 1.65, so H_0 is accepted.

15. (a) $r = \dfrac{.30\sqrt{9}}{\sqrt{1 - .30^2}} = .94;$ if $D = 9$, $\alpha = .05$, then $t = 1.83;$ H_0 is accepted.

(b) $t = \dfrac{.30\sqrt{100}}{\sqrt{1 - .30^2}} = 3.14;$ if $D = 100$, $\alpha = .05$, then $t (= z) = 1.65;$ H_0 is rejected.

17. $\pm1.96 = \dfrac{.203 - \mu_z}{1/\sqrt{225}}$, $\mu_z = .072$ or $.334$, $\rho = .07$ to $.32.$

chapter seventeen

nonparametric tests

A nonparametric test is one in which (a) it is not necessary to assume the population is normally distributed (or any other very strong assumption about the population distribution; for this reason nonparametric tests are also called "distribution-free tests") or (b) categorical or ranked data are used. You will find the computations quite simple; both the mean and the standard deviation are meaningless for ranked data.

17.1 INTRODUCTION

For many of the tests you have learned to use, it has been necessary to assume that the population is normally distributed. If it is normal, the error made in making inferences about the population from sample data can be estimated. But if it is not normal, the error may be large and cannot be estimated. It is therefore valuable to know some tests in which such a strong assumption as normality doesn't need to be made about the population distribution; these are called **nonparametric** tests. You have already studied one of the most important: χ^2 test. There are many others, but only four of them are included here.

The word "nonparametric" is confusing. It was originally applied when **no** assumption needed to be made about the population distribution (as in the χ^2 tests of Chapter 13). Almost always, however, some assumption needs to be made; in some of the tests in this chapter, for example, the population distribution is assumed to be continuous so that comparative rankings can be given to scores in different samples (the third score in sample 1 is higher than the third score in sample 2). So "nonparametric" is now used to mean that a less stringent requirement than normality is made for the population distribution. Also, the meaning of the phrase "nonparametric test" has been extended to include any test using categorical or ranked data. Thus, a comparison of the heights of two samples of women is made with a nonparametric test if, instead of measuring each person's height, all the women are simply arranged in order of height in a row (the two tallest come from sample 1, the third tallest comes from sample 2, and so on). Then, because the data are ranked and not metric, a nonparametric test is used—even if it is known that the heights in both populations from which the samples were chosen are normally distributed.

Nonparametric tests have many advantages: You avoid the error caused by assuming a population is normally distributed when it is not, the computations that need to be made are often very simple, and the data may be easier to collect (almost certainly so when categorical or ranked rather than metric). Why are nonparametric tests not always used, then? The answer is that you don't get something for nothing. If the population distribution is normal so that, say, a *t* test or a nonparametric test may be chosen, the former will generally give a smaller value of β than the latter for a given, fixed value of α; if this is the case, sample size will have to be larger for the nonparametric test if the same limits on α and β are to be attained as with a t test. For the same sample size, then, the t test results usually will be more reliable. Also, the null hypotheses are sometimes more general; rejection may imply "two population distributions are different," but you don't know whether they have different means, different variances, or different distributions (for example, one normal and the other not).

17.2 THE SIGN TEST

The **sign test** is used in a "before-after" type of experiment in which the experimenter discovers only whether or not his experiment has succeeded. For example, 60 drivers chosen at random might be given a test of reflexes before and after drinking 1 ounce of alcohol, and rated $+$ if reaction time improves, 0 if it stays the same, and $-$ if reaction time is less. The numbers of $+$ and of $-$ scores are counted; any individual with a 0 is dropped from the experiment. If we wish to test the claim that drinking alcohol decreases reaction time, then H_0 states that there is no change in reaction time and H_1 that there are more minus than plus signs. If the sample is a random one, then under the null hypothesis the probability that any individual drawn from the population will have a slower reaction time is 1/2; the sign test uses a binomial distribution, then, with $H_0: P = 1/2$.

Example 1

Eleven white children are chosen at random from seventh graders at Robbins Junior High School and tested on their prejudice towards blacks. They are re-tested after a series of eight very open discussions with some black students at the school. Six show less prejudice, 1 shows more, and 4 show no difference. Decide whether you have significant evidence that this type of discussion will lessen prejudice of all white seventh graders at the school ($\alpha = .10$).

H_0: There is no difference in prejudice ($P = .5$, where P is the proportion of $-$ scores).

H_1: There is less prejudice ($P < .5$).

If $P = .5$, the probability that 6 or more of 7 students (note that the 4 students showing no change are ignored) will show less prejudice is $_7C_6(.5)^6(.5)^1 + _7C_7(.5)^7 = .055 + .008 = .063$ (using Table 2, Appendix C). This is less than .10, so H_0 is rejected. At a .10 level of significance, the series of discussions will cause a change in the attitudes of white seventh graders in the school.

Note that, since the binomial distribution is used in the sign test, the binomial assumptions (see page 89) must be satisfied. This means that the sample consists of scores which are independent; the white students do not influence each other.

Remember that the normal distribution can be used as an approximation to the binomial distribution if n is sufficiently large, but that a correction of .5 for continuity must be used (see page 118). Recall also that the mean of a binomial distribution with $P = 1/2$ is $nP = n/2$ and the standard deviation is $\sqrt{nPQ} = \sqrt{n}/2$. So to test $H_0: P = 1/2$

for large n, you will compute $z = \dfrac{X \pm .5 - n/2}{\sqrt{n/2}}$; the .5 term in the numerator is the correction for continuity, and the $-$ sign is chosen if $X > n/2$, the $+$ sign if $X < n/2$. (This choice of the sign reflects conservatism, since H_0 is less likely to be rejected with this choice.)

Example 2

In a test of 60 drivers, it is found that 32 have a slower reaction time, 10 show no difference, and 18 have faster reactions after drinking 1 ounce of of alcohol.

$H_0: P = .5$.

$H_1: P > .5$ (this means P is the proportion of drivers with a slower reaction time).

One-tail test, $\alpha = .05$, so $z = 1.65$.

$$z = \frac{X \pm 1/2 - n/2}{\sqrt{n/2}}$$

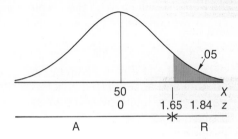

$X = 32 =$ number of drivers with slower reaction time

$n = 32 + 18 = 50$ (Note that the drivers with the same reaction time are not included; $n \ne 60$.)

$$z = \frac{32 - 1/2 - 50/2}{\sqrt{50/2}} = \frac{6.50}{3.54} = 1.84$$

H_0 is rejected; at a 5 per cent level of significance, drivers have a slower reaction time after drinking 1 ounce of alcohol.

In using the sign test, you do not use any information you have about actual scores (and therefore about the size of the differences between scores). It is not surprising, therefore, that a t test for differences of matched pairs (see page 215) is preferred **if** the data are metric and **if** the required assumption that the differences are normally distributed is met. This is always the case: if data are metric, use of the actual data (rather than simply the rankings) will give a more reliable result **if the assumptions about the population distribution that are required for use of the test are met.**

17.3   THE RUNS TEST

The **runs test** is a very simple one that can be used to test the randomness of a sample. The median of the sample is determined, and then each score in turn is marked X if it is below the median or Y if it is above the median. The number of "runs" (that is, clumps of adjacent X's or Y's) is counted.

Example 1

Determine the number of runs for the following scores: 494, 486, 443, 419, 494, 436, 416, 481, 408, 451, 434, 488, 488, 415, 453, 401, 454, 403, 454, 456. The median must first be computed; it is 452 (check this!). Assigning the label X to scores below 452 and Y to those above 452, and crossing out the one score equal to 452, we find 13 runs:

$$\overline{YY}\ \ \underline{XX}\ \ \overline{Y}\ \ \underline{XX}\ \ \overline{Y}\ \ \underline{XXX}\ \ \overline{YY}\ \ \underline{X}\ \ \overline{Y}\ \ \underline{X}\ \ \overline{Y}\ \ \underline{X}\ \ \overline{YY}$$

Each run is indicated by a horizontal line. The first two numbers are above the median and make up run 1, the next two are below the median and make up run 2, and so on.

If the sample is random, then we do not expect to select first all the scores which are above the median, and later all those below it; nor do we expect the scores to gyrate evenly above and below the median. Rather, we expect that the sample scores, considered in the order they are selected for the sample, will have a reasonable number of runs below and above the median. A two-tail test is therefore used for testing randomness of a sample.

The symbols R, n_X, and n_Y will be used for the number of runs, of X's, and of Y's, respectively. Critical values of R for $\alpha = .05$ and for n_X and n_Y between 4 and 20 are given in Table 9, Appendix C. Reject H_0: the sample is random and accept H_1: it is not a random sample unless the number of runs is between the limits given in the table (reject H_0 if the number of runs equals a given limit). Exercise 11, page 322, will give you some idea of how such a table is made up.

Example 2

The scores given in Example 1 are the weights in grams of 20 cans of peaches packed on one machine at the Freestone Canning Co. Use the runs test about the median to test the hypothesis that the sample is random ($\alpha = .05$).

$R = 13$, $n_X = 10$, $n_Y = 10$ (see Example 1). The critical values of R are 6 and 16, so H_0 is accepted. (But the machine had better be repaired quickly!)

What if n_X or $n_Y > 20$ and Table 9 cannot be used? Fortunately, the sampling distribution of R is approximately a normal distribution in this case, with

$$\mu_R = 1 + \frac{2n_X n_Y}{n_X + n_Y}$$

$$\sigma_R = \sqrt{\frac{2n_X n_Y(2n_X n_Y - n_X - n_Y)}{(n_X + n_Y)^2(n_X + n_Y - 1)}},$$

An example will illustrate this, and at the same time show another use of the runs test: A list of numbers can be tested for randomness by, for example, putting even numbers in the X sample and odd numbers in the Y sample. Here again a two-tail test is used, since we expect neither that many odds and then many evens will clump together, nor that odds and evens will always alternate.

Example 3

The following numbers are chosen from the third row in the table of random numbers in appendix C (Table 3):

54222   56179   09833   34227   43897   38517   11617   30338

Use a runs test on odd and even digits to see if these numbers are indeed random (.05 level of significance).

H_0: the numbers are random; H_1: they are not random.

Upon assigning X's to even numbers and Y's to odd, we have 22 runs:

$\overline{Y\underline{XXXX}}$ $\overline{Y\underline{X}YYY}$ $\overline{\underline{X}Y\underline{X}YY}$ $\overline{Y\underline{XXX}Y}$ $\overline{\underline{X}Y\underline{X}YY}$ $\overline{Y\underline{X}YYY}$ $\overline{YY\underline{X}YY}$ $\overline{Y\underline{X}YY\underline{X}}$

$n_X = 16, n_Y = 24;$

$$\mu_R = \frac{2 * 16 * 24}{16 + 24} + 1 = 20.2$$

$$\sigma_R = \sqrt{\frac{2 * 16 * 24(2 * 16 * 24 - 16 - 24)}{40^2(40 - 1)}} = 2.99$$

$$z = \frac{R - \mu_R}{\sigma_R} = \frac{22 - 20.2}{2.99} = .60$$

Since $\alpha = .05$, critical values of z are ± 1.96. H_0 is accepted: The sample is random.

17.4 THE MANN-WHITNEY TEST

This test (also called the Wilcoxon test or the rank-sum test) is used to test whether two random samples are taken from identical populations. It is assumed that the populations have identical shape and variability; do they also have the same median?

It must be possible to lump together the scores in the two samples and rank them. A test statistic U is computed, based on a comparison of the sums of ranks of the separate samples in the pooled set. Table 10, Appendix C, will be used for small samples (each less than 20), and a z test for large samples.

Example 1

At a .05 level of significance, do the following random samples come from the same population?

X: 7   9   10   11   13

Y: 12   13   14   15   16   17   18

H_0: the two samples come from the same population

H_1: they come from different populations

The two samples are lumped together and then ranked in the combined sample:

	Sample											
	X	X	X	X	Y	X	Y	Y	Y	Y	Y	Y
Score	7	9	10	11	12	13	13	14	15	15	17	18
Rank	1	2	3	4	5	6.5	6.5	8	9	10	11	12

Notice that two scores of 13 occur; these would be ranked 6 and 7, so both are given the mean of these two numbers.

Sum of ranks of X scores $= \Sigma R_X = 1 + 2 + 3 + 4 + 6.5 = 16.5$
(Do not confuse R_X here with the number of runs R in the runs test!)

Sum of ranks of Y scores $= \Sigma R_Y = 5 + 6.5 + 8 + 9 + 10 + 11 + 12 = 61.5$

$n_X = 5, n_Y = 7$

As a check, we should have $R_X + R_Y = \dfrac{n(n+1)}{2}$, where $n = n_X + n_Y = 5 + 7$ $= 12$:

$$16.5 + 61.5 = \frac{12 * 13}{2} = 78 \quad ✔$$

Now find

$$U_X = n_X n_Y + \frac{n_X(n_X + 1)}{2} - \Sigma R_X$$

$$U_Y = n_X n_Y + \frac{n_Y(n_Y + 1)}{2} - \Sigma R_Y$$

Check that $U_X + U_Y = n_X n_Y$

Let U equal the smaller of U_X, U_Y

$$U_X = 5 * 7 + \frac{5 * 6}{2} - 16.5 = 33.5$$

$$U_Y = 5 * 7 + \frac{7 * 8}{2} - 61.5 = 1.5$$

$$33.5 + 1.5 = 5 * 7 = 35 \quad ✔$$

$$U = 1.5$$

In Table 10, Appendix C, the critical value of U given for $n_X = 5$, $n_Y = 7$ is 5. H_0 is accepted if the computed value of U is greater than the value given in the table. In this example it is less, so H_0 is rejected; there is a significant difference between the samples and it is concluded that they come from different populations.

Here we used ranked data, but metric scores are known and their means and standard deviations can be computed: $\bar{X} = 10$, $s_X = 2.2$, $\bar{Y} = 15$, $s_Y = 2.2$. Perhaps a t test of difference of means could be used. (Look back at Example 2, page 212.) What are the relative advantages of t and Mann-Whitney tests? For both tests, it is assumed that the two sampled populations have the same shape and dispersion; for the t test, it is also necessary to assume that the shape is that of the normal curve. If there is any cause to doubt this (and the doubt is usually greatest for small samples), then the Mann-Whitney test should be used, since both test whether the measures of center of the two populations are the same. For a given α, the probability of rejecting H_0 when it is false (this is called the **power** of the test, and is measured by $1 - \beta$) is slightly higher with a t test. Sample size must be increased slightly if the same power is to be achieved using a Mann-Whitney test.

If you are sure of your arithmetic, you may shorten the computation of U by finding the sum of ranks for only one of the samples, and then the check formula suggested above is used to find the sum or ranks for the other sample. ($U_Y = n_X n_Y - U_X$, for example.)

What if the sample sizes are larger than those given in Table 10, Appendix C? The sampling distribution of U approximates a normal distribution for samples of size about 20 or larger (and the approximation gets better as sample sizes increase), with mean

$$\mu_U = \frac{n_X n_Y}{2}$$

and with standard error

$$\sigma_U = \sqrt{\frac{n_X n_Y (n + 1)}{12}}$$

Example 2

Two random samples, chosen independently, are ranked as follows:

X: 1, 2, 4, 5, 10, 11, 12, 14, 15, 16, 20, 21, 24, 27, 28, 29, 34, 35, 36
Y: 3, 6, 7, 8, 9, 13, 17, 18, 19, 22, 23, 25, 26, 30, 31, 32, 33, 37, 38, 39, 40, 41, 42, 43

Test, at a .05 level of significance, whether these are random samples from the same population.

H_0: the two samples come from the same population; H_1: they don't.

$$\Sigma R_X = 344, n_X = 19, n_Y = 24, n = 19 + 24 = 43$$

$$\Sigma R_Y = \frac{n(n + 1)}{2} - \Sigma R_X = \frac{43 * 44}{2} - 344 = 602$$

$$U_X = 19 * 24 + \frac{19 * 20}{2} - 344 = 302$$

In this example, $\mu_U = \dfrac{19 * 24}{2} = 228$, $\sigma_U = \sqrt{\dfrac{19 * 24 * 44}{12}} = 40.9$.

The z score for U_X is

$$z = \frac{U_X - \mu_U}{\sigma_U} = \frac{302 - 228}{40.9} = 1.81.$$

Now compute U_Y:

$$U_Y = 19 * 24 + \frac{24 * 25}{2} - 602 = 154$$

The z score for U_Y is $\frac{154 - 228}{40.9} = -1.81$. But this is the negative of the z score for U_X! This will always be the case, so it's only necessary to compute one of them.

With $\alpha = .05$, using a two-tail test, the critical values of z are ± 1.96. H_0 is accepted: The two samples do come from the same population.

17.5 SPEARMAN RANK CORRELATION COEFFICIENT

This coefficient, r_S, measures the correlation between two paired samples of ranked data. For example, two judges may each rank a random sample of n students according to their ability in writing. How well do the judges agree? Or the abilities of one individual in writing, accounting, mechanical skills, and so on may be ranked at the beginning and end of a training course. Does the individual still do best at writing, worst at accounting? As with the Pearson r, if the rankings agree (each member of the sample is ranked the same by both judges, or has the same rank in attributes at different times) we expect a high positive correlation; $r_S = +1$. If one judge thinks worst of the composition which the other judge rates best, and vice versa, then there is a high negative correlation.

The Spearman rank correlation coefficient is the Pearson correlation r applied to the **ranks** in two paired samples (not to the original scores). The customary formula for r could, of course, be used, but an even simpler computation for r_S may be substituted:

1. List the n pairs of ranks: X, Y.

2. Find the differences d between the two sets of ranks.

3. Square these differences and add the squares ($= \Sigma d^2$).

4. Compute r_S:

$$r_S = 1 - \frac{6\Sigma d^2}{n(n^2 - 1)}$$

The Spearman rank correlation coefficient is useful (1) when only ranks are known or (2) when metric scores are known but it's easier to determine r_S than r. But there is one great drawback to use of the Spearman coefficient: even if r_S is close to 1 (or −1), we then know that as X increases, Y increases (or decreases), but we do not know by how much. If the Pearson r is close to 1 (−1), on the other hand, we know there is a linear relationship between X and Y. Whenever X increases by a certain amount, Y increases (or decreases) by a constant amount; the regression line will fit the points (X, Y) almost exactly.

Example 1

Six paintings were ranked as follows by two judges; compute r_s: (a) using the four steps on page 318, (b) using the formula for the Pearson r.

Painting	**X** First Judge	**Y** Second Judge	**(a)** d	d^2	XY	**(b)** X^2	Y^2
A	2	2	0	0	4	4	4
B	1	3	−2	4	3	1	9
C	4	4	0	0	16	16	16
D	5	6	−1	1	30	25	36
E	6	5	1	1	36	36	25
F	3	1	2	4	3	9	1
	(21)	(21)		10	86	91	91

(a) $\Sigma d^2 = 10$, $n = 6$.

$$r_s = 1 - \frac{6 * 10}{6 * (6^2 - 1)} = .71$$

(b) $r = \dfrac{6 * 86 - 21 * 21}{\sqrt{6 * 91 - 21^2}\ \sqrt{6 * 91 - 21^2}} = \dfrac{75}{105} = .71 = r_s.$

What if there are ties in either the X or the Y list? In this case, replace the equal ranks by the mean of the ranks for which they stand. Thus, if a sample of 5 shows 2 scores tied for rank 3 (that is, the ranked scores appear as 1,2,3,3,5), then both of the 3's are replaced by 3.5, the mean of 3 and 4.

Example 2

Compute r_s for the following rankings of 6 individuals on 2 tests:

			Individual			
	A	B	C	D	E	F
Test X	1	2	2	4	5	6
Test Y	3	2	1	4	6	5

There is a tie for second place among the X rankings, so each of those 2's is replaced by 2.5.

X	Y	d	d^2
1	3	−2	4
2.5	2	.5	.25
2.5	1	1.5	2.25
4	4	0	0
5	6	−1	1.
6	5	1	1.
			8.5

$$n = 6$$

$$r_s = 1 - \frac{6 * 8.5}{6 * (6^2 - 1)} = 1 - .24 = .76$$

Example 3

Eight factory workers chosen at random in the Beacon Light Co. plant are ranked as follows on tests of finger dexterity and mechanical ability. Is there a correlation between them?

Worker	Finger Dexterity	Mechanical Ability	X	Y	d	d²	
A	1	2	1	3	−2	4	
B	2	1	2	1	1	1	
C	3	8	4	8	−4	16	$n = 8$
D	3	2	4	3	1	1	
E	3	2	4	3	1	1	
F	6	6	6	6.5	− .5	.25	
G	7	6	7	6.5	.5	.25	
H	8	5	8	5	3	9	
						32.50	

The X column is the Finger Dexterity column adjusted for the triple tie for third place; the Y column is the Mechanical Ability column adjusted for ties for second and sixth places.

$$r_s = 1 - \frac{6 * 32.50}{8 * (8^2 - 1)} = 1 - .39 = .61$$

What about testing hypotheses about the Spearman rank correlation coefficient for the population, ρ_s? If the null hypothesis is that $\rho_s = 0$, a t test can be made of the same type as when testing $H_0: \rho = 0$:

$$t = \frac{r_s\sqrt{n - 2}}{\sqrt{1 - r_s^2}}$$

with $n - 2$ degrees of freedom. However, n should be 10 or larger for this test.

Example 4

Eighteen workers chosen at random in the ABC factory show a Spearman rank correlation of .60 between finger dexterity and mechanical ability. For all workers in this factory, is there a significant correlation? (.05 level of significance.)

$H_0: \rho_s = 0, H_1: \rho_s \neq 0; \alpha = .05.$

$$r_s = .60, n = 18, \text{ so } t = \frac{.6 * \sqrt{18 - 2}}{\sqrt{1 - .6^2}} = 3.00$$

For a two-tailed test with 16 degrees of freedom and $\alpha = .05, t = \pm2.12$. H_0 is rejected: There is a significant correlation.

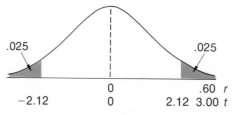

.025   .025

−2.12
                    0   .60 r
                    0   2.12 3.00 t

17.6 VOCABULARY AND SYMBOLS

nonparametric test   Mann-Whitney test $R_X, R_Y, U_X,$
sign test U_Y, μ_U, σ_U
runs test R, μ_R, σ_R Wilcoxon test
run   rank-sum test
                          Spearman rank correlation coefficient r_S, d, ρ_S

17.7 EXERCISES

1.   A foolhardy† agricultural chemist who wants to compare Kepone and Myrex as insecticides for fire ants goes to a large pasture in which there are many nests and randomly chooses 36, which are then paired by size. He puts the same dosage of corn treated with Kepone or Myrex on one of each pair. A week later he returns to the pasture and notes that, in 8 pairs, the nest treated with Myrex has fewer ants; in 6 pairs, all fire ants have died; and in 4 pairs, Kepone has done better. Use the sign test to see whether the two chemicals differ in their effects on fire ants (.10 level of significance).

2.   Two mountain climbers had an argument about whether narrow or wide pitons were more popular, and enlisted the help of several climbing clubs in an experiment. Fifteen ropes (2 climbers each) climbed Class V pitches in Franconia Notch twice, once with narrow pitons and once with wide pitons. Of the 30 climbers, 20 preferred the narrow and 10 the wide pitons. At a .05 level of significance, is there a difference in preference?

3.   Seventy-seven salesmen for the Coolidge Vacuum Cleaner Company are given a week's special training course. Forty of them show substantially better sales the following month, 28 show no change, and 9 don't sell as much as in the month before the course. At a .02 level of significance, have sales increased after the training course?

4.   Twelve children in the same class at school are ranked for popularity (a) by group preference and (b) by rating on a social attitudes test. The results are as follows:

						Child						
	A	B	C	D	E	F	G	H	I	J	K	L
Group rating	6	2	7	3	4	11	1	5	10	8	9	12
Test rating	7	1	8	4	6	3	2	5	10	9	11	12

Use a sign test on differences in rank to see whether the two methods of ranking differ at a .05 level of significance.

5.   Thirty-five children in a class are ranked for popularity (a) by group preference and (b) by score on a social attitudes test. Nineteen have higher rankings by group preference, eleven have lower, and five have the same ranking. Use a sign test on differences of rank to decide whether rankings by the two methods differ significantly ($\alpha = .05$).

6.   The open discussions with black students discussed in Example 1, page 312, are carried out with 261 white children at Robbins Junior High School. In re-testing after the discussions, 113 white children show less prejudice, 83 show more, and 65 show no change. Do you have significant evidence that this type of discussion will lessen prejudice of all white children at the school? ($\alpha = .05$.)

---

†Fire ants sting, and can kill a horse.

7.   The following numbers are taken from the first line, columns 3−6, of Table 3, Appendix C:

| 81536 | 09686 | 26743 | 87001 |

Use the runs test, counting runs above and below the median, to test whether these numbers are random ($\alpha = .05$).

8.   Use the runs test, counting runs of odd and even numbers, to test whether the numbers in Exercise 7 are random ($\alpha = .05$).

9.   The median of the numbers in Example 3, page 320, is 3.5. Use the runs test about the median to test whether or not these numbers are random ($\alpha = .05$).

10.   Among cases chosen from the files at the Union County Court House, the prison terms given were 3, 1, 1/2, 2, 20, 5, 1, 6, 3, 1, 1/2, 1/6, 4, 2, 15, 7, 1, 1/2, 2, 3, 1, 4, 1, 2, 3, 1, 2, 18, 8, 1, 2, 3, 2, 5, 1, 2, 1, 7, 12, 9, 2, 1, 3, 4, 2, 1 1/2, 1/2, 6, 1/2, 3, 1, 5, 2, 7, 2, 1, 1, 4, 1, 1 years. Is this a random sample, at a .10 level of significance?

11.   (a)   If $n_X = 2$, $n_Y = 1$, and X and Y are ranked at random in a combined list, what is the probability that (a) $R = 3$? (b) $R = 2$? [Hint: (a) $p(XYX) = ?$ (b) $P(XXY$ or $YXX) = ?$]
             (b)   If $n_X = 3$, $n_Y = 2$, and X and Y are ranked at random in a combined list, what is the probability that (i) $R = 2$? (ii) $R = 3$? (iii) $R = 4$? (iv) $R = 5$?
             (c)   If $\alpha = .40$, for what values of R would H_0 be rejected on a two-tail runs test if $N_X = 3$, $n_Y = 2$?
     This exercise should give you a faint idea of how the critical values of R for the runs test are determined. A technique other than listing is needed for larger values of n_X and n_Y, but for this you need a separate course in probability.

12.   At a .05 level of significance, do the following random samples come from identical populations?

First sample *(X):*   6, 9, 5, 8, 6, 4
Second sample *(Y):*   3, 6, 10, 7, 4, 9, 11, 7

13.   A random sample of married men is chosen, and the men are asked whether they are satisfied or discontented with their marriages; then each man is asked how long he has been married. The results of this survey are:

Number of years married

Satisfied	1, 3, 3, 2, 7, 10
Not satisfied	16, 4, 9, 20, 16, 7, 4, 6, 18

At a .05 level of significance, is there a difference in the two groups?

14.   Six courses, chosen at random, offered at Blackwood University were evaluated 1 to 6, with 6-"excellent," by the students in the courses for appropriateness of work expected for the number of credits given and for reasonableness of cost of textbooks. The mean scores of the six courses were as follows:

	Course					
	A	B	C	D	E	F
Appropriateness of work expected	3.1	2.4	5.0	3.0	1.8	4.2
Reasonableness of textbook cost	4.0	5.7	5.4	3.4	1.2	5.0

(a) Use a t test to check, at a .05 level of significance, the hypothesis that all courses in the University have the same mean score on the two questions rated above.

(b) Use a Mann-Whitney test to see whether all courses in the University have the same distribution on these two questions on the evaluation form ($\alpha = .05$).

(c) If the mean scores on the six courses are ranked, find the Spearman rank correlation between the two questions.

15.   Six students are chosen at random from those studying both Economics 101 and Psychology 24, the former with Professor A and the latter with Professor B. The students rate the professors as follows on teaching ability, on a 1–6 scale on which 6 is "excellent":

			Student			
	A	B	C	D	E	F
Professor A	5	4	6	4	4	5
Professor B	4	3	5	2	4	3

There is no reason to assume that ratings given by all students are normally distributed. Is there a significant difference in the ratings of the two professors? ($\alpha = .05$.)

16.   Fifty American scientists are chosen at random, and for each the number of published scientific articles is counted. Of the 50, 27 are the oldest child in the family. Based on the following rankings, is there a difference in the number of publications of first-born and other children who become scientists? ($\alpha = .01$.)

Rank (based on number of publications)

Oldest child	2, 2, 2, 4, 7, 7, 7, 11, 11, 11, 16, 16, 16, 16, 16, 22, 22, 22, 28, 31, 33, 33, 33, 38, 38, 38, 44
Not oldest child	5, 9, 13, 19, 20, 25, 25, 25, 27, 29.5, 29.5, 35, 36, 41, 41, 41, 43, 47.5, 47.5, 47.5, 47.5, 47.5, 47.5

17.   Goodland Tire Co. believes that a chemical additive to the rubber used in its tires will improve wear. Random samples, each of 25 tires, are tested under the same conditions with the following results:

Wear (in thousandths of an inch)

Without additive	23, 14, 12, 11, 13, 18, 15, 16, 13, 10, 15, 20, 14, 9, 7, 20, 14, 18, 24, 19, 13, 16, 15, 20, 15
With additive	14, 19, 8, 10, 16, 12, 11, 9, 18, 12, 7, 6, 8, 11, 17, 13, 13, 15, 10, 8, 14, 7, 5, 13, 14

(a) Use a t test to see whether there is significantly less wear of tires made with rubber to which the chemical has been added ($\alpha = .05$). Assume that variances are the same for wear measurements on both types of tire, and that the measurements are normally distributed for both. $\overline{X} = 15.4$, $s_x^2 = 17.4$, $\overline{Y} = 11.6$, $s_y^2 = 14.5$.

(b) Use a Mann-Whitney test to decide whether the chemical additive has improved the wear of tires, at a .05 level of significance.

18.   Giovanni Chianti's seven movies were ranked by the film department students of Henry Fisk College on character development and on suspenseful ending. Find the correlation between their rankings.

	Movie						
	A	B	C	D	E	F	G
Character development	1	2	2	4	5	6	6
Suspenseful ending	3	4	2	1	7	5	6

19.   Ten dressmakers chosen at random in the Princess Dress Co. are ranked as follows for craftsmanship and speed. Find the correlation between these two rankings.

	Dressmaker									
	A	B	C	D	E	F	G	H	I	J
Craftsmanship	10	2	4	5	9	5	1	3	7	8
Speed	10	5	2	4	8	6	2	1	8	7

20.   For the following paired scores, (a) determine r, (b) determine r_S, and (c) draw a scatter plot, and then explain the difference between r and r_S.

$$X:\ 3\quad 5\quad 0\quad 2\quad 4\quad 1$$
$$Y:\ 3\quad 25\quad 0\quad 2\quad 16\quad 1$$

21.   Fifty-one factory workers show a Spearman rank correlation of .55 between absenteeism and difficulty of family problems. For all workers, is there any significant correlation? (.05 level of significance.)

22.   Refer to Exercise 14.
    (a) For all courses in the University, is there a significant correlation between ratings on the two questions? ($\alpha = .05$; determine ρ_S.)
    (b) Suppose that 100 courses, instead of 6, were evaluated, but that the computed Spearman correlation coefficient was unchanged. Is there a significant correlation between ratings on the two questions? ($\alpha = .05$.)

ANSWERS

1. $P =$ proportion in which Myrex does better. $H_0: P = .5$, $H_1: P > .5$, $n = 12$. (Ignore the 6 in which both were effective.) $X = 8$. $p(8$ or more in 12 trials with $P = .5) = .121 + .054 + .016 + .003 = .154$; this is greater than .10, so H_0 is accepted; there is no difference.

3. $H_0: P = .5$, $H_1: P > .5$, $n = 49$, $\mu = 24.5$, $\sigma = \sqrt{49 * .5 * .5} = 3.5$, $z = \dfrac{40 - .5 - 24.5}{3.5} = 4.29$. $\alpha = .02$, so the critical value of z is 2.05. Reject H_0; sales have increased.

5. $n = 30$, $\mu = 15$, $\sigma = 2.7$, $z = \dfrac{19 - .5 - 15}{2.7} = 1.30$. $\alpha = .05$, the critical value of z is 1.64. Accept $H_0: P = .5$ and reject $H_1: P \neq .5$.

7. $Md = 5.5$; $\underline{8}\overline{15}\underline{36}\ \overline{0}\underline{9}\overline{68}\underline{6}\ \overline{26}\underline{7}\overline{43}\ \underline{87}\overline{001}$; $R = 10$, $n_X = 12$, $n_Y = 8$. $\alpha = .05$, the critical values of R are 6 and 16. Accept H_0; the numbers are random.

9. $\overline{54}\underline{222}\ \overline{56}\underline{1}\overline{79}\ \underline{09}\overline{833}\ \underline{3}\overline{4}\underline{22}\overline{7}\ \underline{43}\overline{897}\ \underline{38}\overline{51}\underline{7}\ \underline{116}\overline{1}\underline{7}\ \underline{30}\overline{338}$; $R = 23$, $n_X = n_Y = 20$; the critical values of R are 14 and 28. Accept H_0; the numbers are random.

11.   (a) (i) $\dfrac{2}{3} * \dfrac{1}{2} * \dfrac{1}{1} = \dfrac{1}{3}$; (ii) $\dfrac{2}{3} * \dfrac{1}{2} * \dfrac{1}{1} + \dfrac{1}{3} * \dfrac{2}{2} * \dfrac{1}{1} = \dfrac{2}{3}$.

(b)   There are 5! permutations of 5 letters, and 3! * 2! of 3 X's and 2 Y's, so the probability of any particular permutation is 3! * 2!/5! = 1/10. (i) $p(XXXYY \text{ or } YYXXX) = .2$; (ii) $p(YXXXY \text{ or } XYYXX \text{ or } XXYYX) = .3$; (iii) $p(XYXXY \text{ or } XXYXY \text{ or } YXXYX \text{ or } YXYXX) = .4$; (iv) $p(XYXYX) = .1$.

(c)   Reject if $R = 2$ or 5.

13.   Score: $\bar{1}$ $\bar{2}$ $\bar{3}$ $\bar{3}$ $\underline{4}$ $\underline{4}$ $\underline{6}$ $\underline{7}$ $\bar{7}$ $\underline{9}$ $\overline{10}$ $\underline{16}$ $\underline{16}$ $\underline{18}$ $\underline{20}$
      Rank: $\bar{1}$ $\bar{2}$ $\bar{3}$ $\bar{4}$ $\underline{5}$ $\underline{6}$ $\underline{7}$ 8.5 8.5 $\underline{10}$ $\overline{11}$ $\underline{12}$ $\underline{13}$ $\underline{14}$ $\underline{15}$

$n_X = 6, n_Y = 9, R_X = 29.5, U_X = 6 * 9 + 6 * 7/2 - 29.5 = 45.5, U_Y = 6 * 9 - 45.5 = 8.5 = U$. The critical value of U is 10, so H_0 is rejected; the two groups differ.

15.   Score: $\underline{2}$ $\underline{3}$ $\underline{3}$ $\underline{4}$ $\underline{4}$ $\bar{4}$ $\bar{4}$ $\bar{4}$ $\bar{5}$ $\bar{5}$ $\underline{5}$ $\bar{6}$
      Rank: $\underline{1}$ $\underline{2}$ $\underline{3}$ $\underline{6}$ $\underline{6}$ $\bar{6}$ $\bar{6}$ $\bar{6}$ $\overline{10}$ $\overline{10}$ $\underline{10}$ $\overline{12}$

$n_X = n_Y = 6, R_X = 28, U_X = 6 * 6 + 6 * 7/2 - 28 = 29, U_Y = 6 * 6 - 29 = 7. \alpha = .05$, the critical value of U is 5, so H_0 is accepted.

17.   (a) $H_0 : \mu_X - \mu_Y = 0, H_1 : \mu_X - \mu_Y > 0$ (X is *without additive*), $\sigma_{(\bar{X} - \bar{Y})} = \sqrt{\dfrac{24 * 17.4 + 24 * 14.5}{48}} * \dfrac{2}{25} = 1.13, t = \dfrac{15.4 - 11.6}{1.13} = 3.01. D = 48$, critical $t (= z) = 1.64$; H_0 is rejected.

(b)   Score: 5 6 7 8 9 10 11 12 13 14 15 16 17 18 19 20 23 24
      Frequency: X 0 0 1 0 1 1 1 1 1 3 3 4 2 0 2 1 3 1 1
                  Y 1 1 2 3 1 2 2 2 2 3 3 1 1 1 1 1 0 0 0

X ranks: 4, 9.5, 12, 15, 18, 22.5, 22.5, 22.5, 28.5, 28.5, 28.5, 34, 34, 34, 34, 38, 38, 42, 42, 44.5, 46, 47, 48, 49, 50

Y ranks: 1, 2, 4, 4, 6, 7, 8, 9.5, 12, 12, 14, 15, 18, 22.5, 22.5, 22.5, 28.5, 28.5, 28.5, 34, 38, 39, 40, 42, 44.5

$R_Y = 483, U_Y = 25 * 25 + 25 * 26/2 - 483 = 467, U_X = 25^2 - 467 = 158 = U, \mu_U = 25^2/2 = 312.5, \sigma_U = \sqrt{25 * 25 * 51/2} = 126, z = \dfrac{158 - 312.5}{126} = -1.23; \alpha = .05$, the critical value of z is -1.64. Accept H_0.

19. d: 0, −3, 2, 1, 1, −1, −1, 2, −1, 1; $\Sigma d^2 = 23, r_s = 1 - \dfrac{6 * 23}{10 * 99} = .86$.

21. $H_0 : \rho_s = 0, H_1 : \rho_s \neq 0, t = \dfrac{.50\sqrt{49}}{\sqrt{1 - .50^2}} = 4.04, D = 4$, the critical values of t are ± 1.96. Reject H_0; there is a significant correlation.

appendix a

symbols

Numbers in parentheses indicate pages on which symbols are introduced.

a	Y-intercept of the line which best fits sample data (272)	$_nC_X$	number of combinations of n things taken X at a time (88)
A	an event; a subset of a sample space (62)	d	difference in value between two paired scores (215)
A	Y-intercept of the line which fits population data (296)	d	difference in rankings of scores in paired samples (318)
$\overline{A}$	complement of the set A; the event "A does not occur" (70)	D, D_X, D_Y	number of degrees of freedom (209, 241)
$A(z)$	area under the normal curve between 0 and z (118)	E	error (163)
		E	expected frequency (223)
ANOVA	analysis of variance (247)	f	frequency (16)
		$F(D_X, D_Y)$	F score (242)
α (alpha)	probability of a Type I error; level of significance (180)	$F_\alpha(D_X, D_Y)$	critical value of F for a one-tail test, level of significance $= \alpha$ (249)
b	slope of the line which best fits sample data (272)	H_0	null hypothesis (177)
B	slope of the line which fits population data (296)	H_1	alternative hypothesis (179)
		k	number of samples (250)
β (beta)	probability of a Type II error (180)		
		Md	median (30)
$1 - \beta$	power of a statistical test (317)	Mo	mode (29)

μ (mu)	population mean (83)	$p(A)$	probability of the event A (60)
μ_d	mean of paired differences in the population (215)	$p(B/A)$	probability of event B, given that event A occurs (73)
μ_p	mean of the sampling distribution of proportions (150)	P	proportion of the time that an event occurs in a population (150)
$\mu_{p_X-p_Y}$	mean of the sampling distribution of differences of proportions (160)	P_a	ath percentile (37)
μ_R	mean of the sampling distribution of number of runs (314)	$_nP_X$	number of permutations of n things taken X at a time (85)
$\mu_{\bar{X}}$	mean of the sampling distribution of means (133)	q	probability of failure in one or more independent trials $(q = 1 - p)$ (90)
$\mu_{\bar{X}-\bar{Y}}$	mean of the sampling distribution of differences of means (155)	q	proportion of the time that an event does not occur in a sample (150)
μ_{Y/X_0}	mean of population Y values for $X = X_0$ (299)	Q	semi-interquartile range or quartile deviation (46)
n	sample size (31)	Q	proportion of the time that an event does not occur in a population (150)
$n!$	n factorial (87)		
N	population size (109)		
O	observed frequency (223)	Q_1	first quartile (37)
		Q_3	third quartile (37)
p	probability of success in 1 of n independent trials (90)	r	linear correlation coefficient for samples (276)
p	proportion of the time that an event occurs in a sample (150)	r^2	coefficient of determination (281)
$p(a < X < b)$	probability that X has a value between a and b (101)	r_s	Spearman rank correlation coefficient (318)
		R	number of runs (314)

Symbol	Description	Symbol	Description
R_X $[R_Y]$	sum of ranks of X $[Y]$ scores (316)	$\hat{\sigma}^2$ *within*	estimate of σ^2 based on variability within samples (250)
ρ (rho)	linear correlation coefficient for populations (303)	σ_p	standard error of proportions (150)
s	standard deviation of a sample (47)	$\sigma_{(p_X - p_Y)}$	standard error of differences of proportions (160)
s	number of successful outcomes among n equally likely outcomes (60)	σ_R	standard error of the number of runs (314)
s^2	variance of a sample (47)	$\sigma_{\bar{X}}$	standard error of the mean (133)
s_b	standard error of the slope of the line of best fit (296)	$\sigma_{(\bar{X} - \bar{Y})}$	standard error of differences of means (155)
s_d	standard deviation of differences in paired samples (215)	Σ (capital sigma)	summation notation (4)
s_e	standard error of estimate (218)	t	Student's t distribution or t score (206)
s_{Y_p}	standard deviation of the Y values predicted for $X = X_0$ (302)	U	See Mann-Whitney test, page (316)
		X	score (16)
s_{Y/X_0}	standard error of the mean of Y values corresponding to X_0 (299)	X	midpoint of a class (17)
		$\bar{X}$	mean of a sample (31)
S	sample space (62)		
S	See Scheffé's test, page 255.	Y_c	value of Y computed from the line of best fit for samples (272)
σ (sigma)	standard deviation of a population (83)	Y_p	value of Y predicted in a population for $X = X_0$ (302)
σ^2	variance of a population (83)		
$\hat{\sigma}^2$ *between*	estimate of σ^2 based on variability between samples (249)	z	score in standard deviation units (111)

Z	Fisher Z score (306)	$\leq$	less than or equal to (70)
$\neq$	does not equal	$\approx$	approximately equals (19)
$*$	times (1)	∞	infinity (18)
$>$	greater than	$\lvert a \rvert$	absolute value of a (46)
$<$	less than		

appendix b

formulas

Numbers in parentheses indicate pages.

Mean

of a sample, ungrouped data $\qquad$ $\overline{X} = \dfrac{\Sigma X}{n}$ $\qquad$ (31)

of a sample, grouped data $\qquad$ $\overline{X} = \dfrac{\Sigma fX}{n}$ $\qquad$ (34)

of a population, ungrouped data $\qquad$ $\mu = \dfrac{\Sigma X}{N}$ $\qquad$ (109)

of a probability distribution $\qquad$ $\mu = \Sigma X p(X)$ $\qquad$ (82)

of a binomial distribution $\qquad$ $\mu = np$ $\qquad$ (93)

Variation ratio $\qquad$ $V = \dfrac{n - f_{Mo}}{n}$ $\qquad$ (44)

Semi-interquartile range $\qquad$ $Q = \dfrac{Q_3 - Q_1}{2}$ $\qquad$ (46)

Variance

of a sample, ungrouped data $\qquad$ $s^2 = \dfrac{\Sigma (X - \overline{X})^2}{n - 1}$ $\qquad$ (47)

$s^2 = \dfrac{\Sigma X^2 - (\Sigma X)^2 / n}{n - 1}$ $\qquad$ (48)

of a sample, grouped data $\qquad$ $s^2 = \dfrac{\Sigma f(X - \overline{X})^2}{n - 1}$ $\qquad$ (51)

$s^2 = \dfrac{\Sigma fX^2 - (\Sigma fX)^2 / n}{n - 1}$ $\qquad$ (52)

of a population, ungrouped data $\qquad$ $\sigma^2 = \dfrac{\Sigma (X - \mu)^2}{N}$ $\qquad$ (109)

$\sigma^2 = \dfrac{\Sigma X^2 - (\Sigma X)^2 / N}{N}$

of a population, grouped data	$\sigma^2 = \dfrac{\Sigma fX^2 - (\Sigma fX)^2/N}{N}$	(134)
of a probability distribution	$\sigma^2 = \Sigma(X - \mu)^2 p(X)$	(83)
of a binomial distribution	$\sigma^2 = npq$	(93)

Standard deviation

of a sample	$s = \sqrt{s^2}$	(47)
of a population	$\sigma = \sqrt{\sigma^2}$	(109)

Skewness	$\text{Skewness} = \dfrac{3(\overline{X} - Md)}{s}$	(54)

Probability

if outcomes are equally likely	$p(A) = \dfrac{s}{n}$	(60)
	$p(A \text{ or } B)$ $= p(A) + p(B) - p(A \text{ and } B)$	(71)
if A and B are mutually exclusive events	$p(A \text{ or } B) = p(A) + p(B)$	(69)
if A and B are independent	$p(A \text{ and } B) = p(A)p(B)$	(74)
if A and B are not independent	$p(A \text{ and } B) = p(A)p(B/A)$	(74)

Slope of a straight line	$\text{slope} = \dfrac{\text{change in } Y \text{ values}}{\text{change in } X \text{ values}}$	(266)
Straight line	$Y = a + bX$	(269)
Regression equation	$Y_c = a + bX$	(272)
	$b = \dfrac{n\Sigma XY - (\Sigma X)(\Sigma Y)}{n\Sigma X^2 - (\Sigma X)^2}$	(273)
	$a = \dfrac{\Sigma Y}{n} - b\dfrac{\Sigma X}{n}$	(273)
Standard error of estimate	$s_e = \sqrt{\dfrac{\Sigma(Y - Y_c)^2}{n - 2}}$	(278)
	$\Sigma(Y - Y_c)^2 = \Sigma Y^2 - a\Sigma Y - b\Sigma XY$	
Error variation	$\Sigma(Y - Y_c)^2$	(279)
Explained variation	$(\Sigma Y_c - \overline{Y})^2$	(279)
Total variation	$\Sigma(Y - \overline{Y})^2$	(280)
Coefficient of determination	$r^2 = \dfrac{\text{Explained variation}}{\text{Total variation}}$	(281)

Coefficient of (linear) correlation
$$r = (\text{sign of } b) \sqrt{r^2} \tag{281}$$

$$r = \frac{n\Sigma XY - (\Sigma X)(\Sigma Y)}{\sqrt{n\Sigma X^2 - (\Sigma X)^2}\ \sqrt{n\Sigma Y^2 - (\Sigma Y)^2}} \tag{282}$$

$$r = b\frac{s_X}{s_Y} \tag{283}$$

Rough estimate
$$r = \pm\left(1 - \frac{W}{L}\right) \tag{277}$$

Coefficient of correlation between ranks
(Spearman rank correlation coefficient)
$$r_s = 1 - \frac{6\Sigma d^2}{n(n^2 - 1)} \tag{318}$$

Formulas for Statistical Inference

Inference about	Formula	Standard error†	
Mean			
If σ is known **and** population is normal or $n \geq 30$	$z = \dfrac{\overline{X} - \mu}{\sigma_{\overline{X}}}$	$\sigma_{\overline{X}} = \dfrac{\sigma}{\sqrt{n}}$	(141)
If σ is unknown **and** population is normal	$t = \dfrac{\overline{X} - \mu}{s/\sqrt{n}}, \; D = n - 1$		(210)
Proportion			
If P is known	$z = \dfrac{p - P}{\sigma_p}$	$\sigma_p = \sqrt{\dfrac{PQ}{n}}$	(185)
If P is unknown	$z = \dfrac{p - P}{\sigma_p}$	$\sigma_p = \sqrt{\dfrac{pq}{n}}$	(153)
Differences of means, two independent samples			
If σ_X, σ_Y are known **and** both populations are normal or $n_X \geq 30$, $n_Y \geq 30$	$z = \dfrac{(\overline{X} - \overline{Y}) - (\mu_X - \mu_Y)}{\sigma_{(\overline{X} - \overline{Y})}}$	$\sigma_{(\overline{X} - \overline{Y})} = \sqrt{\sigma_X^2 + \sigma_Y^2}$	(157)
If σ_X, σ_Y are unknown but equal **and** both populations are normal	$\begin{cases} t = \dfrac{(\overline{X} - \overline{Y}) - (\mu_X - \mu_Y)}{\sigma_{(\overline{X} - \overline{Y})}} \\ D = n_X + n_Y - 2 \end{cases}$	$\sigma_{(\overline{X} - \overline{Y})} = \sqrt{\dfrac{(n_X - 1)s_X^2 + (n_Y - 1)s_Y^2}{n_X + n_Y - 2}\left(\dfrac{1}{n_X} + \dfrac{1}{n_Y}\right)}$	(211)

If σ_X, σ_Y are unknown and unequal **and** both populations are normal (213)

$$t = \frac{(\bar{X} - \bar{Y}) - (\mu_X - \mu_Y)}{\sigma_{(\bar{X}-\bar{Y})}}$$

$$D = \frac{(s_X^2/n_X + s_Y^2/n_Y)^2}{\left(\dfrac{s_X^2}{n_X}\right)^2 \dfrac{1}{n_X + 1} + \left(\dfrac{s_Y^2}{n_Y}\right)^2 \dfrac{1}{n_Y + 1}} - 2$$

$$\sigma_{(\bar{X}-\bar{Y})} = \sqrt{\frac{s_X^2}{n_X} + \frac{s_Y^2}{n_Y}}$$

Differences of means, two or more independent samples

If all populations are normal and have the same variance (251)

$$F = \frac{\hat{\sigma}^2 \text{ between}}{\hat{\sigma}^2 \text{ within}}$$

$$\hat{\sigma}^2 \text{ between} = \frac{A - B^2/n_T}{k - 1}, \quad \hat{\sigma}^2 \text{ within} = \frac{C - A}{n_T - k}$$

(See page 251 for A, B, and C.)

Which means are different? (255)

Compare $|\bar{X}_i - \bar{X}_j|$ and $S\sqrt{1/n_i + 1/n_j}$

(Scheffé's test)

$$S = \sqrt{(k - 1)F_\alpha(k - 1, \, n - k)(\hat{\sigma}^2 \text{ within})}$$

Difference of means, paired samples

If differences are normal (215)

$$t = \frac{d - \mu_d}{s_d}, \quad D = n - 1$$

Sign test (nonparametric) (313)

$$z = \frac{X \pm .5 - n/2}{\sqrt{n}/2}$$

Difference of proportions (162)

$$z = \frac{(p_X - p_Y) - (P_X - P_Y)}{\sigma_{(p_X - p_Y)}}$$

$$\sigma_{(p_X - p_Y)} = \sqrt{\sigma_{p_X}^2 + \sigma_{p_Y}^2}$$

†The factor $\sqrt{\dfrac{N - n}{N - 1}}$ has been omitted from certain formulas. See pages 135 and 150.

Inference about	Formula	Standard error	
Variances			
If both populations are normal	$F = \dfrac{s_X^{\,2}}{s_Y^{\,2}},\ D_X = n_X - 1,\ D_Y = n_Y - 1$		(243)
Goodness of fit			
$D > 1$	$\left\{\begin{array}{l} \chi_P^{\,2} = \Sigma\ \dfrac{(O-E)^2}{E} \\[2mm] D = (\text{number of } E \text{ entries}) - 1 \end{array}\right.$		(223)
$D = 1$	$\chi_P^{\,2} = \Sigma\ \dfrac{(|O-E| - .5)^2}{E}$		(235)
Are two variables independent?			
$D > 1$	$\left\{\begin{array}{l} \chi_P^{\,2} = \Sigma\ \dfrac{(O-E)^2}{E} \\[2mm] D = (n-1)(m-1) \end{array}\right.$		(230)
$D = 1$	$\chi_P^{\,2} = \Sigma\ \dfrac{(|O-E| - .5)^2}{E}$		(235)
Regression line			
If, for each X, Y scores in a population are normal with the same variance:			
Slope B of regression line	$t = \dfrac{b - B}{s_b},\ D = n - 2$	$s_b = \sqrt{\dfrac{n(\Sigma Y^2 - a\Sigma Y - b\Sigma XY)}{(n-2)[n\Sigma X^2 - (\Sigma X)^2]}}$	(296)
Mean of Y scores corresponding to X_0	$\left\{\begin{array}{l} t = \dfrac{Y_c - \mu_{Y/X_0}}{s_{Y/X_0}} \\[2mm] D = n - 2 \end{array}\right.$	$s_{Y/X_0} = s_e\sqrt{\dfrac{1}{n} + \dfrac{n(X_0 - \overline{X})^2}{n\Sigma X^2 - (\Sigma X)^2}}$ $s_e = \sqrt{\dfrac{\Sigma Y^2 - a\Sigma Y - b\Sigma XY}{n-2}}$	(299)
Prediction interval of Y for a given X_0 (**not** statistical inference)	$\left\{\begin{array}{l} t = \dfrac{Y_c - Y_p}{s_{Y_p}} \\[2mm] D = n - 2 \end{array}\right.$	$s_{Y_p} = s_e\sqrt{1 + \dfrac{1}{n} + \dfrac{n(X_0 - \overline{X})^2}{n\Sigma X^2 - (\Sigma X)^2}}$	(302)

Correlation

If X, Y are normally distributed (have a bivariate normal distribution; see page 000)

$H_0: \rho = 0$

$$t = \frac{r\sqrt{n-2}}{\sqrt{1-r^2}}, D = n - 2$$

(305)

$H_0: \rho = A, A \neq 0$

$$z = \frac{Z - \mu_z}{\sigma_z}$$

$$\sigma_z = \frac{1}{\sqrt{n-3}}$$

(306)

$H_0: \rho_S = 0$ (if $n > 10$)

$$t = \frac{r_S\sqrt{n-2}}{\sqrt{1-r_S^2}}, D = n - 2$$

(320)

Randomness of a sample

Runs test (nonparametric)

$$\mu_R = 1 + \frac{2n_X n_Y}{n_X + n_Y}$$

$$\sigma_R = \sqrt{\frac{2n_X n_Y(2n_X n_Y - n_X - n_Y)}{(n_X + n_Y)^2(n_X + n_Y - 1)}}$$

(314)

Are two populations the same?

Mann-Whitney test: nonparametric (if two populations have the same shape and variance, do they have the same median?)

$$U_X = n_X n_Y + \frac{n_X(n_X + 1)}{2} - \Sigma R_X$$
$$= n_X n_Y - U_Y$$

(316)

$$U_Y = n_X n_Y + \frac{n_Y(n_Y + 1)}{2} - \Sigma R_Y$$

$$= n_X n_Y - U_X$$

$U =$ smaller of U_X, U_Y

$$\mu_U = \frac{n_X n_Y}{2}$$

$$\sigma_U = \sqrt{\frac{n_X n_Y(n_X + n_Y + 1)}{12}}$$

(317)

appendix C

tables

TABLE 1 SQUARES AND SQUARE ROOTS

If n is between	SQUARE Multiply n² by	SQUARE ROOT Use column headed	Multiply by
.000100 and .000999	.00000001	$\sqrt{n}$	.01
.00100 and .00999	.000001	$\sqrt{10n}$	.01
.0100 and .0999	.0001	$\sqrt{n}$	.1
.100 and .999	.01	$\sqrt{10n}$	.1
1.00 and 9.99	1	$\sqrt{n}$	1
10.0 and 99.9	100	$\sqrt{10n}$	1
100 and 999	10,000	$\sqrt{n}$	10
1000 and 9,990	1,000,000	$\sqrt{10n}$	10
10,000 and 99,900	100,000,000	$\sqrt{n}$	100
100,000 and 999,000	10,000,000,000	$\sqrt{10n}$	100
1,100,000 and 9,990,000	10,000,000,000	$\sqrt{n}$	1000

n	n^2	$\sqrt{n}$	$\sqrt{10n}$	n	n^2	$\sqrt{n}$	$\sqrt{10n}$	n	n^2	$\sqrt{n}$	$\sqrt{10n}$
1.00	1.0000	1.000	3.162	1.25	1.5625	1.118	3.536	1.50	2.2500	1.225	3.873
1.01	1.0201	1.005	3.178	1.26	1.5876	1.122	3.550	1.51	2.2801	1.229	3.886
1.02	1.0404	1.010	3.194	1.27	1.6129	1.127	3.564	1.52	2.3104	1.233	3.899
1.03	1.0609	1.015	3.209	1.28	1.6384	1.131	3.578	1.53	2.3409	1.237	3.912
1.04	1.0816	1.020	3.225	1.29	1.6641	1.136	3.592	1.54	2.3716	1.241	3.924
1.05	1.1025	1.025	3.240	1.30	1.6900	1.140	3.606	1.55	2.4025	1.245	3.937
1.06	1.1236	1.030	3.256	1.31	1.7161	1.145	3.619	1.56	2.4336	1.249	3.950
1.07	1.1449	1.034	3.271	1.32	1.7424	1.149	3.633	1.57	2.4649	1.253	3.962
1.08	1.1664	1.039	3.286	1.33	1.7689	1.153	3.647	1.58	2.4964	1.257	3.975
1.09	1.1881	1.044	3.302	1.34	1.7956	1.158	3.661	1.59	2.5281	1.261	3.987
1.10	1.2100	1.049	3.317	1.35	1.8225	1.162	3.674	1.60	2.5600	1.266	4.000
1.11	1.2321	1.054	3.332	1.36	1.8496	1.166	3.688	1.61	2.5921	1.269	4.012
1.12	1.2544	1.058	3.347	1.37	1.8769	1.170	3.701	1.62	2.6244	1.273	4.025
1.13	1.2769	1.063	3.362	1.38	1.9044	1.175	3.715	1.63	2.6569	1.277	4.037
1.14	1.2996	1.068	3.376	1.39	1.9321	1.179	3.728	1.64	2.6896	1.281	4.050
1.15	1.3225	1.072	3.391	1.40	1.9600	1.183	3.742	1.65	2.7225	1.285	4.062
1.16	1.3456	1.077	3.406	1.41	1.9881	1.187	3.755	1.66	2.7556	1.288	4.074
1.17	1.3689	1.082	3.421	1.42	2.0164	1.192	3.768	1.67	2.7889	1.292	4.087
1.18	1.3924	1.086	3.435	1.43	2.0449	1.196	3.782	1.68	2.8224	1.296	4.099
1.19	1.4161	1.091	3.450	1.44	2.0736	1.200	3.795	1.69	2.8561	1.300	4.111
1.20	1.4400	1.095	3.464	1.45	2.1025	1.204	3.808	1.70	2.8900	1.304	4.123
1.21	1.4641	1.100	3.479	1.46	2.1316	1.208	3.821	1.71	2.9241	1.308	4.135
1.22	1.4884	1.105	3.493	1.47	2.1609	1.212	3.834	1.72	2.9584	1.311	4.147
1.23	1.5129	1.109	3.507	1.48	2.1904	1.217	3.847	1.73	2.9929	1.315	4.159
1.24	1.5376	1.114	3.521	1.49	2.2201	1.221	3.860	1.74	3.0276	1.319	4.171

TABLE 1 *Continued* SQUARES AND SQUARE ROOTS

n	n²	√n	√10n	n	n²	√n	√10n	n	n²	√n	√10n
1.75	3.0625	1.323	4.183	2.30	5.2900	1.517	4.796	2.85	8.1225	1.688	5.339
1.76	3.0976	1.327	4.195	2.31	5.3361	1.520	4.806	2.86	8.1796	1.691	5.348
1.77	3.1329	1.330	4.207	2.32	5.3824	1.523	4.817	2.87	8.2369	1.694	5.357
1.78	3.1684	1.334	4.219	2.33	5.4289	1.526	4.827	2.88	8.2944	1.697	5.367
1.79	3.2041	1.338	4.231	2.34	5.4756	1.530	4.837	2.89	8.3521	1.700	5.376
1.80	3.2400	1.342	4.243	2.35	5.5225	1.533	4.848	2.90	8.4100	1.703	5.385
1.81	3.2761	1.345	4.254	2.36	5.5696	1.536	4.858	2.91	8.4681	1.706	5.394
1.82	3.3124	1.349	4.266	2.37	5.6169	1.539	4.868	2.92	8.5264	1.709	5.404
1.83	3.3489	1.353	4.278	2.38	5.6644	1.543	4.879	2.93	8.5849	1.712	5.413
1.84	3.3856	1.356	4.290	2.39	5.7121	1.546	4.889	2.94	8.6436	1.715	5.422
1.85	3.4225	1.360	4.301	2.40	5.7600	1.549	4.899	2.95	8.7025	1.718	5.431
1.86	3.4596	1.364	4.313	2.41	5.8081	1.552	4.909	2.96	8.7616	1.720	5.441
1.87	3.4969	1.367	4.324	2.42	5.8564	1.556	4.919	2.97	8.8209	1.723	5.450
1.88	3.5344	1.371	4.336	2.43	5.9049	1.559	4.930	2.98	8.8804	1.726	5.459
1.89	3.5721	1.375	4.347	2.44	5.9536	1.562	4.990	2.99	8.9401	1.729	5.468
1.90	3.6100	1.378	4.359	2.45	6.0025	1.565	4.950	3.00	9.0000	1.732	5.477
1.91	3.6481	1.382	4.370	2.46	6.0516	1.568	4.960	3.01	9.0601	1.735	5.486
1.92	3.6864	1.386	4.382	2.47	6.1009	1.572	4.970	3.02	9.1204	1.738	5.495
1.93	3.7249	1.389	4.393	2.48	6.1504	1.575	4.980	3.03	9.1809	1.741	5.505
1.94	3.7636	1.393	4.405	2.49	6.2001	1.578	4.990	3.04	9.2416	1.744	5.514
1.95	3.8025	1.396	4.416	2.50	6.2500	1.581	5.000	3.05	9.3025	1.746	5.523
1.96	3.8416	1.400	4.427	2.51	6.3001	1.584	5.010	3.06	9.3636	1.749	5.532
1.97	3.8809	1.404	4.438	2.52	6.3504	1.587	5.020	3.07	9.4249	1.752	5.541
1.98	3.9204	1.407	4.450	2.53	6.4009	1.591	5.030	3.08	9.4864	1.755	5.550
1.99	3.9601	1.411	4.461	2.54	6.4516	1.594	5.040	3.09	9.5481	1.758	5.559
2.00	4.0000	1.414	4.472	2.55	6.5025	1.597	5.050	3.10	9.6100	1.761	5.568
2.01	4.0401	1.418	4.483	2.56	6.5536	1.600	5.060	3.11	9.6721	1.764	5.577
2.02	4.0804	1.421	4.494	2.57	6.6049	1.603	5.070	3.12	9.7344	1.766	5,586
2.03	4.1209	1.425	4.506	2.58	6.6564	1.606	5.079	3.13	9.7969	1.769	5.595
2.04	4.1616	1.428	4.517	2.59	6.7081	1.609	5.089	3.14	9.8596	1.772	5.604
2.05	4.2025	1.432	4.528	2.60	6.7600	1.612	5.099	3.15	9.9225	1.775	5.612
2.06	4.2436	1.435	4.539	2.61	6.8121	1.616	5.109	3.16	9.9856	1.778	5.621
2.07	4.2849	1.439	4.550	2.62	6.8644	1.619	5.119	3.17	10.0489	1.780	5.630
2.08	4.3264	1.442	4.561	2.63	6.9169	1.622	5.128	3.18	10.1124	1.783	5.639
2.09	4.3681	1.446	4.572	2.64	6.9696	1.625	5.138	3.19	10.1761	1.786	5.648
2.10	4.4100	1.449	4.583	2.65	7.0225	1.628	5.148	3.20	10.2400	1.789	5.657
2.11	4.4521	1.453	4.593	2.66	7.0756	1.631	5.158	3.21	10.3041	1.792	5.666
2.12	4.4944	1.456	4.604	2.67	7.1289	1.634	5.167	3.22	10.3684	1.794	5.674
2.13	4.5369	1.459	4.615	2.68	7.1824	1.637	5.177	3.23	10.4329	1.797	5.683
2.14	4.5796	1.463	4.626	2.69	7.2361	1.640	5.187	3.24	10.4976	1.800	5.692
2.15	4.6225	1.466	4.632	2.70	7.2900	1.643	5.196	3.25	10.5625	1.803	5.701
2.16	4.6656	1.470	4.648	2.71	7.3441	1.646	5.206	3.26	10.6276	1.806	5.710
2.17	4.7089	1.473	4.658	2.72	7.3984	1.649	5.215	3.27	10.6929	1.808	5.718
2.18	4.7524	1.476	4.669	2.73	7.4529	1.652	5.225	3.28	10.7584	1.811	5.727
2.19	4.7961	1.480	4.680	2.74	7.5076	1.655	5.235	3.29	10.8241	1.814	5.736
2.20	4.8400	1.483	4.690	2.75	7.5625	1.658	5.244	3.30	10.8900	1.817	5.745
2.21	4.8841	1.487	4.701	2.76	7.6176	1.661	5.254	3.31	10.9561	1.819	5.753
2.22	4.9284	1.490	4.712	2.77	7.6729	1.664	5.263	3.32	11.0224	1.822	5.762
2.23	4.9729	1.493	4.722	2.78	7.7284	1.667	5.273	3.33	11.0889	1.825	5.771
2.24	5.0176	1.497	4.733	2.79	7.7841	1.670	5.282	3.34	11.1556	1.828	5.779
2.25	5.0625	1.500	4.743	2.80	7.8400	1.673	5.292	3.35	11.2225	1.830	5.788
2.26	5.1076	1.503	4.754	2.81	7.8961	1.676	5.301	3.36	11.2896	1.833	5.797
2.27	5.1529	1.507	4.764	2.82	7.9524	1.679	5.310	3.37	11.3569	1.836	5.805
2.28	5.1984	1.510	4.775	2.83	8.0089	1.682	5.320	3.38	11.4244	1.838	5.814
2.29	5.2441	1.513	4.785	2.84	8.0656	1.685	5.329	3.39	11.4921	1.841	5.822

TABLE 1 *Continued* SQUARES AND SQUARE ROOTS

n	n^2	$\sqrt{n}$	$\sqrt{10n}$	n	n^2	$\sqrt{n}$	$\sqrt{10n}$	n	n^2	$\sqrt{n}$	$\sqrt{10n}$
3.40	11.5600	1.844	5.831	3.97	15.7609	1.992	6.301	4.54	20.6116	2.131	6.738
3.41	11.6281	1.847	5.840	3.98	15.8408	1.995	6.309	4.55	20.7025	2.133	6.745
3.42	11.6964	1.849	5.848	3.99	15.9201	1.997	6.317	4.56	20.7936	2.135	6.753
3.43	11.7649	1.852	5.857	4.00	16.0000	2.000	6.325	4.57	20.8849	2.138	6.760
3.44	11.8336	1.855	5.865	4.01	16.0801	2.002	6.332	4.58	20.9764	2.140	6.768
3.45	11.9025	1.857	5.874	4.02	16.1604	2.005	6.340	4.59	21.0681	2.142	6.775
3.46	11.9716	1.860	5.882	4.03	16.2409	2.007	6.348	4.60	21.1600	2.145	6.782
3.47	12.0409	1.863	5.891	4.04	16.3216	2.010	6.356	4.61	21.2521	2.147	6.790
3.48	12.1104	1.865	5.899	4.05	16.4025	2.012	6.364	4.62	21.3444	2.149	6.797
3.49	12.1801	1.868	5.908	4.06	16.4836	2.015	6.372	4.63	21.4369	2.152	6.804
3.50	12.2500	1.871	5.916	4.07	16.5649	2.017	6.380	4.64	21.5296	2.154	6.812
3.51	12.3201	1.874	5.925	4.08	16.6464	2.020	6.387	4.65	21.6225	2.156	6.819
3.52	12.3904	1.876	5.933	4.09	16.7281	2.022	6.395	4.66	21.7156	2.159	6.826
3.53	12.4609	1.879	5.941	4.10	16.8100	2.025	6.403	4.67	21.8089	2.161	6.834
3.54	12.5316	1.881	5.950	4.11	16.8921	2.027	6.411	4.68	21.9024	2.163	6.841
3.55	12.6025	1.884	5.958	4.12	16.9744	2.030	6.419	4.69	21.9961	2.166	6.848
3.56	12.6736	1.887	5.967	4.13	17.0569	2.032	6.427	4.70	22.0900	2.168	6.856
3.57	12.7449	1.889	5.975	4.14	17.1396	2.035	6.434	4.71	22.1841	2.170	6.863
3.58	12.8164	1.892	5.983	4.15	17.2225	2.037	6.442	4.72	22.2784	2.173	6.870
3.59	12.8881	1.895	5.992	4.16	17.3056	2.040	6.450	4.73	22.3729	2.175	6.878
3.60	12.9600	1.897	6.000	4.17	17.3889	2.042	6.458	4.74	22.4676	2.177	6.885
3.61	13.0321	1.900	6.008	4.18	17.4724	2.044	6.465	4.75	22.5625	2.179	6.892
3.62	13.1044	1.903	6.017	4.19	17.5561	2.047	6.473	4.76	22.6576	2.182	6.899
3.63	13.1769	1.905	6.025	4.20	17.6400	2.049	6.481	4.77	22.7529	2.184	6.907
3.64	13.2496	1.908	6.033	4.21	17.7241	2.052	6.488	4.78	22.8484	2.186	6.914
3.65	13.3225	1.911	6.042	4.22	17.8084	2.054	6.496	4.79	22.9441	2.189	6.921
3.66	13.3956	1.913	6.050	4.23	17.8929	2.057	6.504	4.80	23.0400	2.191	6.928
3.67	13.4689	1.916	6.058	4.24	17.9776	2.059	6.512	4.81	23.1361	2.193	6.935
3.68	13.5424	1.918	6.066	4.25	18.0625	2.062	6.519	4.82	23.2324	2.195	6.943
3.69	13.6161	1.921	6.075	4.26	18.1476	2.064	6.527	4.83	23.3289	2.198	6.950
3.70	13.6900	1.924	6.083	4.27	18.2329	2.066	6.535	4.84	23.4256	2.200	6.957
3.71	13.7641	1.926	6.091	4.28	18.3184	2.069	6.542	4.85	23.5225	2.202	6.964
3.72	13.8384	1.929	6.099	4.29	18.4041	2.071	6.550	4.86	23.6196	2.205	6.971
3.73	13.9129	1.931	6.107	4.30	18.4900	2.074	6.557	4.87	23.7169	2.207	6.979
3.74	13.9876	1.934	6.116	4.31	18.5761	2.076	6.565	4.88	23.8144	2.209	6.986
3.75	14.0625	1.936	6.124	4.32	18.6624	2.078	6.573	4.89	23.9121	2.211	6.993
3.76	14.1376	1.939	6.132	4.33	18.7489	2.081	6.580	4.90	24.0100	2.214	7.000
3.77	14.2129	1.942	6.140	4.34	18.8356	2.083	6.588	4.91	24.1081	2.216	7.007
3.78	14.2884	1.944	6.148	4.35	18.9225	2.086	6.595	4.92	24.2064	2.218	7.014
3.79	14.3641	1.947	6.156	4.36	19.0096	2.088	6.603	4.93	24.3049	2.220	7.021
3.80	14.4400	1.949	6.164	4.37	19.0969	2.090	6.611	4.94	24.4036	2.223	7.029
3.81	14.5161	1.952	6.173	4.38	19.1844	2.093	6.618	4.95	24.5025	2.225	7.036
3.82	14.5924	1.954	6.181	4.39	19.2721	2.095	6.626	4.96	24.6016	2.227	7.043
3.83	14.6689	1.957	6.189	4.40	19.3600	2.098	6.633	4.97	24.7009	2.229	7.050
3.84	14.7456	1.960	6.197	4.41	19.4481	2.100	6.641	4.98	24.8004	2.232	7.057
3.85	14.8225	1.962	6.205	4.42	19.5364	2.102	6.648	4.99	24.9001	2.234	7.064
3.86	14.8996	1.965	6.213	4.43	19.6249	2.105	6.656	5.00	25.0000	2.236	7.071
3.87	14.9769	1.967	6.221	4.44	19.7136	2.107	6.663	5.01	25.1001	2.238	7.078
3.88	15.0544	1.970	6.229	4.45	19.8025	2.110	6.671	5.02	25.2004	2.241	7.085
3.89	15.1321	1.972	6.237	4.46	19.8916	2.112	6.678	5.03	25.3009	2.243	7.092
3.90	15.2100	1.975	6.245	4.47	19.9809	2.114	6.686	5.04	25.4016	2.245	7.099
3.91	15.2881	1.977	6.253	4.48	20.0704	2.117	6.693	5.05	25.5025	2.247	7.106
3.92	15.3664	1.980	6.261	4.49	20.1601	2.119	6.701	5.06	25.6036	2.249	7.113
3.93	15.4449	1.982	6.269	4.50	20.2500	2.121	6.708	5.07	25.7049	2.252	7.120
3.94	15.5236	1.985	6.277	4.51	20.3401	2.124	6.716	5.08	25.8064	2.254	7.127
3.95	15.6025	1.987	6.285	4.52	20.4304	2.126	6.723	5.09	25.9081	2.256	7.134
3.96	15.6816	1.990	6.293	4.53	20.5209	2.128	6.731	5.10	26.0100	2.258	7.141

TABLE 1 *Continued* SQUARES AND SQUARE ROOTS

n	n²	√n	√10n	n	n²	√n	√10n	n	n²	√n	√10n
5.11	26.1121	2.261	7.148	5.69	32.3761	2.385	7.543	6.27	39.3129	2.504	7.918
5.12	26.2144	2.263	7.155	5.70	32.4900	2.387	7.550	6.28	39.4384	2.506	7.925
5.13	26.3169	2.265	7.162	5.71	32.6041	2.391	7.556	6.29	39.5641	2.508	7.931
5.14	26.4196	2.267	7.169	5.72	32.7184	2.392	7.563	6.30	39.6900	2.510	7.937
5.15	26.5225	2.269	7.176	5.73	32.8329	2.394	7.570	6.31	39.8161	2.512	7.944
5.16	26.6256	2.272	7.183	5.74	32.9476	2.396	7.576	6.32	39.9424	2.514	7.950
5.17	26.7289	2.274	7.190	5.75	33.0625	2.398	7.583	6.33	40.0689	2.516	7.956
5.18	26.8324	2.276	7.197	5.76	33.1776	2.400	7.589	6.34	40.1956	2.518	7.962
5.19	26.9361	2.278	7.204	5.77	33.2929	2.402	7.596	6.35	40.3225	2.520	7.969
5.20	27.0400	2.280	7.211	5.78	33.4084	2.404	7.603	6.36	40.4496	2.522	7.975
5.21	27.1441	2.283	7.218	5.79	33.5241	2.406	7.609	6.37	40.5769	2.524	7.981
5.22	27.2484	2.285	7.225	5.80	33.6400	2.408	7.616	6.38	40.7044	2.526	7.987
5.23	27.3529	2.287	7.232	5.81	33.7561	2.410	7.622	6.39	40.8321	2.528	7.994
5.24	27.4576	2.289	7.239	5.82	33.8724	2.412	7.629	6.40	40.9600	2.530	8.000
5.25	27.5625	2.291	7.246	5.83	33.9889	2.415	7.635	6.41	41.0881	2.532	8.006
5.26	27.6676	2.293	7.253	5.84	34.1056	2.417	7.642	6.42	41.2164	2.534	8.013
5.27	27.7729	2.296	7.259	5.85	34.2225	2.419	7.648	6.43	41.3449	2.536	8.019
5.28	27.8784	2.298	7.266	5.86	34.3396	2.421	7.655	6.44	41.4736	2.538	8.025
5.29	27.9841	2.300	7.273	5.87	34.4569	2.423	7.662	6.45	41.6025	2.540	8.031
5.30	28.0900	2.302	7.280	5.88	34.5744	2.425	7.668	6.46	41.7316	2.542	8.037
5.31	28.1961	2.304	7.287	5.89	34.6921	2.427	7.675	6.47	41.8609	2.544	8.044
5.32	28.3024	2.307	7.294	5.90	34.8100	2.429	7.681	6.48	41.9904	2.546	8.050
5.33	28.4089	2.309	7.301	5.91	34.9281	2.431	7.688	6.49	42.1201	2.548	8.056
5.34	28.5156	2.311	7.308	5.92	35.0464	2.433	7.694	6.50	42.2500	2.550	8.062
5.35	28.6225	2.313	7.314	5.93	35.1649	2.435	7.701	6.51	42.3801	2.551	8.068
5.36	28.7296	2.315	7.321	5.94	35.2836	2.437	7.707	6.52	42.5104	2.553	8.075
5.37	28.8369	2.317	7.328	5.95	35.4025	2.439	7.714	6.53	42.6409	2.555	8.081
5.38	28.9444	2.319	7.335	5.96	35.5216	2.441	7.720	6.54	42.7716	2.557	8.087
5.39	29.0521	2.322	7.342	5.97	35.6409	2.443	7.727	6.55	42.9025	2.559	8.093
5.40	29.1600	2.324	7.348	5.98	35.7604	2.445	7.733	6.56	43.0336	2.561	8.099
5.41	29.2681	2.326	7.355	5.99	35.8801	2.447	7.740	6.57	43.1649	2.563	8.106
5.42	29.3764	2.328	7.362	6.00	36.0000	2.449	7.746	6.58	43.2964	2.565	8.112
5.43	29.4849	2.330	7.369	6.01	36.1201	2.452	7.752	6.59	43.4281	2.567	8.118
5.44	29.5936	2.332	7.376	6.02	36.2404	2.454	7.759	6.60	43.5600	2.569	8.124
5.45	29.7025	2.334	7.382	6.03	36.3609	2.456	7.765	6.61	43.6921	2.571	8.130
5.46	29.8116	2.337	7.389	6.04	36.4816	2.458	7.772	6.62	43.8244	2.573	8.136
5.47	29.9209	2.339	7.396	6.05	36.6025	2.460	7.778	6.63	43.9569	2.575	8.142
5.48	30.0304	2.341	7.403	6.06	36.7236	2.462	7.785	6.64	44.0896	2.577	8.149
5.49	30.1401	2.343	7.408	6.07	36.8449	2.464	7.791	6.65	44.2225	2.579	8.155
5.50	30.2500	2.345	7.416	6.08	36.9664	2.466	7.797	6.66	44.3556	2.581	8.161
5.51	30.3601	2.347	7.423	6.09	37.0881	2.468	7.804	6.67	44.4889	2.583	8.167
5.52	30.4704	2.349	7.430	6.10	37.2100	2.470	7.810	6.68	44.6224	2.585	8.173
5.53	30.5809	2.352	7.436	6.11	37.3321	2.472	7.817	6.69	44.7561	2.587	8.179
5.54	30.6916	2.354	7.443	6.12	37.4544	2.474	7.823	6.70	44.8900	2.588	8.185
5.55	30.8025	2.356	7.450	6.13	37.5769	2.476	7.829	6.71	45.0241	2.590	8.191
5.56	30.9136	2.358	7.457	6.14	37.6996	2.478	7.836	6.72	45.1584	2.592	8.198
5.57	31.0249	2.360	7.463	6.15	37.8225	2.480	7.842	6.73	45.2929	2.594	8.204
5.58	31.1364	2.362	7.470	6.16	37.9456	2.482	7.849	6.74	45.4276	2.596	8.210
5.59	31.2481	2.364	7.477	6.17	38.0689	2.484	7.853	6.75	45.5625	2.598	8.216
5.60	31.3600	2.366	7.483	6.18	38.1924	2.486	7.861	6.76	45.6976	2.600	8.222
5.61	31.4721	2.369	7.490	6.19	38.3161	2.488	7.868	6.77	45.8329	2.602	8.228
5.62	31.5844	2.371	7.497	6.20	38.4400	2.490	7.874	6.78	45.9684	2.604	8.234
5.63	31.6969	2.373	7.503	6.21	38.5641	2.492	7.880	6.79	46.1041	2.606	8.240
5.64	31.8096	2.375	7.510	6.22	38.6884	2.494	7.887	6.80	46.2400	2.608	8.246
5.65	31.9225	2.377	7.517	6.23	38.8129	2.496	7.893	6.81	46.3761	2.610	8.252
5.66	32.0356	2.379	7.523	6.24	38.9376	2.498	7.899	6.82	46.5124	2.612	8.258
5.67	32.1489	2.381	7.530	6.25	39.0625	2.500	7.906	6.83	46.6489	2.613	8.264
5.68	32.2624	2.383	7.537	6.26	39.1876	2.502	7.912	6.84	46.7856	2.615	8.270

TABLE 1 *Continued* SQUARES AND SQUARE ROOTS

n	n²	√n	√10n	n	n²	√n	√10n	n	n²	√n	√10n
6.85	46.9225	2.617	8.276	7.43	55.2049	2.726	8.620	8.01	64.1601	2.830	8.950
6.86	47.0596	2.619	8.283	7.44	55.3536	2.728	8.626	8.02	64.3204	2.832	8.955
6.87	47.1969	2.621	8.289	7.45	55.5025	2.729	8.631	8.03	64.4809	2.834	8.961
6.88	47.3344	2.623	8.295	7.46	55.6516	2.731	8.637	8.04	64.6416	2.835	8.967
6.89	47.4721	2.625	8.301	7.47	55.8009	2.733	8.643	8.05	64.8025	2.837	8.972
6.90	47.6100	2.627	8.307	7.48	55.9504	2.735	8.649	8.06	64.9636	2.839	8.978
6.91	47.7481	2.629	8.313	7.49	56.1001	2.737	8.654	8.07	65.1249	2.841	8.983
6.92	47.8864	2.631	8.319	7.50	56.2500	2.739	8.660	8.08	65.2864	2.843	8.989
6.93	48.0249	2.632	8.325	7.51	56.4001	2.740	8.666	8.09	65.4481	2.844	8.994
6.94	48.1636	2.634	8.331	7.52	56.5504	2.742	8.672	8.10	65.6100	2.846	9.000
6.95	48.3025	2.636	8.337	7.53	56.7009	2.744	8.678	8.11	65.7721	2.848	9.006
6.96	48.4416	2.638	8.343	7.54	56.8516	2.746	8.688	8.12	65.9344	2.850	9.011
6.97	48.5809	2.640	8.349	7.55	57.0025	2.748	8.689	8.13	66.0969	2.851	9.017
6.98	48.7204	2.642	8.355	7.56	57.1536	2.750	8.695	8.14	66.2596	2.853	9.022
6.99	48.8601	2.644	8.361	7.57	57.3049	2.751	8.701	8.15	66.4225	2.855	9.028
7.00	49.0000	2.646	8.367	7.58	57.4564	2.753	8.706	8.16	66.5856	2.857	9.033
7.01	49.1401	2.648	8.373	7.59	57.6081	2.755	8.712	8.17	66.7489	2.858	9.039
7.02	49.2804	2.650	8.379	7.60	57.7600	2.757	8.718	8.18	66.9124	2.860	9.044
7.03	49.4209	2.651	8.385	7.61	57.9121	2.759	8.724	8.19	67.0761	2.862	9.050
7.04	49.5616	2.653	8.390	7.62	58.0644	2.760	8.729	8.20	67.2400	2.864	9.055
7.05	49.7025	2.655	8.396	7.63	58.2169	2.762	8.735	8.21	67.4041	2.865	9.061
7.06	49.8436	2.657	8.402	7.64	58.3696	2.764	8.741	8.22	67.5684	2.867	9.066
7.07	49.9849	2.659	8.408	7.65	58.5225	2.766	8.746	8.23	67.7329	2.869	9.072
7.08	50.1264	2.661	8.414	7.66	58.6756	2.768	8.752	8.24	67.8976	2.871	9.077
7.09	50.2681	2.663	8.420	7.67	58.8289	2.769	8.758	8.25	68.0625	2.872	9.083
7.10	50.4100	2.665	8.426	7.68	58.9824	2.771	8.764	8.26	68.2276	2.874	9.088
7.11	50.5521	2.666	8.432	7.69	59.1361	2.773	8.769	8.27	68.3929	2.876	9.094
7.12	50.6944	2.668	8.438	7.70	59.2900	2.775	8.775	8.28	68.5584	2.878	9.010
7.13	50.8369	2.670	8.444	7.71	59.4441	2.777	8.781	8.29	68.7241	2.879	9.105
7.14	50.9796	2.672	8.450	7.72	59.5984	2.778	8.786	8.30	68.8900	2.881	9.110
7.15	51.1225	2.674	8.456	7.73	59.7529	2.780	8.792	8.31	69.0561	2.883	9.116
7.16	51.2656	2.676	8.462	7.74	59.9076	2.782	8.798	8.32	69.2224	2.884	9.121
7.17	51.4089	2.678	8.468	7.75	60.0625	2.784	8.803	8.33	69.3889	2.886	9.127
7.18	51.5524	2.680	8.473	7.76	60.2176	2.786	8.809	8.34	69.5556	2.888	9.132
7.19	51.6961	2.681	8.479	7.77	60.3729	2.787	8.815	8.35	69.7225	2.890	9.138
7.20	51.8400	2.683	8.485	7.78	60.5284	2.789	8.820	8.36	69.8896	2.891	9.143
7.21	51.9841	2.685	8.491	7.79	60.6841	2.791	8.826	8.37	70.0569	2.893	9.149
7.22	52.1284	2.687	8.497	7.80	60.8400	2.793	8.832	8.38	70.2244	2.895	9.154
7.23	52.2729	2.689	8.503	7.81	60.9961	2.795	8.837	8.39	70.3921	2.897	9.160
7.24	52.4176	2.691	8.509	7.82	61.1524	2.796	8.843	8.40	70.5600	2.898	9.165
7.25	52.5625	2.693	8.515	7.83		2.798	8.849	8.41	70.7281	2.900	9.167
7.26	52.7076	2.694	8.521	7.84	61.4656	2.800	8.854	8.42	70.8964	2.902	9.176
7.27	52.8529	2.696	8.526	7.85	61.6225	2.802	8.860	8.43	71.0649	2.903	9.182
7.28	52.9984	2.698	8.532	7.86	61.7796	2.804	8.866	8.44	71.2336	2.905	9.187
7.29	53.1441	2.700	8.538	7.87	61.9369	2.805	8.871	8.45	71.4025	2.907	9.192
7.30	53.2900	2.702	8.544	7.88	62.0944	2.807	8.877	8.46	71.5716	2.909	9.198
7.31	53.4361	2.704	8.550	7.89	62.2521	2.809	8.883	8.47	71.7409	2.910	9.203
7.32	53.5824	2.706	8.556	7.90	62.4100	2.811	8.888	8.48	71.9104	2.912	9.209
7.33	53.7289	2.707	8.562	7.91	62.5681	2.812	8.894	8.49	72.0801	2.914	9.214
7.34	53.8756	2.709	8.567	7.92	62.7264	2.814	8.899	8.50	72.2500	2.915	9.220
7.35	54.0225	2.711	8.573	7.93	62.8849	2.816	8.905	8.51	72.4201	2.917	9.225
7.36	54.1696	2.713	8.579	7.94	63.0436	2.818	8.911	8.52	72.5904	2.919	9.230
7.37	54.3169	2.715	8.585	7.95	63.2025	2.820	8.916	8.53	72.7609	2.921	9.236
7.38	54.4644	2.717	8.591	7.96	63.3616	2.821	8.922	8.54	72.9316	2.922	9.241
7.39	54.6121	2.718	8.597	7.97	63.5209	2.823	8.927	8.55	73.1025	2.924	9.247
7.40	54.7600	2.720	8.602	7.98	63.6804	2.825	8.933	8.56	73.2736	2.926	9.252
7.41	54.9081	2.722	8.608	7.99	63.8401	2.827	8.939	8.57	73.4449	2.927	9.257
7.42	55.0564	2.724	8.614	8.00	64.0000	2.828	8.944	8.58	73.6164	2.929	9.263

TABLE 1 *Continued* SQUARES AND SQUARE ROOTS

n	n²	$\sqrt{n}$	$\sqrt{10n}$	n	n²	$\sqrt{n}$	$\sqrt{10n}$	n	n²	$\sqrt{n}$	$\sqrt{10n}$
8.59	73.7881	2.931	9.268	9.11	82.9921	3.018	9.545	9.63	92.7369	3.103	9.813
8.60	73.9600	2.933	9.274	9.12	83.1744	3.020	9.550	9.64	92.9296	3.105	9.818
8.61	74.1321	2.934	9.279	9.13	83.3569	3.022	9.555	9.65	93.1225	3.106	9.823
8.62	74.3044	2.936	9.284	9.14	83.5396	3.023	9.560	9.66	93.3156	3.108	9.829
8.63	74.4769	2.938	9.290	9.15	83.7225	3.025	9.566	9.67	93.5089	3.110	9.834
8.64	74.6496	2.939	9.295	9.16	83.9056	3.027	9.571	9.68	93.7024	3.111	9.839
8.65	74.8225	2.941	9.301	9.17	84.0889	3.028	9.576	9.69	93.8961	3.113	9.844
8.66	74.9956	2.943	9.306	9.18	84.2724	3.030	9.581	9.70	94.0900	3.114	9.849
8.67	75.1689	2.944	9.311	9.19	84.4561	3.032	9.586	9.71	94.2841	3.116	9.854
8.68	75.3424	2.946	9.317	9.20	84.6400	3.033	9.592	9.72	94.4784	3.118	9.859
8.69	75.5161	2.948	9.322	9.21	84.8241	3.035	9.597	9.73	94.6729	3.119	9.864
8.70	75.6900	2.950	9.327	9.22	85.0084	3.036	9.602	9.74	94.8676	3.121	9.869
8.71	75.8641	2.951	9.333	9.23	85.1929	3.038	9.607	9.75	95.0625	3.123	9.874
8.72	76.0384	2.953	9.338	9.24	85.3776	3.040	9.612	9.76	95.2576	3.124	9.879
8.73	76.2129	2.955	9.343	9.25	85.5625	3.041	9.618	9.77	95.4529	3.126	9.884
8.74	76.3876	2.956	9.349	9.26	85.7476	3.043	9.623	9.78	95.6484	3.127	9.889
8.75	76.5625	2.958	9.354	9.27	85.9329	3.045	9.628	9.79	95.8441	3.129	9.894
8.76	76.7376	2.960	9.359	9.28	86.1184	3.046	9.633	9.80	96.0400	3.131	9.899
8.77	76.9129	2.961	9.365	9.29	86.3041	3.048	9.638	9.81	96.2361	3.132	9.905
8.78	77.0884	2.963	9.370	9.30	86.4900	3.050	9.644	9.82	96.4324	3.134	9.910
8.79	77.2641	2.965	9.376	9.31	86.6761	3.051	9.649	9.83	96.6289	3.135	9.915
8.80	77.4400	2.966	9.381	9.32	86.8624	3.053	9.654	9.84	96.8256	3.137	9.920
8.81	77.6161	2.968	9.386	9.33	87.0489	3.055	9.659	9.85	97.0225	3.138	9.925
8.82	77.7924	2.970	9.391	9.34	87.2356	3.056	9.664	9.86	97.2196	3.140	9.930
8.83	77.9689	2.972	9.397	9.35	87.4225	3.058	9.670	9.87	97.4169	3.142	9.935
8.84	78.1456	2.973	9.402	9.36	87.6096	3.059	9.675	9.88	97.6144	3.143	9.940
8.85	78.3225	2.975	9.407	9.37	87.7969	3.061	9.680	9.89	97.8121	3.145	9.945
8.86	78.4996	2.977	9.413	9.38	87.9844	3.063	9.685	9.90	98.0100	3.146	9.950
8.87	78.6769	2.978	9.418	9.39	88.1721	3.064	9.690	9.91	98.2081	3.148	9.955
8.88	78.8544	2.980	9.423	9.40	88.3600	3.066	9.695	9.92	98.4064	3.150	9.960
8.89	79.0321	2.982	9.429	9.41	88.5481	3.068	9.701	9.93	98.6049	3.151	9.965
8.90	79.2100	2.983	9.434	9.42	88.7364	3.069	9.706	9.94	98.8036	3.153	9.970
8.91	79.3881	2.985	9.439	9.43	88.9249	3.071	9.711	9.95	99.0025	3.154	9.975
8.92	79.5664	2.987	9.445	9.44	89.1136	3.072	9.716	9.96	99.2016	3.156	9.980
8.93	79.7449	2.988	9.450	9.45	89.3025	3.074	9.721	9.97	99.4009	3.157	9.985
8.94	79.9236	2.990	9.455	9.46	89.4916	3.076	9.726	9.98	99.6004	3.159	9.990
8.95	80.1025	2.992	9.460	9.47	89.6809	3.077	9.731	9.99	99.8001	3.161	9.995
8.96	80.2816	2.993	9.466	9.48	89.8704	3.079	9.737				
8.97	80.4609	2.995	9.471	9.49	90.0601	3.081	9.742				
8.98	80.6404	2.997	9.476	9.50	90.2500	3.082	9.747				
8.99	80.8201	2.998	9.482	9.51	90.4401	3.084	9.752				
9.00	81.0000	3.000	9.487	9.52	90.6304	3.085	9.757				
9.01	81.1801	3.001	9.492	9.53	90.8209	3.087	9.762				
9.02	81.3604	3.003	9.497	9.54	91.0116	3.089	9.767				
9.03	81.5409	3.005	9.503	9.55	91.2025	3.090	9.772				
9.04	81.7216	3.007	9.508	9.56	91.3936	3.092	9.778				
9.05	81.9025	3.008	9.513	9.57	91.5849	3.093	9.783				
9.06	82.0836	3.010	9.518	9.58	91.7764	3.095	9.788				
9.07	82.2649	3.012	9.524	9.59	91.9681	3.097	9.793				
9.08	82.4464	3.013	9.529	9.60	92.1600	3.098	9.798				
9.09	82.6281	3.015	9.534	9.61	92.3521	3.100	9.803				
9.10	82.8100	3.017	9.539	9.62	92.5444	3.102	9.808				

TABLE 2 BINOMIAL PROBABILITIES

$$_nC_X p^X q^{n-X}$$

n	X	.01	.05	.10	.20	.30	.40	.50	.60	.70	.80	.90	.95	.99
2	0	980	902	810	640	490	360	250	160	090	040	010	002	0+
	1	020	095	180	320	420	480	500	480	420	320	180	095	020
	2	0+	002	010	040	090	160	250	360	490	640	810	902	980
3	0	970	857	729	512	343	216	125	.064	027	008	001	0+	0+
	1	029	135	243	384	441	432	375	288	189	096	027	007	0+
	2	0+	007	027	096	189	288	375	432	441	384	243	135	029
	3	0+	0+	001	008	027	064	125	216	343	512	729	857	970
4	0	961	815	656	410	240	130	062	026	008	002	0+	0+	0+
	1	039	171	292	410	412	346	250	154	076	026	004	0+	0+
	2	001	014	049	154	265	346	375	346	265	154	049	014	001
	3	0+	0+	004	026	076	154	250	346	412	410	292	171	039
	4	0+	0+	0+	002	008	026	062	130	240	410	656	815	961
5	0	951	774	590	328	168	078	031	010	002	0+	0+	0+	0+
	1	048	204	328	410	360	259	156	077	028	006	0+	0+	0+
	2	001	021	073	205	309	346	312	230	132	051	008	001	0+
	3	0+	001	008	051	132	230	312	346	309	205	073	021	001
	4	0+	0+	0+	006	028	077	156	259	360	410	328	204	048
	5	0+	0+	0+	0+	002	010	031	078	168	328	590	774	951
6	0	941	735	531	262	118	047	016	004	001	0+	0+	0+	0+
	1	057	232	354	393	303	187	094	037	010	002	0+	0+	0+
	2	001	031	098	246	324	311	234	138	060	015	001	0+	0+
	3	0+	002	015	082	185	276	312	276	185	082	015	002	0+
	4	0+	0+	001	015	060	138	234	311	324	246	098	031	001
	5	0+	0+	0+	002	010	037	094	187	303	393	354	232	057
	6	0+	0+	0+	0+	001	004	016	047	118	262	531	735	941
7	0	932	698	478	210	082	028	008	002	0+	0+	0+	0+	0+
	1	066	257	372	367	247	131	055	017	004	0+	0+	0+	0+
	2	002	041	124	275	318	261	164	077	025	004	0+	0+	0+
	3	0+	004	023	115	227	290	273	194	097	029	003	0+	0+
	4	0+	0+	003	029	097	194	273	290	227	115	023	004	0+
	5	0+	0+	0+	004	025	077	164	261	318	275	124	041	002
	6	0+	0+	0+	0+	004	017	055	131	247	367	372	257	066
	7	0+	0+	0+	0+	0+	002	008	028	082	210	478	698	932
8	0	923	663	430	168	058	017	004	001	0+	0+	0+	0+	0+
	1	075	279	383	336	198	090	031	008	001	0+	0+	0+	0+
	2	003	051	149	294	296	209	109	041	010	001	0+	0+	0+
	3	0+	005	033	147	254	279	219	124	047	009	0+	0+	0+
	4	0+	0+	005	046	136	232	273	232	136	046	005	0+	0+
	5	0+	0+	0+	009	047	124	219	279	254	147	033	005	0+
	6	0+	0+	0+	001	010	041	109	209	296	294	149	051	003
	7	0+	0+	0+	0+	001	008	031	090	198	336	383	279	075
	8	0+	0+	0+	0+	0+	001	004	017	058	168	430	663	923
9	0	914	630	387	134	040	010	002	0+	0+	0+	0+	0+	0+
	1	083	299	387	302	156	060	018	004	0+	0+	0+	0+	0+
	2	003	063	172	302	267	161	070	021	004	0+	0+	0+	0+
	3	0+	008	045	176	267	251	164	074	021	003	0+	0+	0+
	4	0+	001	007	066	172	251	246	167	074	017	001	0+	0+
	5	0+	0+	001	017	074	167	246	251	172	066	007	001	0+
	6	0+	0+	0+	003	021	074	164	251	267	176	045	008	0+
	7	0+	0+	0+	0+	004	021	070	161	267	302	172	063	003

344

TABLE 2 *Continued* BINOMIAL PROBABILITIES

n	X	.01	.05	.10	.20	.30	.40	p .50	.60	.70	.80	.90	.95	.99
9	8	0+	0+	0+	0+	0+	004	018	060	156	302	387	299	083
	9	0+	0+	0+	0+	0+	0+	002	010	040 *	134	387	630	914
10	0	904	599	349	107	028	006	001	0+	0+	0+	0+	0+	0+
	1	091	315	387	268	121	040	010	002	0+	0+	0+	0+	0+
	2	004	075	194	302	233	121	044	011	001	0+	0+	0+	0+
	3	0+	010	057	201	267	215	117	042	009	001	0+	0+	0+
	4	0+	001	011	088	200	251	205	111	037	006	0+	0+	0+
	5	0+	0+	001	026	103	201	246	201	103	026	001	0+	0+
	6	0+	0+	0+	006	037	111	205	251	200	088	011	001	0+
	7	0+	0+	0+	001	009	042	117	215	267	201	057	010	0+
	8	0+	0+	0+	0+	001	011	044	121	233	302	194	075	004
	9	0+	0+	0+	0+	0+	002	010	040	121	268	387	315	091
	10	0+	0+	0+	0+	0+	0+	001	006	028	107	349	599	904
11	0	895	569	314	086	020	004	0+	0+	0+	0+	0+	0+	0+
	1	099	329	384	236	093	027	005	001	0+	0+	0+	0+	0+
	2	005	087	213	295	200	089	027	005	001	0+	0+	0+	0+
	3	0+	014	071	221	257	177	081	023	004	0+	0+	0+	0+
	4	0+	001	016	111	220	236	161	070	017	002	0+	0+	0+
	5	0+	0+	002	039	132	221	226	147	057	010	0+	0+	0+
	6	0+	0+	0+	010	057	147	226	221	132	039	002	0+	0+
	7	0+	0+	0+	002	017	070	161	236	220	111	016	001	0+
	8	9+	0+	0+	0+	004	023	081	177	257	221	071	014	0+
	9	0+	0+	0+	0+	001	005	027	089	200	295	213	087	005
	10	0+	0+	0+	0+	0+	001	005	027	093	236	384	329	099
	11	0+	0+	0+	0+	0+	0+	0+	004	020	086	314	569	895
12	0	886	540	282	069	014	002	0+	0+	0+	0+	0+	0+	0+
	1	107	341	377	206	071	017	003	0+	0+	0+	0+	0+	0+
	2	006	099	230	283	168	064	016	002	0+	0+	0+	0+	0+
	3	0+	017	085	236	240	142	054	012	001	0+	0+	0+	0+
	4	0+	002	021	133	231	213	121	042	008	001	0+	0+	0+
	5	0+	0+	004	053	158	227	193	101	029	003	0+	0+	0+
	6	0+	0+	0+	016	079	177	226	177	079	016	0+	0+	0+
	7	0+	0+	0+	003	029	101	193	227	158	053	004	0+	0+
	8	0+	0+	0+	001	008	042	121	213	231	133	021	002	0+
	9	0+	0+	0+	0+	001	012	054	142	240	236	085	017	0+
	10	0+	0+	0+	0+	0+	002	016	064	168	283	230	099	006
	11	0+	0+	0+	0+	0+	0+	003	017	071	206	377	341	107
	12	0+	0+	0+	0+	0+	0+	0+	002	014	069	282	540	886
13	0	878	513	254	055	010	001	0+	0+	0+	0+	0+	0+	0+
	1	115	351	367	179	054	011	002	0+	0+	0+	0+	0+	0+
	2	007	111	245	268	139	045	010	001	0+	0+	0+	0+	0+
	3	0+	021	100	246	218	111	035	006	001	0+	0+	0+	0+
	4	0+	003	028	154	234	184	087	024	003	0+	0+	0+	0+
	5	0+	0+	006	069	180	221	157	066	014	001	0+	0+	0+
	6	0+	0+	001	023	103	197	209	131	044	006	0+	0+	0+
	7	0+	0+	0+	006	044	131	209	197	103	023	001	0+	0+
	8	0+	0+	0+	001	014	066	157	221	180	069	006	0+	0+
	9	9+	0+	0+	0+	003	024	087	184	234	154	028	003	0+
	10	0+	0+	0+	0+	001	006	035	111	218	246	100	021	0+
	11	0+	0+	0+	0+	0+	001	010	045	139	268	245	111	007
	12	0+	0+	0+	0+	0+	0+	002	011	054	179	367	351	115
	13	0+	0+	0+	0+	0+	0+	0+	001	010	055	254	513	878
14	0	869	488	229	044	007	001	0+	0+	0+	0+	0+	0+	0+
	1	123	359	356	154	041	007	001	0+	0+	0+	0+	0+	0+
	2	008	123	257	250	113	032	006	001	0+	0+	0+	0+	0+

TABLE 2 *Continued* BINOMIAL PROBABILITIES

n	X	.01	.05	.10	.20	.30	.40	p .50	.60	.70	.80	.90	.95	.99
14	3	0+	026	114	250	194	085	022	003	0+	0+	0+	0+	0+
	4	0+	004	035	172	229	155	061	014	001	0+	0+	0+	0+
	5	0+	0+	008	086	196	207	122	041	007	0+	0+	0+	0+
	6	0+	0+	001	032	126	207	183	092	023	002	0+	0+	0+
	7	0+	0+	0+	009	062	157	209	157	062	009	0+	0+	0+
	8	0+	0+	0+	002	023	092	183	207	126	032	001	0+	0+
	9	0+	0+	0+	0+	007	041	122	207	196	086	008	0+	0+
	10	0+	0+	0+	0+	001	014	061	155	229	172	035	004	0+
	11	0+	0+	0+	0+	0+	003	022	085	194	250	114	026	0+
	12	0+	0+	0+	0+	0+	001	006	032	113	250	257	123	008
	13	0+	0+	0+	0+	0+	0+	001	007	041	154	356	359	123
	14	0+	0+	0+	0+	0+	0+	0+	001	007	044	229	488	869
15	0	860	463	206	035	005	0+	0+	0+	0+	0+	0+	0+	0+
	1	130	366	343	132	031	005	0+	0+	0+	0+	0+	0+	0+
	2	009	135	267	231	092	022	003	0+	0+	0+	0+	0+	0+
	3	0+	031	129	250	170	063	014	002	0+	0+	0+	0+	0+
	4	0+	005	043	188	219	127	042	007	001	0+	0+	0+	0+
	5	0+	001	010	103	206	186	092	024	003	0+	0+	0+	0+
	6	0+	0+	002	043	147	207	153	061	012	001	0+	0+	0+
	7	0+	0+	0+	014	081	177	196	118	035	003	0+	0+	0+
	8	0+	0+	0+	003	035	118	196	177	081	014	0+	0+	0+
	9	0+	0+	0+	001	012	061	153	207	147	043	002	0+	0+
	10	0+	0+	0+	0+	003	024	092	186	206	103	010	001	0+
	11	0+	0+	0+	0+	001	007	042	127	219	188	043	005	0+
	12	0+	0+	0+	0+	0+	002	014	063	170	250	129	031	0+
	13	0+	0+	0+	0+	0+	0+	003	022	092	231	267	135	009
	14	0+	0+	0+	0+	0+	0+	0+	005	031	132	343	366	130
	15	0+	0+	0+	0+	0+	0+	0+	0+	005	035	206	463	860
16	0	851	440	185	028	003	0+	0+	0+	0+	0+	0+	0+	0+
	1	138	371	329	113	023	003	0+	0+	0+	0+	0+	0+	0+
	2	010	146	274	211	073	015	002	0+	0+	0+	0+	0+	0+
	3	0+	036	142	246	146	047	008	001	0+	0+	0+	0+	0+
	4	0+	006	051	200	204	101	028	004	0+	0+	0+	0+	0+
	5	0+	001	014	120	210	162	067	014	001	0+	0+	0+	0+
	6	0+	0+	003	055	165	198	122	039	006	0+	0+	0+	0+
	7	0+	0+	0+	020	101	189	175	084	018	001	0+	0+	0+
	8	0+	0+	0+	006	049	142	196	142	049	006	0+	0+	0+
	9	0+	0+	0+	001	018	084	175	189	101	020	0+	0+	0+
	10	0+	0+	0+	0+	006	039	122	198	165	055	003	0+	0+
	11	0+	0+	0+	0+	001	014	067	162	210	120	014	001	0+
	12	0+	0+	0+	0+	0+	004	028	101	204	200	051	006	0+
	13	0+	0+	0+	0+	0+	001	008	047	146	246	142	036	0+
	14	0+	0+	0+	0+	0+	0+	002	015	073	211	274	146	010
	15	0+	0+	0+	0+	0+	0+	0+	003	023	113	329	371	138
	16	0+	0+	0+	0+	0+	0+	0+	0+	003	028	185	440	851
17	0	843	418	167	022	002	0+	0+	0+	0+	0+	0+	0+	0+
	1	145	374	315	096	017	002	0+	0+	0+	0+	0+	0+	0+
	2	012	158	280	191	058	010	001	0+	0+	0+	0+	0+	0+
	3	001	042	156	239	124	034	005	0+	0+	0+	0+	0+	0+
	4	0+	008	060	209	187	080	018	002	0+	0+	0+	0+	0+
	5	0+	001	018	136	208	138	047	008	001	0+	0+	0+	0+
	6	0+	0+	004	068	178	184	094	024	003	0+	0+	0+	0+
	7	0+	0+	001	027	120	193	148	057	010	0+	0+	0+	0+
	8	0+	0+	0+	008	064	161	186	107	028	002	0+	0+	0+
	9	0+	0+	0+	002	028	107	186	161	064	008	0+	0+	0+
	10	0+	0+	0+	0+	010	057	148	193	120	027	001	0+	0+
	11	0+	0+	0+	0+	003	024	094	184	178	068	004	0+	0+
	12	0+	0+	0+	0+	001	008	047	138	208	136	018	001	0+
	13	0+	0+	0+	0+	0+	002	018	080	187	209	060	008	0+
	14	0+	0+	0+	0+	0+	0+	005	034	124	239	156	042	001
	15	0+	0+	0+	0+	0+	0+	001	010	058	191	280	158	012

TABLE 2 *Continued* BINOMIAL PROBABILITIES

n	X	.01	.05	.10	.20	.30	.40	p = θ .50	.60	.70	.80	.90	.95	.99
17	16	0+	0+	0+	0+	0+	0+	0+	002	017	096	315	374	145
	17	0+	0+	0+	0+	0+	0+	0+	0+	002	022	167	418	843
18	0	835	397	150	018	002	0+	0+	0+	0+	0+	0+	0+	0+
	1	152	376	300	081	013	001	0+	0+	0+	0+	0+	0+	0+
	2	013	168	284	172	046	007	001	0+	0+	0+	0+	0+	0+
	3	001	047	168	230	105	025	003	0+	0+	0+	0+	0+	0+
	4	0+	009	070	215	168	061	012	001	0+	0+	0+	0+	0+
	5	0+	001	002	151	202	115	033	004	0+	0+	0+	0+	0+
	6	0+	0+	005	082	187	166	071	014	001	0+	0+	0+	0+
	7	0+	0+	001	035	138	189	121	037	005	0+	0+	0+	0+
	8	0+	0+	0+	012	081	173	167	077	015	001	0+	0+	0+
	9	0+	0+	0+	003	038	128	186	128	038	003	0+	0+	0+
	10	0+	0+	0+	001	015	077	167	173	081	012	0+	0+	0+
	11	0+	0+	0+	0+	005	037	121	189	138	035	001	0+	0+
	12	0+	0+	0+	0+	001	014	071	166	187	082	005	0+	0+
	13	0+	0+	0+	0+	0+	004	033	115	202	151	022	001	0+
	14	0+	0+	0+	0+	0+	001	012	061	168	215	070	009	0+
	15	0+	0+	0+	0+	0+	0+	003	025	105	230	168	097	001
	16	0+	0+	0+	0+	0+	0+	001	007	046	172	284	168	013
	17	0+	0+	0+	0+	0+	0+	0+	001	013	081	300	376	152
	18	0+	0+	0+	0+	0+	0+	0+	0+	002	018	150	397	835
19	0	826	377	135	014	001	0+	0+	0+	0+	0+	0+	0+	0+
	1	159	377	285	069	009	001	0+	0+	0+	0+	0+	0+	0+
	2	014	179	285	154	036	005	0+	0+	0+	0+	0+	0+	0+
	3	001	053	180	218	087	018	002	0+	0+	0+	0+	0+	0+
	4	0+	011	080	218	149	047	007	0+	0+	0+	0+	0+	0+
	5	0+	002	027	164	192	093	022	002	0+	0+	0+	0+	0+
	6	0+	0+	007	096	192	145	052	008	0+	0+	0+	0+	0+
	7	0+	0+	001	044	152	180	096	024	002	0+	0+	0+	0+
	8	0+	0+	0+	017	098	180	144	053	008	0+	0+	0+	0+
	9	0+	0+	0+	005	051	146	176	098	022	001	0+	0+	0+
	10	0+	0+	0+	001	022	098	176	146	051	005	0+	0+	0+
	11	0+	0+	0+	0+	008	053	144	180	098	017	0+	0+	0+
	12	0+	0+	0+	0+	002	024	096	180	152	044	001	0+	0+
	13	0+	0+	0+	0+	0+	008	052	145	192	096	007	0+	0+
	14	0+	0+	0+	0+	0+	002	022	093	192	164	027	002	0+
	15	0+	0+	0+	0+	0+	0+	007	047	149	218	080	011	0+
	16	0+	0+	0+	0+	0+	0+	002	018	087	218	180	053	001
	17	0+	0+	0+	0+	0+	0+	0+	005	036	154	285	179	014
	18	0+	0+	0+	0+	0+	0+	0+	001	009	069	285	377	159
	19	0+	0+	0+	0+	0+	0+	0+	0+	001	014	135	377	826
20	0	818	358	122	012	001	0+	0+	0+	0+	0+	0+	0+	0+
	1	165	377	270	058	007	0+	0+	0+	0+	0+	0+	0+	0+
	2	016	189	285	137	028	003	0+	0+	0+	0+	0+	0+	0+
	3	001	060	190	205	072	012	001	0+	0+	0+	0+	0+	0+
	4	0+	013	090	218	130	035	005	0+	0+	0+	0+	0+	0+
	5	0+	002	032	175	179	075	015	001	0+	0+	0+	0+	0+
	6	0+	0+	009	109	192	124	037	005	0+	0+	0+	0+	0+
	7	0+	0+	002	054	164	166	074	015	001	0+	0+	0+	0+
	8	0+	0+	0+	022	114	180	120	036	004	0+	0+	0+	0+
	9	0+	0+	0+	007	065	160	160	071	012	0+	0+	0+	0+
	10	0+	0+	0+	002	031	117	176	117	031	002	0+	0+	0+
	11	0+	0+	0+	0+	012	071	160	160	065	007	0+	0+	0+
	12	0+	0+	0+	0+	004	036	120	180	114	022	0+	0+	0+
	13	0+	0+	0+	0+	001	015	074	166	164	054	002	0+	0+
	14	0+	0+	0+	0+	0+	005	037	124	192	109	009	0+	0+
	15	0+	0+	0+	0+	0+	001	015	075	179	175	032	002	0+
	16	0+	0+	0+	0+	0+	0+	005	035	130	218	090	013	0+
	17	0+	0+	0+	0+	0+	0+	001	012	072	205	190	060	001
	18	0+	0+	0+	0+	0+	0+	0+	003	028	137	285	189	016
	19	0+	0+	0+	0+	0+	0+	0+	0+	007	058	270	377	165
	20	0+	0+	0+	0+	0+	0+	0+	0+	001	012	122	358	818

TABLE 3 RANDOM NUMBERS

	1	2	3	4	5	6	7	8
1	04479	44211	81536	09686	26743	87001	62392	59946
2	87019	90503	16034	07862	19701	85949	85876	58188
3	54222	56179	09833	34227	43897	38517	11617	30338
4	17929	24021	50932	89349	08012	37925	59003	95503
5	56399	82269	69443	62020	03365	82164	01356	24871
6	79242	52682	36255	74168	28636	93043	65454	36152
7	88869	22489	50467	14964	93146	51852	32408	22545
8	83970	03473	42981	83127	98774	74392	12218	91841
9	52754	85751	92705	70949	24331	42672	04885	44251
10	27011	69215	90920	96218	81127	67792	08377	60773
11	67952	43155	65547	50055	97940	38833	08745	69207
12	93376	38289	89474	22350	84982	85224	29969	42745
13	13674	24899	60602	33203	91953	48635	43938	08285
14	19345	11394	09241	72723	09052	76987	89854	48849
15	91609	18375	16171	30692	37389	51879	29556	51315
16	68537	17630	70322	26128	15645	91691	81064	58083
17	10038	17181	93964	41122	13020	98243	46447	28675
18	57023	28928	73917	94774	62542	30536	14777	72360
19	70791	39030	11261	76783	31184	38669	95862	99067
20	88033	42447	17815	75351	40853	83513	88714	09887
21	71858	65129	60871	04586	90651	93207	85501	83600
22	40357	80097	82138	61279	70478	49731	94154	50436
23	03339	05350	61895	46420	81433	61995	16654	91274
24	43084	21898	98854	70139	31516	29990	40919	05125
25	77223	58612	93223	12495	12628	43715	88010	03080
26	87135	86620	56893	82220	33968	13380	38087	74056
27	82521	26025	67975	81512	85227	39786	82990	38936
28	63590	66694	35357	19452	67724	10912	58569	66929
29	38133	07569	71030	75769	89240	48888	27184	78014
30	70369	48709	65114	69725	42994	22584	18455	52022
31	16168	91235	17509	72148	34676	61011	03681	21135
32	38541	45056	27395	13139	57487	57389	10764	62267
33	35508	90052	94492	83678	11316	98396	20893	87494
34	27147	55333	29880	81775	05384	86224	43487	86643
35	95351	12900	12689	07330	29470	39802	79928	68896
36	48047	70852	63798	62452	83695	38200	17414	13151
37	86417	48099	72299	46033	88948	93459	89657	52339
38	10361	07412	48001	57271	13210	04328	23855	65719
39	29998	88220	63213	98976	78720	61138	90709	50003
40	13008	59213	55737	68130	74358	74687	79519	29409
41	92882	31482	05651	53952	00915	43967	62276	47818
42	57149	14046	02876	79221	76700	68078	67712	98230
43	87080	09985	68303	23068	73514	39328	56046	98785
44	33504	84019	91220	05463	19500	66509	87209	71293
45	44702	70429	73468	16316	87536	49921	00239	37743
46	40473	76124	12097	56736	84635	77172	47155	77306
47	31727	64165	28937	14805	22863	62154	87637	80982
48	98855	63471	83278	00131	90229	02976	49485	67541
49	56067	73922	05810	24125	09603	99539	04848	57223
50	87152	73758	86758	77787	47126	31822	72088	00927

Draw Any Straight = .975k/1

$$\frac{52}{52} \quad \frac{16}{51} \quad \frac{12}{50} \quad \frac{8}{49} \quad \frac{4}{48}$$

Draw Straight Flush = 250k/1

$$\frac{52}{52} \quad \frac{4}{51} \quad \frac{3}{50} \quad \frac{2}{49} \quad \frac{1}{48}$$

Draw 3 of a kind = .425k/1

$$\frac{52}{52} \quad \frac{3}{51} \quad \frac{2}{50}$$

Draw 4 of a kind = 20.8k/1

$$54/52 \quad 3/51 \quad 2/50 \quad 1/49$$

Draw Full Hse. = 6.8k/1

$$\frac{52}{52} \quad \frac{3}{51} \quad \frac{2}{50} \quad \frac{49}{49} \quad \frac{3}{48}$$

Draw Flush = .505k/1

$$\frac{52}{52} \cdot \frac{12}{51} \cdot \frac{11}{50} \cdot \frac{10}{49} \quad \frac{9}{48}$$

NB
Adjust ODDS FIRST CARD FROM (52/52)
to (13/52) for particular Suit
or (4/52) "   "   "   CARD

THINGS I MUST DO
TODAY

DATE _____

1. _____
2. _____
3. _____
4. _____
5. _____
6. _____
7. _____
8. _____
9. _____
10. _____
11. _____
12. _____

smoot associates, inc.

p.o. box t
norwell, massachusetts 02061
(617) 826-4482

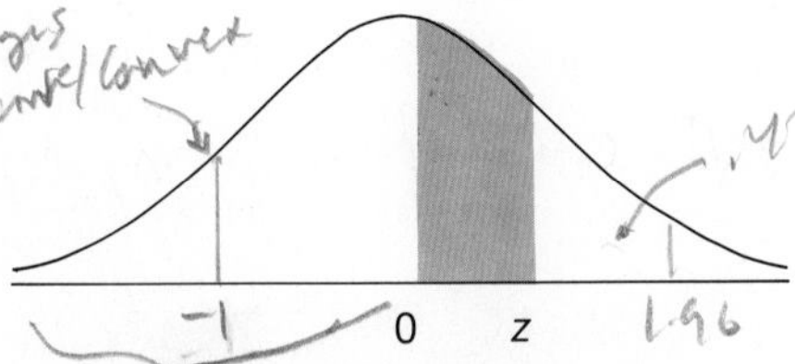

Handwritten notes (top):
- Area under curve = Probability
- Std Curve: M = 0; σ = 1

See p.111 + convert 'x' to 'z'
$z = \frac{x - x̄}{σ}$
'z' score 2.3 = 2.3 σ above the mean

Handwritten annotations on the curve: Changes Concave/Convex; -1; 0; z; 1.96; Area = .5; .475 x 2 = 95% Confidence

TABLE 4 AREAS UNDER THE <u>NORMAL</u> CURVE *(handwritten: Most under -3 > z < 3)*

z	.00	.01	.02	.03	.04	.05	.06	.07	.08	.09
0.0	.0000	.0040	.0080	.0120	.0160	.0199	.0239	.0279	.0319	.0359
0.1	.0398	.0438	.0478	.0517	.0557	.0596	.0636	.0675	.0714	.0753
0.2	.0793	.0832	.0871	.0910	.0948	.0987	.1026	.1064	.1103	.1141
0.3	.1179	.1217	.1255	.1293	.1331	.1368	.1406	.1443	.1480	.1517
0.4	.1554	.1591	.1628	.1664	.1700	.1736	.1772	.1808	.1844	.1879
0.5	.1915	.1950	.1985	.2019	.2054	.2088	.2123	.2157	.2190	.2224
0.6	.2257	.2291	.2324	.2357	.2389	.2422	.2454	.2486	.2518	.2549
0.7	.2580	.2611	.2642	.2673	.2704	.2734	.2764	.2794	.2823	.2852
0.8	.2881	.2910	.2939	.2967	.2995	.3023	.3051	.3078	.3106	.3133
0.9	.3159	.3186	.3212	.3238	.3264	.3289	.3315	.3340	.3365	.3389
1.0	.34134	.34375	.34614	.34850	.35083	.35314	.35543	.35769	.35993	.36214
1.1	.36433	.36650	.36864	.37076	.37286	.37493	.37698	.37900	.38100	.38298
1.2	.38493	.38686	.38877	.39065	.39251	.39435	.39617	.39796	.39973	.40147
1.3	.40320	.40490	.40658	.40824	.40988	.41149	.41309	.41466	.41621	.41774
1.4	.41924	.42073	.42220	.42364	.42507	.42647	.42786	.42922	.43056	.43189
1.5	.43319	.43448	.43574	.43699	.43822	.43943	.44062	.44179	.42295	.44408
1.6	.44520	.44630	.44738	.44845	.44950	.45053	.45154	.45254	.45352	.45449
1.7	.45543	.45637	.45728	.45818	.45907	.45994	.46080	.46164	.46246	.46327
1.8	.46407	.46485	.46562	.46638	.46712	.46784	.46856	.46926	.46995	.47062
1.9	.47128	.47193	.47257	.47320	.47381	.47441	.47500	.47558	.47615	.47670
2.0	.47725	.47778	.47831	.47882	.47932	.47982	.48030	.48077	.48124	.48169
2.1	.48214	.48257	.48300	.48341	.48382	.48422	.48461	.48500	.48537	.48574
2.2	.48610	.48645	.48679	.48713	.48745	.48778	.48809	.48840	.48870	.48899
2.3	.48928	.48956	.48983	.49010	.49036	.49061	.49086	.49111	.49134	.49158
2.4	.49180	.49202	.49224	.49245	.49266	.49286	.49305	.49324	.49343	.49361
2.5	.49379	.49396	.49413	.49430	.49446	.49461	.49477	.49492	.49506	.49520
2.6	.49534	.49547	.49560	.49573	.49585	.49598	.49609	.49621	.49632	.49643
2.7	.49653	.49664	.49674	.49683	.49693	.49702	.49711	.49720	.49728	.49736
2.8	.49744	.49752	.49760	.49767	.49774	.49781	.49788	.49795	.49801	.49807
2.9	.49813	.49819	.49825	.49831	.49836	.49841	.49846	.49851	.49856	.49861
3.0	.49865	.49869	.49874	.49878	.49882	.49886	.49889	.49893	.49897	.49900
3.1	.49903	.49906	.49910	.49913	.49916	.49918	.49921	.49924	.49926	.49929
3.2	.49931	.49934	.49936	.49938	.49940	.49942	.49944	.49946	.49948	.49950
3.3	.49952									
3.4	.49966									
3.5	.49977									
3.6	.49984									
3.7	.49989									
3.8	.49993									
3.9	.49995									
4.0	.49997									

Handwritten annotations within table: (.90) near 1.5/.05; (.95) near 1.9/.05; (.99) near 2.5/.08

Don't Know Std. Deviation (only w/sample)

appendix c *(Used for Sample ≤ 30)*

*problem will tell you to infer
if one or two tail, particularly
one tail w/one random sample*

$t\ score = t = \dfrac{\bar{x}}{6s/\sqrt{n}}$

Sample mean ≥ Mean

or

*for confidence Intervals,
Sample Mean ≠ Mean*

TABLE 5 STUDENT'S *t* DISTRIBUTIONS

Degrees of freedom

One-tail:	.10	.05	.025	.01	.005
Two-tail:	.20	.10	.05	.02	.01
$D=(n-1)$					
1	3.08	6.31	12.7	31.8	63.7
2	1.89	2.92	4.30	6.97	9.92
3	1.64	2.35	3.18	4.54	5.84
4	1.53	2.13	2.78	3.75	4.60
5	1.48	2.02	2.57	3.37	4.03
6	1.44	1.94	2.45	3.14	3.71
7	1.42	1.90	2.37	3.00	3.50
8	1.40	1.86	2.31	2.90	3.36
9	1.38	1.83	2.26	2.82	3.25
10	1.37	1.81	2.23	2.76	3.17
11	1.36	1.80	2.20	2.72	3.10
12	1.36	1.78	2.18	2.68	3.06
13	1.35	1.77	2.16	2.65	3.01
14	1.35	1.76	2.15	2.62	2.98
15	1.34	1.75	2.13	2.60	2.95
16	1.34	1.75	2.12	2.58	2.92
17	1.33	1.74	2.11	2.57	2.90
18	1.33	1.73	2.10	2.55	2.88
19	1.33	1.73	2.09	2.54	2.86
20	1.33	1.73	2.09	2.53	2.85
21	1.32	1.72	2.08	2.52	2.83
22	1.32	1.72	2.07	2.51	2.82
23	1.32	1.71	2.07	2.50	2.81
24	1.32	1.71	2.06	2.49	2.80
25	1.32	1.71	2.06	2.49	2.79
26	1.32	1.71	2.06	2.48	2.78
27	1.31	1.70	2.05	2.47	2.77
28	1.31	1.70	2.05	2.47	2.76
29	1.31	1.70	2.05	2.46	2.76
30	1.31	1.70	2.04	2.46	2.75
40	1.30	1.68	2.02	2.42	2.70
∞	1.28	1.65	1.96	2.33	2.58

Closer to Normal Distrib. as n→30

u u
3 scores z
probabilities .80 .90 .95 .98 .99

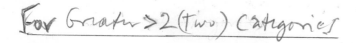

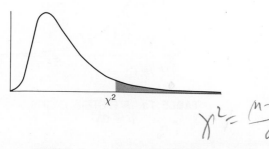

$$\chi^2 = \frac{(n-1)s^2}{\sigma^2}$$

Dgy of Freedom

TABLE 6 χ^2 DISTRIBUTIONS

$D = n-1$	.10	.05	.025	.01
1	2.71	3.84	5.02	6.63
2	4.61	5.99	7.38	9.21
3	6.25	7.81	9.35	11.3
4	7.78	9.49	11.1	13.3
5	9.24	11.1	12.8	15.1
6	10.6	12.6	14.4	16.8
7	12.0	14.1	16.0	18.5
8	13.4	15.5	17.5	20.1
9	14.7	16.9	19.0	21.7
10	16.0	18.3	20.5	23.2
11	17.3	19.7	21.9	24.7
12	18.5	21.0	23.3	26.2
13	19.8	22.4	24.7	27.7
14	21.1	23.7	26.1	29.1
15	22.3	25.0	27.5	30.6
16	23.5	26.3	28.8	32.0
17	24.8	27.6	30.2	33.4
18	26.0	28.9	31.5	34.8
19	27.2	30.1	32.9	36.2
20	28.4	31.4	34.2	37.6
21	29.6	32.7	35.5	38.9
22	30.8	33.9	36.8	40.3
23	32.0	35.2	38.1	41.6
24	33.2	36.4	39.4	43.0
25	34.4	37.7	40.6	44.3
26	35.6	38.9	41.9	45.6
27	36.7	40.1	43.2	47.0
28	37.9	41.3	44.5	48.3
29	39.1	42.6	45.7	49.6
30	40.3	43.8	47.0	50.9
40	51.8	55.8	59.3	63.7
50	63.2	67.5	71.4	76.2
60	74.4	79.1	83.3	88.4
70	85.5	90.5	95.0	100.4
80	96.6	101.9	106.6	112.3

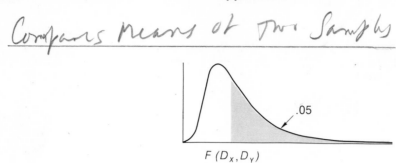

Comparing Means of Two Samples

$$F (D_X, D_Y)$$

TABLE 7a F DISTRIBUTIONS
$\alpha = .05$

D_X

D_Y	1	2	3	4	5	6	7	8	9
1	161	200	216	225	230	234	237	239	241
2	18.5	19.0	19.2	19.3	19.3	19.3	19.4	19.4	19.4
3	10.1	9.55	9.28	9.12	9.01	8.94	8.88	8.85	8.81
4	7.71	6.94	6.59	6.39	6.26	6.16	6.09	6.04	6.00
5	6.61	5.79	5.41	5.19	5.05	4.95	4.88	4.82	4.77
6	5.99	5.14	4.76	4.53	4.39	4.28	4.21	4.15	4.10
7	5.59	4.74	4.35	4.12	3.97	3.87	3.79	3.73	3.68
8	5.32	4.46	4.07	3.84	3.69	3.58	3.50	3.44	3.39
9	5.12	4.26	3.86	3.63	3.48	3.37	3.29	3.23	3.18
10	4.96	4.10	3.71	3.48	3.33	3.22	3.14	3.07	3.02
11	4.84	3.98	3.59	3.36	3.20	3.09	3.01	2.95	2.90
12	4.75	3.89	3.49	3.26	3.11	3.00	2.91	2.85	2.80
13	4.67	3.81	3.41	3.18	3.03	2.92	2.83	2.77	2.71
14	4.60	3.74	3.34	3.11	2.96	2.85	2.76	2.70	2.65
15	4.54	3.68	3.29	3.06	2.90	2.79	2.71	2.64	2.59
16	4.49	3.63	3.24	3.01	2.85	2.74	2.66	2.59	2.54
17	4.45	3.59	3.20	2.96	2.81	2.70	2.61	2.55	2.49
18	4.41	3.55	3.16	2.93	2.77	2.66	2.58	2.51	2.46
19	4.38	3.52	3.13	2.90	2.74	2.63	2.54	2.48	2.42
20	4.35	3.49	3.10	2.87	2.71	2.60	2.51	2.45	2.39
21	4.32	3.47	3.07	2.84	2.68	2.57	2.49	2.42	2.37
22	4.30	3.44	3.05	2.82	2.66	2.55	2.46	2.40	2.34
23	4.28	3.42	3.03	2.80	2.64	2.53	2.44	2.37	2.32
24	4.26	3.40	3.01	2.78	2.62	2.51	2.42	2.36	2.30
25	4.24	3.39	2.99	2.76	2.60	2.49	2.40	2.34	2.28
26	4.23	3.37	2.98	2.74	2.59	2.47	2.39	2.32	2.27
27	4.21	3.35	2.96	2.73	2.57	2.46	2.37	2.31	2.25
28	4.20	3.34	2.95	2.71	2.56	2.45	2.36	2.29	2.24
29	4.18	3.33	2.93	2.70	2.55	2.43	2.35	2.28	2.22
30	4.17	3.32	2.92	2.69	2.53	2.42	2.33	2.27	2.21
40	4.08	3.23	2.84	2.61	2.45	2.34	2.25	2.18	2.12
60	4.00	3.15	2.76	2.53	2.37	2.25	2.17	2.10	2.04
80	3.96	3.11	2.72	2.48	2.32	2.21	2.12	2.05	1.99
100	3.94	3.09	2.70	2.46	2.30	2.19	2.10	2.03	1.97
120	3.92	3.07	2.68	2.45	2.29	2.17	2.09	2.02	1.96
∞	3.84	3.00	2.60	2.37	2.21	2.10	2.01	1.94	1.88

TABLE 7a *Continued F* DISTRIBUTIONS
$\alpha = .05$

D_X

D_Y	12	14	16	20	24	30	40	60	100
1	244	245	246	248	249	250	251	252	253
2	19.4	19.4	19.4	19.5	19.5	19.5	19.5	19.5	19.5
3	8.74	8.71	8.69	8.66	8.64	8.62	8.59	8.57	8.56
4	5.91	5.87	5.84	5.80	5.77	5.75	5.72	5.69	5.66
5	4.68	4.64	4.60	4.56	4.53	4.50	4.46	4.43	4.40
6	4.00	3.96	3.92	3.87	3.84	3.81	3.77	3.74	3.71
7	3.57	3.52	3.49	3.44	3.41	3.38	3.34	3.30	3.28
8	3.28	3.23	3.20	3.15	3.12	3.08	3.04	3.01	2.98
9	3.07	3.02	2.98	2.94	2.90	2.86	2.83	2.79	2.76
10	2.91	2.87	2.82	2.77	2.74	2.70	2.66	2.62	2.59
11	2.79	2.74	2.70	2.65	2.61	2.57	2.53	2.49	2.45
12	2.69	2.64	2.60	2.54	2.51	2.47	2.43	2.38	2.35
13	2.60	2.55	2.51	2.46	2.42	2.38	2.34	2.30	2.26
14	2.53	2.48	2.44	2.39	2.35	2.31	2.27	2.22	2.19
15	2.48	2.43	2.39	2.33	2.29	2.25	2.20	2.16	2.12
16	2.42	2.37	2.33	2.28	2.24	2.19	2.15	2.11	2.07
17	2.38	2.33	2.29	2.23	2.19	2.15	2.10	2.06	2.02
18	2.34	2.29	2.25	2.19	2.15	2.11	2.06	2.02	1.98
19	2.31	2.26	2.21	2.16	2.11	2.07	2.03	1.98	1.94
20	2.28	2.23	2.18	2.12	2.08	2.04	1.99	1.95	1.90
21	2.25	2.20	2.15	2.10	2.05	2.01	1.96	1.92	1.87
22	2.23	2.18	2.13	2.07	2.03	1.98	1.94	1.89	1.84
23	2.20	2.14	2.10	2.05	2.01	1.96	1.91	1.86	1.82
24	2.18	2.13	2.09	2.03	1.98	1.94	1.89	1.84	1.80
25	2.16	2.11	2.06	2.01	1.96	1.92	1.87	1.82	1.77
26	2.15	2.10	2.05	1.99	1.95	1.90	1.85	1.80	1.76
27	2.13	2.08	2.03	1.97	1.93	1.88	1.84	1.79	1.74
28	2.12	2.06	2.02	1.96	1.91	1.87	1.82	1.77	1.72
29	2.10	2.05	2.00	1.94	1.90	1.85	1.81	1.75	1.71
30	2.09	2.04	1.99	1.93	1.89	1.84	1.79	1.74	1.69
40	2.00	1.95	1.90	1.84	1.79	1.74	1.69	1.64	1.59
60	1.92	1.86	1.81	1.75	1.70	1.65	1.59	1.53	1.48
80	1.88	1.82	1.77	1.70	1.65	1.60	1.54	1.47	1.42
100	1.85	1.79	1.75	1.68	1.63	1.57	1.51	1.44	1.39
120	1.83	1.77	1.73	1.66	1.61	1.55	1.50	1.43	1.37
∞	1.75	1.69	1.64	1.57	1.52	1.46	1.39	1.32	1.24

354 appendix c

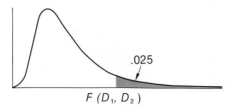

$F(D_1, D_2)$

TABLE 7b F DISTRIBUTIONS
$\alpha = .025$

D_X

D_Y	1	2	3	4	5	6	7	8	9
1	648	800	864	900	922	937	948	957	963
2	38.5	39.0	39.2	39.2	39.3	39.3	39.4	39.4	39.4
3	17.4	16.0	15.4	15.1	14.9	14.7	14.6	14.5	14.5
4	12.2	10.6	9.98	9.60	9.36	9.20	9.07	8.98	8.90
5	10.0	8.43	7.76	7.39	7.15	6.98	6.85	6.76	6.68
6	8.81	7.26	6.60	6.23	5.99	5.82	5.70	5.60	5.52
7	8.07	6.54	5.89	5.52	5.29	5.12	4.99	4.90	4.82
8	7.57	6.06	5.42	5.05	4.82	4.65	4.53	4.43	4.36
9	7.21	5.71	5.08	4.72	4.48	4.32	4.20	4.10	4.03
10	6.94	5.46	4.83	4.47	4.24	4.07	3.95	3.85	3.78
11	6.72	5.26	4.63	4.28	4.04	3.88	3.76	3.66	3.59
12	6.55	5.10	4.47	4.12	3.89	3.73	3.61	3.51	3.44
13	6.41	4.97	4.35	4.00	3.77	3.60	3.48	3.39	3.31
14	6.30	4.86	4.24	3.89	3.66	3.50	3.37	3.28	3.21
15	6.20	4.77	4.15	3.80	3.58	3.41	3.29	3.20	3.12
16	6.12	4.69	4.08	3.73	3.50	3.34	3.22	3.12	3.05
17	6.04	4.62	4.01	3.66	3.44	3.28	3.16	3.06	2.98
18	5.98	4.56	3.95	3.61	3.38	3.22	3.10	3.01	2.93
19	5.92	4.51	3.90	3.56	3.33	3.17	3.05	2.96	2.88
20	5.87	4.46	3.86	3.51	3.29	3.13	3.01	2.91	2.84
21	5.83	4.42	3.82	3.48	3.25	3.09	2.97	2.87	2.80
22	5.79	4.38	3.78	3.44	3.22	3.05	2.93	2.84	2.76
23	5.75	4.35	3.75	3.41	3.18	3.02	2.90	2.81	2.73
24	5.72	4.32	3.72	3.38	3.15	2.99	2.87	2.78	2.70
25	5.69	4.29	3.69	3.35	3.13	2.97	2.85	2.75	2.68
26	5.66	4.26	3.66	3.32	3.10	2.94	2.82	2.72	2.65
27	5.63	4.23	3.64	3.30	3.08	2.92	2.80	2.70	2.63
28	5.61	4.21	3.62	3.28	3.06	2.90	2.78	2.68	2.61
29	5.59	4.19	3.60	3.26	3.04	2.88	2.76	2.66	2.59
30	5.57	4.18	3.59	3.25	3.03	2.87	2.75	2.65	2.57
40	5.42	4.05	3.46	3.13	2.90	2.74	2.62	2.53	2.45
60	5.29	3.93	3.34	3.01	2.79	2.63	2.51	2.41	2.33
80	5.20	3.85	3.28	2.94	2.72	2.57	2.45	2.37	2.25
100	5.17	3.82	3.24	2.90	2.68	2.53	2.41	2.31	2.21
120	5.15	3.80	3.23	2.89	2.67	2.52	2.39	2.30	2.22
∞	5.02	3.69	3.12	2.79	2.57	2.41	2.29	2.19	2.11

TABLE 7b *Continued F* DISTRIBUTIONS
$\alpha = .025$

D_1

D_2	12	14	16	20	24	30	40	60	100
1	977	984	989	993	997	1000	1006	1010	1013
2	39.4	39.4	39.4	39.4	39.5	39.5	39.5	39.5	39.5
3	14.3	14.3	14.3	14.2	14.1	14.1	14.0	14.0	13.9
4	8.75	8.69	8.64	8.56	8.51	8.46	8.41	8.36	8.32
5	6.52	6.45	6.41	6.33	6.28	6.23	6.18	6.12	6.08
6	5.37	5.28	5.25	5.17	5.12	5.07	5.01	4.96	4.92
7	4.67	4.59	4.55	4.47	4.42	4.36	4.31	4.25	4.21
8	4.20	4.13	4.08	4.00	3.95	3.89	3.84	3.78	3.74
9	3.87	3.80	3.75	3.67	3.62	3.56	3.51	3.45	3.40
10	3.62	3.55	3.50	3.42	3.37	3.31	3.26	3.20	3.15
11	3.43	3.36	3.31	3.23	3.17	3.12	3.06	3.00	2.95
12	3.28	3.20	3.15	3.07	3.02	2.96	2.91	2.85	2.80
13	3.15	3.08	3.03	2.95	2.89	2.84	2.78	2.72	2.67
14	3.05	2.98	2.92	2.84	2.79	2.73	2.67	2.61	2.56
15	2.96	2.89	2.84	2.76	2.70	2.64	2.59	2.52	2.47
16	2.89	2.82	2.76	2.68	2.63	2.57	2.51	2.45	2.40
17	2.82	2.75	2.70	2.62	2.56	2.50	2.44	2.38	2.33
18	2.77	2.70	2.64	2.56	2.50	2.44	2.38	2.32	2.27
19	2.72	2.65	2.59	2.51	2.45	2.39	2.33	2.27	2.22
20	2.68	2.60	2.54	2.46	2.41	2.35	2.29	2.22	2.17
21	2.64	2.57	2.50	2.42	2.37	2.31	2.25	2.18	2.13
22	2.60	2.53	2.47	2.39	2.33	2.27	2.21	2.14	2.09
23	2.57	2.50	2.44	2.36	2.30	2.24	2.18	2.11	2.06
24	2.54	2.47	2.41	2.33	2.27	2.21	2.15	2.08	2.03
25	2.51	2.44	2.38	2.30	2.24	2.18	2.12	2.05	2.00
26	2.48	2.41	2.35	2.27	2.21	2.15	2.09	2.02	1.97
27	2.46	2.39	2.33	2.25	2.19	2.13	2.07	2.00	1.95
28	2.44	2.37	2.31	2.23	2.17	2.11	2.05	1.98	1.93
29	2.42	2.35	2.29	2.21	2.15	2.09	2.03	1.96	1.91
30	2.41	2.34	2.28	2.20	2.14	2.07	2.01	1.94	1.89
40	2.29	2.21	2.15	2.07	2.01	1.94	1.88	1.80	1.76
60	2.17	2.09	2.03	1.94	1.88	1.82	1.74	1.67	1.61
80	2.10	2.03	1.96	1.87	1.81	1.75	1.67	1.59	1.52
100	2.07	1.99	1.93	1.84	1.78	1.71	1.63	1.55	1.48
120	2.05	1.98	1.92	1.82	1.76	1.69	1.61	1.53	1.46
∞	1.94	1.86	1.81	1.71	1.64	1.57	1.48	1.39	1.30

appendix c

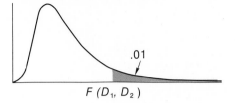

$$F(D_1, D_2)$$

TABLE 7c F DISTRIBUTIONS
$\alpha = .01$

D_X

D_Y	1	2	3	4	5	6	7	8	9
1	4052	4999	5403	5625	5764	5859	5928	5981	6022
2	98.5	99.0	99.2	99.3	99.3	99.3	99.3	99.4	99.4
3	34.1	30.8	29.5	28.7	28.2	27.9	27.7	27.5	27.3
4	21.2	18.0	16.7	16.0	15.5	15.2	15.0	14.8	14.7
5	16.3	13.3	12.1	11.4	11.0	10.7	10.5	10.3	10.2
6	13.7	10.9	9.78	9.15	8.75	8.47	8.26	8.10	7.98
7	12.3	9.55	8.45	7.85	7.46	7.19	7.00	6.84	6.71
8	11.3	8.65	7.59	7.01	6.63	6.37	6.19	6.03	5.91
9	10.6	8.02	6.99	6.42	6.06	5.80	5.62	5.47	5.35
10	10.0	7.56	6.55	5.99	5.64	5.39	5.21	5.06	4.95
11	9.65	7.20	6.22	5.67	5.32	5.07	4.88	4.74	4.63
12	9.33	6.93	5.95	5.41	5.06	4.82	4.65	4.50	4.39
13	9.07	6.70	5.74	5.20	4.86	4.62	4.44	4.30	4.19
14	8.86	6.51	5.56	5.03	4.69	4.46	4.28	4.14	4.03
15	8.68	6.36	5.42	4.89	4.56	4.32	4.14	4.00	3.89
16	8.53	6.23	5.29	4.77	4.44	4.20	4.03	3.89	3.78
17	8.40	6.11	5.18	4.67	4.34	4.10	3.93	3.79	3.68
18	8.28	6.01	5.09	4.58	4.25	4.01	3.85	3.71	3.60
19	8.18	5.93	5.01	4.50	4.17	3.94	3.77	3.63	3.52
20	8.10	5.85	4.94	4.43	4.10	3.87	3.71	3.56	3.45
21	8.02	5.78	4.87	4.37	4.04	3.81	3.65	3.51	3.40
22	7.94	5.72	4.82	4.31	3.99	3.76	3.59	3.45	3.35
23	7.88	5.66	4.76	4.26	3.94	3.71	3.54	3.40	3.30
24	7.82	5.61	4.72	4.22	3.90	3.67	3.50	3.36	3.25
25	7.77	5.57	4.68	4.18	3.86	3.63	3.46	3.32	3.21
26	7.72	5.53	4.64	4.14	3.82	3.59	3.42	3.29	3.17
27	7.68	5.49	4.60	4.11	3.79	3.56	3.39	3.26	3.14
28	7.64	5.45	4.57	4.08	3.76	3.53	3.36	3.23	3.11
29	7.60	5.42	4.54	4.05	3.73	3.50	3.33	3.20	3.08
30	7.56	5.39	4.51	4.02	3.70	3.47	3.30	3.17	3.06
40	7.31	5.18	4.31	3.83	3.51	3.29	3.12	2.99	2.88
60	7.08	4.98	4.13	3.65	3.34	3.12	2.95	2.82	2.72
80	6.96	4.88	4.04	3.56	3.25	3.04	2.87	2.74	2.64
100	6.90	4.82	3.98	3.51	3.20	2.99	2.82	2.69	2.59
120	6.85	4.79	3.95	3.48	3.17	2.96	2.79	2.66	2.56
∞	6.63	4.60	3.78	3.32	3.02	2.80	2.64	2.51	2.41

TABLE 7c *Continued F* DISTRIBUTIONS
$$\alpha = .01$$

D_X

D_Y	12	14	16	20	24	30	40	100
1	6110	6140	6170	6210	6230	6260	6290	6330
2	99.4	99.4	99.5	99.5	99.5	99.5	99.5	99.5
3	27.1	26.9	26.8	26.7	26.6	26.5	26.4	26.2
4	14.4	14.2	14.1	14.0	13.9	13.8	13.7	13.6
5	9.89	9.77	9.68	9.55	9.47	9.38	9.29	9.13
6	7.72	7.60	7.52	7.39	7.31	7.23	7.14	6.99
7	6.47	6.35	6.27	6.15	6.07	5.98	5.90	5.75
8	5.67	5.56	5.48	5.36	5.28	5.20	5.11	4.90
9	5.11	5.00	4.92	4.80	4.73	4.64	4.56	4.41
10	4.71	4.60	4.52	4.40	4.33	4.25	4.17	4.01
11	4.40	4.29	4.21	4.10	4.02	3.94	3.86	3.70
12	4.16	4.05	3.98	3.86	3.78	3.70	3.61	3.46
13	3.96	3.85	3.78	3.67	3.59	3.51	3.42	3.27
14	3.80	3.70	3.62	3.51	3.43	3.34	3.26	3.11
15	3.67	3.56	3.48	3.36	3.29	3.20	3.12	2.97
16	3.55	3.45	3.37	3.25	3.18	3.10	3.01	2.86
17	3.45	3.35	3.27	3.15	3.08	3.00	2.92	2.76
18	3.37	3.27	3.19	3.07	3.00	2.92	2.83	2.68
19	3.30	3.19	3.12	3.00	2.92	2.84	2.76	2.60
20	3.23	3.13	3.05	2.94	2.86	2.77	2.69	2.53
21	3.17	3.07	2.99	2.88	2.80	2.72	2.63	2.47
22	3.12	3.02	2.94	2.83	2.75	2.67	2.58	2.42
23	3.07	2.97	2.89	2.78	2.70	2.62	2.53	2.37
24	3.03	2.93	2.85	2.74	2.66	2.58	2.49	2.33
25	2.99	2.89	2.81	2.70	2.62	2.54	2.45	2.29
26	2.96	2.86	2.77	2.66	2.58	2.50	2.41	2.25
27	2.93	2.83	2.74	2.63	2.55	2.47	2.38	2.21
28	2.90	2.80	2.71	2.60	2.52	2.44	2.35	2.18
29	2.87	2.77	2.68	2.57	2.49	2.41	2.32	2.15
30	2.84	2.74	2.66	2.55	2.46	2.38	2.29	2.13
40	2.66	2.56	2.49	2.37	2.29	2.20	2.11	1.97
60	2.50	2.40	2.32	2.20	2.12	2.03	1.93	1.74
80	2.41	2.32	2.24	2.11	2.03	1.94	1.84	1.65
100	2.36	2.26	2.19	2.06	1.98	1.89	1.79	1.59
120	2.34	2.24	2.16	2.03	1.95	1.86	1.76	1.56
∞	2.18	2.07	1.99	1.87	1.79	1.70	1.59	

TABLE 8   TRANSFORMATION OF r TO Z AND ρ TO μ_z VALUES

r	Z	r	Z	r	Z	r	Z
.00	.000						
.01	.010	.26	.266	.51	.563	.76	.996
.02	.020	.27	.277	.52	.576	.77	1.020
.03	.030	.28	.288	.53	.590	.78	1.045
.04	.040	.29	.299	.54	.604	.79	1.071
.05	.050	.30	.310	.55	.618	.80	1.099
.06	.060	.31	.321	.56	.633	.81	1.127
.07	.070	.32	.332	.57	.648	.82	1.157
.08	.080	.33	.343	.58	.663	.83	1.188
.09	.090	.34	.354	.59	.678	.84	1.221
.10	.100	.35	.365	.60	.693	.85	1.256
.11	.110	.36	.377	.61	.709	.86	1.293
.12	.121	.37	.388	.62	.725	.87	1.333
.13	.131	.38	.400	.63	.741	.88	1.376
.14	.141	.39	.412	.64	.758	.89	1.422
.15	.151	.40	.424	.65	.775	.90	1.472
.16	.161	.41	.436	.66	.793	.91	1.528
.17	.172	.42	.448	.67	.811	.92	1.589
.18	.182	.43	.460	.68	.829	.93	1.658
.19	.192	.44	.472	.69	.848	.94	1.738
.20	.203	.45	.485	.70	.867	.95	1.832
.21	.213	.46	.497	.71	.887	.96	1.946
.22	.224	.47	.510	.72	.908	.97	2.092
.23	.234	.48	.523	.73	.929	.98	2.298
.24	.245	.49	.536	.74	.950	.99	2.647
.25	.255	.50	.549	.75	.973		

TABLE 9 DISTRIBUTION OF R IN THE RUNS TEST

Any value of R which is equal to or smaller than that shown in A or equal to or larger than that shown in B is significant at the .05 level.

A

n_Y \ n_X	2	3	4	5	6	7	8	9	10	11	12	13	14	15	16	17	18	19	20
2											2	2	2	2	2	2	2	2	2
3					2	2	2	2	2	2	2	2	2	3	3	3	3	3	3
4				2	2	2	3	3	3	3	3	3	3	3	4	4	4	4	4
5			2	2	3	3	3	3	3	4	4	4	4	4	4	4	5	5	5
6		2	2	3	3	3	3	4	4	4	4	5	5	5	5	5	5	6	6
7		2	2	3	3	3	4	4	5	5	5	5	5	6	6	6	6	6	6
8		2	3	3	3	4	4	5	5	5	6	6	6	6	6	7	7	7	7
9		2	3	3	4	4	5	5	5	6	6	6	7	7	7	7	8	8	8
10		2	3	3	4	5	5	5	6	6	7	7	7	7	8	8	8	8	9
11		2	3	4	4	5	5	6	6	7	7	7	8	8	8	9	9	9	9
12	2	2	3	4	4	5	6	6	7	7	7	8	8	8	9	9	9	10	10
13	2	2	3	4	5	5	6	6	7	7	8	8	9	9	9	10	10	10	10
14	2	2	3	4	5	5	6	7	7	8	8	9	9	9	10	10	10	11	11
15	2	3	3	4	5	6	6	7	7	8	8	9	9	10	10	11	11	11	12
16	2	3	4	4	5	6	6	7	8	8	9	9	10	10	11	11	11	12	12
17	2	3	4	4	5	6	7	7	8	9	9	10	10	11	11	11	12	12	13
18	2	3	4	5	5	6	7	8	8	9	9	10	10	11	11	12	12	13	13
19	2	3	4	5	6	6	7	8	8	9	10	10	11	11	12	12	13	13	13
20	2	3	4	5	6	6	7	8	9	9	10	10	11	12	12	13	13	13	14

B

n_Y \ n_X	2	3	4	5	6	7	8	9	10	11	12	13	14	15	16	17	18	19	20
2											6	6	6	6	6	6	6	6	6
3					8	8	8	8	8	8	8	8	8	8	8	8	8	8	8
4				9	9	10	10	10	10	10	10	10	10	10	10	10	10	10	10
5			9	10	10	11	11	12	12	12	12	12	12	12	12	12	12	12	12
6		8	9	10	11	12	12	13	13	13	13	14	14	14	14	14	14	14	14
7		8	10	11	12	13	13	14	14	14	14	15	15	15	16	16	16	16	16
8		8	10	11	12	13	14	14	15	15	16	16	16	16	17	17	17	17	17
9		8	10	12	13	14	14	15	16	16	16	17	17	18	18	18	18	18	18
10		8	10	12	13	14	15	16	16	17	17	18	18	18	19	19	19	20	20
11		8	10	12	13	14	15	16	17	17	18	19	19	19	20	20	20	21	21
12	6	8	10	12	13	14	16	16	17	18	19	19	20	20	21	21	21	22	22
13	6	8	10	12	14	15	16	17	18	19	19	20	20	21	21	22	22	23	23
14	6	8	10	12	14	15	16	17	18	19	20	20	21	22	22	23	23	23	24
15	6	8	10	12	14	15	16	18	18	19	20	21	22	22	23	23	24	24	25
16	6	8	10	12	14	16	17	18	19	20	21	21	22	23	23	24	25	25	25
17	6	8	10	12	14	16	17	18	19	20	21	22	23	23	24	25	25	26	26
18	6	8	10	12	14	16	17	18	19	20	21	22	23	24	25	25	26	26	27
19	6	8	10	12	14	16	17	18	20	21	22	23	23	24	25	26	26	27	27
20	6	8	10	12	14	16	17	18	20	21	22	23	24	25	25	26	27	27	28

TABLE 10   DISTRIBUTION OF U IN THE MANN-WHITNEY TEST
$\alpha = .05$

Any value of U which is equal to or smaller than that shown is significant at the .05 level for a two-tail test.

n of larger sample	2	3	4	5	6	7	8	9	10	11	12	13	14	15	16	17	18	19	20
4	—	—	0																
5	—	0	1	2															
6	—	1	2	3	5														
7	—	1	3	5	6	8													
8	0	2	4	6	8	10	13												
9	0	2	4	7	10	12	15	17											
10	0	3	5	8	11	14	17	20	23										
11	0	3	6	9	13	16	19	23	26	30									
12	1	4	7	11	14	18	22	26	29	33	37								
13	1	4	8	12	16	20	24	28	33	37	41	45							
14	1	5	9	13	17	22	26	31	36	40	45	50	55						
15	1	5	10	14	19	24	29	34	39	44	49	54	59	64					
16	1	6	11	15	21	26	31	37	42	47	53	59	64	70	75				
17	2	6	11	17	22	28	34	39	45	51	57	63	69	75	81	87			
18	2	7	12	18	24	30	36	42	48	55	61	67	74	80	86	93	99		
19	2	7	13	19	25	32	38	45	52	58	65	72	78	85	92	99	106	113	
20	2	8	14	20	27	34	41	48	55	62	69	76	83	90	98	105	112	119	127
21	3	8	15	22	29	36	43	50	58	65	73	80	88	96	103	111	119	126	134
22	3	9	16	23	30	38	45	53	61	69	77	85	93	101	109	117	125	133	141
23	3	9	17	24	32	40	48	56	64	73	81	89	98	106	115	123	132	140	149
24	3	10	17	25	33	42	50	59	67	76	85	94	102	111	120	129	138	147	156
25	3	10	18	27	35	44	53	62	71	80	89	98	107	117	126	135	145	154	163
26	4	11	19	28	37	46	55	64	74	83	93	102	112	122	132	141	151	161	171
27	4	11	20	29	38	48	57	67	77	87	97	107	117	127	137	147	158	168	178
28	4	12	21	30	40	50	60	70	80	90	101	111	122	132	143	154	164	175	186
29	4	13	22	32	42	52	62	73	83	94	105	116	127	138	149	160	171	182	193
30	5	13	23	33	43	54	65	76	87	98	109	120	131	143	154	166	177	189	200
31	5	14	24	34	45	56	67	78	90	101	113	125	136	148	160	172	184	196	208
32	5	14	24	35	46	58	69	81	93	105	117	129	141	153	166	178	190	203	215
33	5	15	25	37	48	60	72	84	96	108	121	133	146	159	171	184	197	210	222
34	5	15	26	38	50	62	74	87	99	112	125	138	151	164	177	190	203	217	230
35	6	16	27	39	51	64	77	89	103	116	129	142	156	169	183	196	210	224	237
36	6	16	28	40	53	66	79	92	106	119	133	147	161	174	188	202	216	231	245
37	6	17	29	41	55	68	81	95	109	123	137	151	165	180	194	209	223	238	252
38	6	17	30	43	56	70	84	98	112	127	141	156	170	185	200	215	230	245	259
39	7	18	31	44	58	72	86	101	115	130	145	160	175	190	206	221	236	252	267
40	7	18	31	45	59	74	89	103	119	134	149	165	180	196	211	227	243	258	274

index

See also Appendix A, Symbols (pages 327–330),
Appendix B, Formulas (pages 331–337), and list
of tables (back end-paper).